# Fault Tolerant Attitude Estimation for Small Satellites

# Fault Tolerant Attitude Estimation for Small Satellites

Chingiz Hajiyev
and
Halil Ersin Soken

**CRC Press**
Taylor & Francis Group
Boca Raton  London  New York

CRC Press is an imprint of the
Taylor & Francis Group, an **informa** business

MATLAB® is a trademark of The MathWorks, Inc. and is used with permission. The MathWorks does not warrant the accuracy of the text or exercises in this book. This book's use or discussion of MATLAB® software or related products does not constitute endorsement or sponsorship by The MathWorks of a particular pedagogical approach or particular use of the MATLAB® software.

First edition published 2021
by CRC Press
6000 Broken Sound Parkway NW, Suite 300, Boca Raton, FL 33487-2742

and by CRC Press
2 Park Square, Milton Park, Abingdon, Oxon, OX14 4RN

© 2021 Taylor & Francis Group, LLC

CRC Press is an imprint of Taylor & Francis Group, LLC

ISBN: 978-0-8153-6981-3 (hbk)
ISBN: 978-1-351-24883-9 (ebk)

Typeset in Times
by SPi Global, India

*To*
*Ulviye, Deniz and Sema*
*and in memory of Galib Hajiyev*

*C.H.*

*To*
*my wife Berna*

*H.E.S.*

# Contents

Preface......................................................................................................xv
Author Biographies..................................................................................xxi

**Chapter 1**    Attitude Parameters ...................................................................1

     1.1    Euler Angles ................................................................................2
           1.1.1    Euler Angles for Vector Transformation.................2
           1.1.2    Propagation of Euler Angles by Time ...................3
     1.2    Quaternions.................................................................................4
           1.2.1    Vector Transformation by Quaternions .................5
           1.2.2    Propagation of Quaternions by Time.....................6
           1.2.3    Euler Angles – Quaternions Relationship .............8
     1.3    Gibbs Vector ...............................................................................9
     1.4    Modified Rodrigues Parameters .............................................10
     1.5    Summary....................................................................................11
     References ............................................................................................11

**Chapter 2**    Mathematical Models for Small Satellite Attitude Dynamics and Kinematics ......................................................................13

     2.1    Coordinate Frames.....................................................................13
     2.2    Satellite Attitude Dynamics .....................................................16
     2.3    Disturbance Torques for a Small Satellite ..............................16
           2.3.1    Gravity Gradient Torque.......................................17
           2.3.2    Aerodynamic (Atmospheric) Torque....................18
           2.3.3    Solar Radiation Torque .........................................18
           2.3.4    Magnetic Torque....................................................19
     2.4    Satellite Kinematics..................................................................21
     2.5    Summary....................................................................................22
     References ............................................................................................22

**Chapter 3**    Attitude Sensors ........................................................................25

     3.1    Magnetometers ..........................................................................25
           3.1.1    Search-coil Magnetometer ...................................26
           3.1.2    Fluxgate Magnetometer........................................27
           3.1.3    MEMS Magnetometer ..........................................28
     3.2    Sun Sensors ...............................................................................30
           3.2.1    Analog Sun Sensors ..............................................31
           3.2.2    Sun Presence Sensors ...........................................34
           3.2.3    Digital Sun Sensors ..............................................35

3.3     Earth Horizon Sensors..........................................................37
        3.3.1    Scanning Earth Horizon Sensors................................38
        3.3.2    Static Earth Horizon Sensors.....................................40
3.4     Star Trackers.......................................................................41
3.5     Gyroscopes........................................................................44
        3.5.1    Fiber Optic Gyros.......................................................45
        3.5.2    MEMS Gyros..............................................................47
3.6     Auxiliary Attitude Sensors for Small Satellites....................48
3.7     Summary.............................................................................49
References........................................................................................49

**Chapter 4**  Attitude Sensor Measurement Models.....................................53
4.1     Magnetometer Models.........................................................53
        4.1.1    Magnetometer Measurement Model..........................53
        4.1.2    Models for the Earth's Magnetic Field in
                 the Reference Frame...................................................54
4.2     Sun Sensor Models..............................................................56
        4.2.1    Sun Sensor Measurement Model................................56
        4.2.2    Models for the Sun Direction Vector in
                 the Reference Frame...................................................58
4.3     Earth Horizon Sensor Models..............................................59
        4.3.1    Earth Horizon Sensor Measurement Model..............59
        4.3.2    Models for the Earth Direction Vector in
                 the Reference Frame...................................................61
4.4     Star Tracker Measurement Model........................................61
4.5     Gyro Measurement Model...................................................63
4.6     Summary.............................................................................64
Notes...............................................................................................64
References........................................................................................64

**Chapter 5**  Attitude Determination Using Two Vector Measurements –
              TRIAD Method ...................................................................67
5.1     TRIAD Method....................................................................67
        5.1.1    TRIAD Algorithm Using Magnetometer and Sun
                 Sensor Measurements................................................69
        5.1.2    Quaternion Estimates from the Attitude Matrix........70
        5.1.3    Using TRIAD Algorithm for More Than Two
                 Vector Measurements................................................72
5.2     Analysis of the TRIAD Method Accuracy............................73
5.3     Increasing Accuracy of the TRIAD Method Using
        Redundancy Techniques......................................................77
5.4     Conclusion and Discussion..................................................79
Note................................................................................................80
References........................................................................................80

**Chapter 6**   Statistical Methods for Three-Axis Attitude Determination ............ 81

     6.1   What is Wahba's Problem?...................................................82
     6.2   Davenport's q-Method .........................................................82
     6.3   QUEST Method ....................................................................84
     6.4   SVD Method .........................................................................87
     6.5   A Brief Comparison of Statistical Methods for Small
         Satellite Implementations .....................................................87
     6.6   Attitude Determination Using GNSS Measurements.............88
     6.7   Conclusion and Discussion...................................................90
     Notes................................................................................................91
     References ........................................................................................91

**Chapter 7**   Kalman Filtering ...................................................................93

     7.1   The Optimal Discrete LKF Derivation .................................94
     7.2   Stability of Optimal LKF.....................................................100
     7.3   LKF in Case of Correlated System and
         Measurement Noise .............................................................101
     7.4   Discrete Kalman Filtering when System and
         Measurement Noises are not Zero-mean Processes ............. 104
     7.5   Divergence in the Kalman Filter and the Methods
         Against Divergence..............................................................105
     7.6   Linearized Kalman Filter.....................................................107
     7.7   Extended Kalman Filter.......................................................109
     7.8   Unscented Kalman Filter......................................................111
     7.9   Other Nonlinear Filtering Algorithms .................................113
     7.10  Conclusion and Discussion..................................................114
     References ......................................................................................114

**Chapter 8**   Adaptive Kalman Filtering ..................................................117

     8.1   *A Priori* Uncertainty and Adaptation...................................119
         8.1.1   *A Priori* Uncertainty.................................................119
         8.1.2   Adaptation ..................................................................120
     8.2   Multiple Model Based Adaptive Estimation..........................121
     8.3   Adaptive Kalman Filtering with Noise Covariance
         Estimation............................................................................124
         8.3.1   Innovation Based Adaptive Estimation ....................124
         8.3.2   Innovation Based Adaptive Filtration
              Algorithm for Stationary Systems.........................125
         8.3.3   Innovation Based Adaptive Filtration
              Algorithm with Feedback.......................................127
         8.3.4   Residual Based Adaptive Estimation.......................128
         8.3.5   Drawbacks of Adaptive Noise Covariance
              Estimation Methods................................................128

8.4     Adaptive Kalman Filtering with Noise Covariance
        Scaling ...................................................................................... 129
        8.4.1     Innovation Based Adaptive Scaling ......................... 129
                  8.4.1.1     R-Adaptation ............................................. 129
                  8.4.1.2     Q-Adaptation ............................................. 131
        8.4.2     Residual Based Adaptive Scaling ............................ 132
8.5     Simplified RKF Against Measurement Faults ...................... 132
8.6     Conclusion .................................................................................. 136
References ............................................................................................ 137

**Chapter 9**     Kalman Filtering for Small Satellite Attitude Estimation ................ 139

9.1     Gyro-based and Dynamics-based Attitude Filtering ............. 140
9.2     Attitude Filtering Using Euler Angles .................................... 141
9.3     Attitude Filtering Using Quaternions ..................................... 141
        9.3.1     Method of Quasi-measurement ................................ 142
        9.3.2     Norm-constrained Kalman Filtering ........................ 143
        9.3.3     Multiplicative Extended Kalman Filter ................... 143
        9.3.4     Unscented Attitude Filtering ................................... 148
9.4     Estimation of Additional Dynamics Parameters ................... 152
        9.4.1     Disturbance Torque Estimation ............................... 153
        9.4.2     Residual Magnetic Moment Estimation ................... 155
9.5     Issues Related to the Attitude Filter's
        Computational Load ................................................................ 159
9.6     Conclusion and Discussion ..................................................... 160
Note ..................................................................................................... 160
References ............................................................................................ 160

**Chapter 10**    Integration of Single-Frame Methods with Filtering
               Algorithms for Attitude Estimation ..................................... 163

10.1    Integration When Attitude is Represented Using
        Euler Angles ............................................................................ 165
10.2    Integration When Attitude is Represented Using
        Quaternions ............................................................................. 168
10.3    Conclusion and Discussion ..................................................... 171
References ............................................................................................ 172

**Chapter 11**    Active Fault Tolerant Attitude Estimation ..................................... 175

11.1    The Innovation and Its Statistical Properties ........................ 177
11.2    Innovation Approach Based Sensor FDI ................................ 179
        11.2.1    Fault Detection via Mathematical Expectation
                  Statistics of Spectral Norm of Normalized
                  Innovation Matrix ..................................................... 179

11.2.2  Innovation Based Sensor Fault Isolation .................. 182

11.2.3  Simulation Results of FDI Algorithms ..................... 185

11.3  Kalman Filter Reconfiguration ............................................. 188

11.3.1  Demonstration for EKF Reconfiguration ................. 191

11.4  The Structure of the Fault Tolerant Attitude Estimation
System ......................................................................................... 194

11.5  Conclusion and Discussions .................................................. 195

Note ......................................................................................................... 196

References ............................................................................................. 196


**Chapter 12**  Fault Tolerant Attitude Estimation: R-Adaptation Methods ............. 199

12.1  Robust Unscented Kalman Filter ........................................ 200

12.1.1  Adapting the R-matrix of UKF Using a Single
Scale Factor ............................................................. 201

12.1.2  Adapting the R-matrix of UKF Using Multiple
Scale Factors ........................................................... 202

12.2  Robust Extended Kalman Filter ......................................... 203

12.2.1  Adapting the R-matrix of EKF Using a Single
Scale Factor ............................................................. 203

12.2.2  Adapting the R-matrix of EKF Using Multiple
Scale Factors ........................................................... 204

12.3  Fault Detection ....................................................................... 204

12.4  Remark on Stability of the Robust Kalman Filters ............. 205

12.5  Demonstrations of REKF and RUKF for Attitude
Estimation of a Small Satellite ........................................... 206

12.5.1  Continuous Bias in Measurements ......................... 207

12.5.2  Measurement Noise Increment ............................... 210

12.5.3  Zero-output Failure ................................................ 211

12.5.4  Discussion on the Implementation of Robust
Attitude Filters on Real Small Satellite Missions .... 214

12.5.5  Comparison of Reconfigured UKF and RUKF
in the Presence of Measurement Faults ................... 216

12.6  Conclusion and Discussion .................................................. 217

References ............................................................................................. 218


**Chapter 13**  Fault Tolerant Attitude Estimation: Q-Adaptation Methods ........... 221

13.1  Adaptation of a Gyro-Based Attitude Filter ......................... 222

13.1.1  Intuitive Tuning of an Attitude Filter ..................... 222

13.1.2  Process Noise Covariance Matrix Estimation
for MEKF ................................................................. 224

13.1.3  Process Noise Covariance Matrix Estimation
for UKF ..................................................................... 226

13.1.4   Demonstration of Adaptive UKF for
Augmented States ........................................................228
13.2   Adaptation of a Dynamics-Based Attitude Filter .................231
13.2.1   Adaptive Fading UKF ...........................................232
13.2.2   Demonstrations for an Adaptive Fading UKF .........234
13.2.2.1   Temporary Uncertainty in Dynamics .......235
13.2.2.2   Permanent Uncertainty in Dynamics ........237
13.3   Conclusion and Discussion ................................................239
Notes ..........................................................................................239
References ..................................................................................239

**Chapter 14**   Integration of R- and Q-Adaptation Methods ..................................241

14.1   Integration R- and Q-Adaptation Methods by
Fault Isolation ..................................................................242
14.1.1   Integration of R- and Q Adaptation Methods ...........242
14.1.2   Demonstration of R- and Q-Adaptation Methods ....244
14.2   Simultaneous Q and R Adaptation ........................................247
14.3   Nontraditional Attitude Filtering with Q-Adaptation ............251
14.3.1   Adapting the Q-matrix in Nontraditional Filter .......253
14.3.2   Demonstrations of SVD/AUKF .................................255
14.3.2.1   SVD/AUKF with Continuous Bias
at Measurements .........................................256
14.3.2.2   SVD/AUKF with Measurement
Noise Increment .........................................257
14.3.2.3   SVD/AUKF with System Change ............258
14.4   Conclusion and Discussion ................................................262
Note ...........................................................................................263
References ..................................................................................263

**Chapter 15**   In-Orbit Calibration of Small Satellite Magnetometers:
Batch Calibration Algorithms ..........................................................265

15.1   Requirement for Magnetometer Calibration .........................265
15.2   Magnetometer Errors in Detail .............................................266
15.2.1   Soft Iron Error .........................................................266
15.2.2   Hard Iron and Null-shift Error .................................267
15.2.3   Time-varying Bias ....................................................268
15.2.4   Scaling .....................................................................268
15.2.5   Nonorthogonality .....................................................268
15.2.6   Misalignment ...........................................................269
15.3   Magnetometer Measurement Models ....................................269
15.3.1   General Model .........................................................269
15.3.2   Models Considering Time-varying Errors ................271

15.4 Batch Magnetometer Calibration Algorithms ...................... 271
    15.4.1 The Cost Function ................................................. 272
    15.4.2 The Minimization Algorithm ................................ 273
    15.4.3 Discussion on Batch Magnetometer Calibration...... 273
15.5 Conclusion ........................................................................ 274
References ................................................................................. 275

**Chapter 16** In-Orbit Calibration of Small Satellite Magnetometers:
Recursive Calibration Algorithms ................................................ 277

16.1 Simultaneous Attitude Estimation and Magnetometer
    Calibration ........................................................................ 278
16.2 UKF for Simultaneous Attitude Estimation and
    Magnetometer Calibration ................................................. 278
    16.2.1 Scale Factor Estimation ...................................... 279
    16.2.2 Demonstration of UKF for Simultaneous Attitude
           Estimation and Magnetometer Calibration.............. 280
16.3 Reconfigurable UKF for Simultaneous Attitude
    Estimation and Magnetometer Calibration........................... 282
    16.3.1 Stopping Rule for Bias Estimation......................... 285
         16.3.1.1 Stopping Rule Formation ........................ 285
         16.3.1.2 Computation of the Covariance
                  Matrix of the Discrepancy Between
                  Two Successive Bias Estimates............... 286
    16.3.2 Demonstration of Reconfigured UKF for
         Simultaneous Attitude Estimation and
         Magnetometer Calibration....................................... 287
16.4 Two-stage UKF for Simultaneous Attitude Estimation
    and Magnetometer Calibration ............................................ 289
    16.4.1 Two-Stage Estimation Procedure ........................... 290
         16.4.1.1 Magnetometer Bias Estimation Stage ...... 290
         16.4.1.2 Gyro Bias Estimation Stage .................... 291
         16.4.1.3 Overall Look to Two-Stage Estimation
                  Scheme ................................................... 291
    16.4.2 Demonstration of Two-stage UKF for
         Simultaneous Attitude Estimation and
         Magnetometer Calibration....................................... 292
         16.4.2.1 Simulation Results for Magnetometer
                  Bias Estimation ...................................... 292
         16.4.2.2 Simulation Results for Gyro Bias
                  Estimation............................................... 293
16.5 Magnetometer Calibration with Known Attitude ................. 293
    16.5.1 Magnetometer Bias and Scale Factor Estimation
         Using a Linear KF ................................................. 296

16.5.2   Simulation Results for Magnetometer Calibration
                via LKF ..................................................................299
16.6   TRIAD+UKF Approach for Attitude Estimation and
         Magnetometer Calibration.....................................................302
16.7   Calibration without Attitude.................................................305
16.8   Magnetometer Bias Estimation for a Spinning Small
         Spacecraft ..........................................................................309
16.9   Discussion on Recursive Magnetometer Calibration ............312
16.10 Conclusion...........................................................................313
References ......................................................................................314

**Index**..............................................................................................317

# Preface

Today we are seeking faster progress in space activities. New mission concepts led by cheap and affordable small satellites are expanding the possibility of space research to more people. As a consequence, spacecraft attitude determination and control (ADC) has become an even more attractive research field. Despite the shrinking sensors and actuators, we need to propose solutions for ADC systems that are as accurate as the ones for larger spacecrafts. Interesting problems include, but are not limited to small, highly capable ADC instrumentation enabling the acquisition of high-quality scientific and exploration information, algorithm design to enable higher performance over reasonable mission durations and ADC subsystems and algorithms to operate a swarm of small satellites in constellation.

One of the major challenges for small satellite ADC system is, surely, the vulnerability of the system to the faults. At large, a small satellite is more open to system faults with its small, commercial off-the-shelf equipment, which are used without comprehensive testing procedures. The main aim of this book is to make a contribution to progress in small satellite technology, especially in the sense of fault tolerant attitude estimation of satellite.

Fault tolerant estimation is investigated in two subcategories like fault tolerant control: active and passive. In the passive category, attitude estimation system continues to operate with the same robust and/or adaptive estimator in case of fault and dissimilarly with the active fault tolerant estimation, structure of the estimator is not changed. The active category involves an online redesign of the attitude estimator after failure has occurred and has been detected.

In this book, both the passive and active fault tolerant estimation methods are considered. The attitude and rate estimation, robust and adaptive Kalman filter algorithms, attitude sensor calibration methods and fault detection and isolation methods for small satellites are illustrated by examples and computer simulations performed via the software package MATLAB®. Most of the book is original and contributed by the authors.

This book is a research monograph in which the theory and application of the proposed methods have been discussed together. It is composed of 16 chapters, as follows:

In **Chapter 1**, we present a brief overview for different parameters that can be used to represent the attitude of a spacecraft. For this purpose, four commonly used techniques, which are Euler angles, quaternions (or Euler symmetric parameters), Gibbs vector and modified Rodrigues parameters, are presented. The characteristics of attitude representations of these techniques are compared and advantages and disadvantages of all methods are discussed. Propagation of Euler angles and quaternions by time which are the basis of attitude kinematics are given.

In **Chapter 2**, the mathematical models for small satellite attitude dynamics and kinematics are presented. First the coordinate frames, which may be of interest when studying attitude dynamics for a spacecraft, are introduced. Then the attitude dynamics equations are given in detail. As a part of the attitude dynamics, the disturbance torques are discussed by considering their dominance on small satellite motion and presenting the necessary equations to model the torques. Last, the satellite kinematics equations are given both in terms of attitude quaternions and Euler angles.

**Chapter 3** presents the functioning principles for small satellite attitude sensors. All of the standard attitude sensors, which are magnetometers, Sun sensors, Earth horizon sensors, star trackers and gyroscopes, are discussed in turn enveloping their different types. Specifically, the newer types of these sensors, mostly based on MEMS (Micro-Electro-Mechanical Systems) technologies and COTS (commercial off-the-shelf) equipment are introduced. In this scope, several recent publications on small satellite attitude determination are surveyed. Schematics and figures for measurement geometries are included to help the readers when investigating how the sensor works.

Sections for each sensor also include a discussion regarding the popularity and suitability of the sensor type for small satellite missions. Example sensors, suitable to be used on small satellites are presented in the associated tables. Last section of the chapter discusses the auxiliary attitude sensors, which can be used instead or in addition to the standard sensors, especially for nanosatellite missions.

In **Chapter 4**, the measurement models for the attitude sensors are presented. For each attitude sensor, first the measurement model is given. Then the procedure to compute the reference direction model is described. These algorithms for reference direction calculation are presented considering they are to be used onboard a small satellite and concerning about their computational load as much as the accuracy.

In this chapter, we also discuss the advantages and disadvantages of using each attitude sensor and the associated models. We provide this information considering the general specifications of each sensor.

In **Chapter 5**, the TRIAD method (two-vector algorithm) is introduced. As the TRIAD method is one of the mostly used attitude determination methods, its error analysis is performed in detail.

In the conduced simulations for a small satellite for attitude estimation using the TRIAD algorithm we use Sun sensor, magnetometer and Earth horizon sensor measurements. In order to increase the attitude determination accuracy, the redundant data processing algorithm, based on the Maximum Likelihood Method, is used.

**Chapter 6** presents the statistical methods for three-axis attitude determination. The aim of these methods is to minimize a well-known cost function that was proposed by Wahba. They provide means for computing an optimal three-axis attitude from multiple vector observations. In specific, we discuss Daventport's q-method, QUEST algorithm and SVD method.

In **Chapter 7**, the Kalman filtering algorithms suitable for estimations for linear and nonlinear systems are considered. The optimal discrete Kalman filter stability, correlated system and measurement noise processes, divergence in the Kalman filter, the methods for numerical stabilization of the filters and other filtering problems are discussed. Filtering when the system and measurement noises are not zero-mean processes is covered.

Three of popular approaches to solve nonlinear estimation problems are the linearized Kalman filter, extended Kalman filter (EKF) and unscented Kalman filter (UKF). The linearized Kalman filter and EKF are linear estimators for a nonlinear system derived by linearization of the nonlinear state and observations equations. Among these, the EKF is an efficient extension to the Kalman filter that was introduced regarding the inherent nonlinearity for many real-world systems and since then it has been widely preferred for various applications including the satellite attitude estimation.

To implement the Kalman filter for nonlinear systems without any linearization step, the unscented transform and so UKF can be used. The UKF algorithm is a more recent filtering method which has many advantages over the well-known EKF. Instead of the analytical linearization used in EKF, UKF statistically captures the system's nonlinearities. The UKF avoids the linearization step of the EKF by introducing sigma points to catch higher order statistic of the system. As a result it satisfies both better estimation accuracy and convergence characteristic.

In **Chapter 8**, the Kalman filter is adapted to ensure robustness against changes in process and measurement noise covariances. The Kalman filter approach to the state estimation is quite sensitive to any malfunctions. If the condition of the real system does not correspond to the models, used in the synthesis of the filter, then these changes significantly decrease the performance of the estimation system.

In real conditions, the statistical characteristics of measurement or system noises may change. It means that estimation algorithm should be adaptive. In such cases the Kalman filter can be adapted and adaptive or robust Kalman filters can be used to recover the possible faults.

In this chapter, first we review the existing estimation-based adaptation techniques for the process and measurement noise covariance matrices to build an adaptive Kalman filter algorithms. In specific, we present the Multiple Model Adaptive Estimation (MMAE) algorithms and noise covariance estimation methods.

Next, we introduce novel noise covariance scaling-based methods for the filter adaptation. By the use of defined variables named scale factor and fading factor, procedures for adapting both the measurement and noise covariance matrices are given.

In **Chapter 9**, we briefly review the details of Kalman filtering for small satellite attitude estimation. Specifically, we discuss different approaches such as dynamics and gyro-based attitude filtering and present different methods to ensure the unit-norm constraint of the quaternions when these parameters are used for attitude representation in the filter. The Multiplicative Extended Kalman Filter and Unscented Kalman Filter algorithms, specifically designed to solve this problem, are presented step-by-step to show the rationale of their design. These algorithms are given in their general form and can be easily modified by the designer, for example if there is need for augmenting the states.

In the last part of the chapter, we also introduce straightforward methods for estimating the attitude dynamics parameters for small satellites, especially to improve the attitude pointing accuracy.

In **Chapter 10**, the single-frame and filtering algorithms are integrated to estimate the attitude of a small satellite. The single-frame method is used to pre-process the vector measurements and provide directly Euler angle or quaternion measurements to the filter block. Using this integrated architecture provides several advantages over the filtering-only methods especially for small satellite application.

In general, the integrated single-frame-aided KF algorithms give more accurate results compared to the filtering-only methods. The integrated algorithm has a robust attitude estimation architecture and much quicker convergence than the vector-based filtering algorithm. Moreover, this approach significantly reduces the complexity of filter design by allowing the use of linear measurement equations.

An active fault tolerant attitude estimation method is proposed in **Chapter 11**. The active category involves either an online redesign of the attitude estimator after sensor failure has occurred and has been detected. The active fault-tolerant attitude estimation system consists of two basic subsystems: sensor fault detection and isolation, and attitude estimator reconfiguration.

Fault detection and isolation algorithms for small satellite attitude determination system using an approach for checking the statistical characteristics of innovation of EKF are proposed. Assuming that the effect of the faulty sensor on its channel is more significant than the other channels, a sensor isolation method is presented by transforming $n$-dimensional innovation to one-dimensional $n$ innovations. Consequently, a new structure of a fault tolerant attitude estimation system for small satellites based on innovation approach is proposed. The advantages and drawbacks of the proposed fault tolerant structure are analyzed and discussed.

In **Chapter 12**, various robust Kalman filters for small satellite attitude estimation are presented in scope of passive fault tolerant attitude estimators. EKF and UKF are taken into consideration as different Kalman filtering algorithms for nonlinear satellite dynamics and kinematics and robust versions of these filters are presented.

Two simple methods for the measurement noise covariance matrix scaling are introduced. In order to show the clear effects of disregarding only the data of the faulty sensor, we performed the adaptation by using both single and multiple scale factors which are two different approaches for the same problem.

Throughout the chapter, Robust Kalman filter algorithms with single and multiple scale factors are tested for attitude estimation problem of a small satellite which has only three magnetometers onboard as the attitude sensors. Results of these proposed algorithms are compared and discussed for different types of magnetometer sensor faults.

In **Chapter 13**, we discuss the details of adapting the process noise covariance (Q) matrix of an attitude filter. We first investigate the applicability of common adaptive methods for gyro-based attitude filtering and provided a performance comparison in between intuitively and adaptively tuned attitude filters. We show that adaptively tuning the attitude filter can be specifically advantageous if additional states are being estimated together with the attitude and gyro biases. We demonstrate the performance of adaptive attitude filter when both the EKF and UKF are used as the main filtering algorithm. We specifically analyze the attitude filter's performance when the process noise covariance matrix is tuned for optimal performance or adapted in real-time.

In the second part of this chapter, we show that it is not possible to get precise estimation results by regular UKF if the model for the process noise covariance mismatches with the real value as a result of change occurred somehow. On the other hand, the presented Adaptive Fading UKF algorithm for dynamics-based attitude estimation is not affected from those changes and secures its good estimation characteristic all the time.

In **Chapter 14**, simultaneous adaptation of the process and measurement noise covariance matrices for the traditional and nontraditional attitude filtering algorithms is discussed. By making use of the R- and Q-adaptation methods, three different integration schemes are considered.

First of these integration schemes is making use of a fault isolation block. In this scheme, at first the fault type is identified and then the necessary adaptation (either R or Q) is applied. In this case, it is simply assumed that the faults in the system (dynamics) and measurements are not concurrent. The R- and Q-adaptation takes place in different times.

The second adaptation scheme is adaptively tuning the Q matrix all the time and adapting the R matrix whenever a measurement malfunction is detected. The Q-adaptation method estimates the Q matrix based on the residual covariance and the adaptation method for the R matrix is an innovation covariance-based scaling method, not the direct estimation of the matrix itself. This integrated scheme is especially useful when we adaptively tune the attitude filter (presumably a gyro-based one) and still make it robust against the measurement faults.

In the last integration method, we use the nontraditional approach for attitude filtering. In this case, the attitude filter is inherently robust against the measurement faults. We adapt only the Q matrix for satisfying the system fault tolerance in the filter. We show that the Q-adaptation method is rather simpler for a nontraditional attitude filter since the measurement model becomes linear. A comprehensive discussion on the fault tolerance of the filter to different types of fault is provided.

In **Chapter 15**, we first review the magnetometer errors and present possible extensions to the general magnetometer model for considering the time-varying errors. Then we discuss the batch methods for magnetometer calibration and their capability for calibration of a small satellite magnetometer. A special attention was given to the optimization algorithms, which are required for cost-function minimization in batch calibration.

In **Chapter 16**, we review the recursive magnetometer calibration algorithms which are capable of sensing the time-variation in the estimated error terms. We categorize the recursive magnetometer calibration algorithms into two: attitude-dependent and attitude-independent algorithms. Studies that propose different filtering algorithms (e.g. EKF, UKF) and different approaches for calibration are reviewed.

A reconfigurable UKF-based estimation algorithm for magnetometer biases and scale factors is presented as a part of the attitude estimation scheme of a picosatellite. In this approach, scale factors are not treated together with other parameters as a part of the state vector; they are estimated separately. After satisfying the conditions of the proposed stopping rule for bias estimation, UKF reconfigures itself for estimation of attitude parameters alone.

As another example for simultaneous attitude estimation and magnetometer calibration, a two-stage algorithm is given. At the initial phase, biases of three orthogonally located magnetometers are estimated as well as the attitude and attitude rates of the satellite. During initial period after the orbit injection, gyro measurements are accepted as bias free. At the second phase, estimated constant magnetometer bias components are taken into account and the algorithm is run for the estimation of the gyro biases. As a result, six different bias terms for two different sensors are obtained in two stages where attitude and attitude rates are estimated regularly.

A linear Kalman filter-based algorithm for the estimation of magnetometer biases and scale factors is presented as the magnetometer calibration method with known attitude. For this method, the attitude transformation matrix is needed to transform the modeled magnetic field vector in orbit frame to the body frame. Measurements from other attitude sensors (e.g. star trackers) are thus essential for this bias calibration algorithm.

In TRIAD+UKF approach for attitude estimation and magnetometer calibration, it is shown that even coarse attitude estimates provided by the TRIAD can be used to form a calibration algorithm with known attitude. The proposed approach is used for concurrent magnetometer calibration and attitude estimation.

Last, attitude-independent recursive calibration algorithms are introduced. These algorithms rely on the scalar check for the magnitude difference for the body frame measurements and the modeled magnetic field.

We believe this monograph will be useful for both researchers in academia and professional engineers in the aerospace industry. It can be also an important reference book for graduate research students. Moreover, Chapters 1, 2, 3, 4, 5 and 6 can be used by undergraduate students studying aerospace engineering and Chapters 7 and 8 can be useful for control and systems engineering students.

We hope that this book will be helpful to researchers, students, our colleagues and all the readers.

**Chingiz Hajiyev**

**Halil Ersin Soken**

Istanbul/Ankara, Turkey

MATLAB® is a registered trademark of The MathWorks, Inc. For product information, please contact:

The MathWorks, Inc.
3 Apple Hill Drive
Natick, MA 01760-2098 USA
Tel: 508-647-7000
Fax: 508-647-7001
E-mail: info@mathworks.com
Web: www.mathworks.com

# Author Biographies

**Chingiz Hajiyev** was born in Kalbajar, Azerbaijan Republic. In 1981, he graduated from the Faculty of Automatic Control Systems of Flying Apparatus, Moscow Aviation Institute (Moscow, Russia) with honors diploma in 1981. He received his Ph.D. and D.Sc.(Eng.) degrees in Process Control in 1987 and 1993, respectively.

From 1987 to 1994 he worked as a Scientific Worker, Senior Scientific Worker, Chief of the Information-Measurement Systems Dept. at the Azerbaijanian Scientific and Production Association (ASPA) "Neftgazavtomat". Between 1994 and 1996 he was a Leading Scientific Worker at the Institute of Cybernetics of the Academy of Sciences of Azerbaijan Republic. He was a professor in the Department of Electronically Calculated System Design, Azerbaijan Technical University, where he had been teaching from 1995 to 1996.

Since 1996 he has been with Department of Aeronautics and Astronautics, Istanbul Technical University (Istanbul, Turkey), where he is currently a professor. From December 2016, he is head of the Aeronautical Department of ITU.

He has more than 500 scientific publications including 14 books, 20 book chapters and more than 300 international journal and international conference papers. Eighty-five research papers are published in the Science Citation Index Expanded (SCIE) journals.

He is on the Editorial Boards of a number of international journals. He is a member of the IFAC (International Federation of Automatic Control) Technical Committee on Automatic Control in Aerospace.

His research interest includes system identification, fault detection and isolation, fault tolerant flight control, spacecraft attitude determination and control, Kalman filtering and integrated navigation systems.

**Halil Ersin Soken** was born in Istanbul, Turkey. He graduated *Summa cum Laude* from Astronautical Engineering Department of Istanbul Technical University in 2007. In 2008, he completed his second degree in Aeronautical Engineering. He earned his PhD in Space and Astronautical Science from the Graduate University for Advanced Studies (Japan) in 2013. During his PhD Studies, he was a visiting researcher in Aalborg University (Denmark) for 3 months.

From 2014 to 2018, Dr. Soken was with the Institute of Space and Astronautical Science (ISAS), Japan Aerospace Exploration Agency (JAXA), first as a research associate and then as a system researcher. From 2018 to 2020, he was with TUBITAK Space Technologies Research Institute (Turkey) as a chief researcher and a team member of the Attitude & Orbit Control Group of the institute. He also served as adjunct lecturer at the Aerospace Department of Middle East Technical University (METU). Starting from the end of 2020, he is assistant professor at the Aerospace Department of METU.

His research interests are spacecraft attitude determination and control, spacecraft dynamics and guidance and navigation. He has co-authored 1 book, more than 20 manuscripts, and more than 50 international conference/symposium papers. He is a member of IEEE, AIAA and affiliate of IFAC.

# 1 Attitude Parameters

In his theorem, Leonhard Euler, a Swiss mathematician and physicist, states that "the most general displacement of a rigid body with one fixed point is a rotation about some axis" [1]. To represent this rotation uniquely, at least three parameters are needed. However, there is not a single certain technique to achieve that and one of several representation methods may be used. In many of these techniques, one has to work with more than three parameters.

There are four presented attitude representation methods in this chapter, which are Euler angles, quaternions (or Euler symmetric parameters), Gibbs vector and modified Rodrigues parameters (MRPs). Euler angles and quaternions are commonly used ones for spacecraft attitude representation among them, whereas Gibbs vector and MRPs, which are derived from quaternions, are especially favorable in attitude filtering. Which one of these four representation methods may be preferred for the attitude representation of the small satellite is a question that should be answered, depending on the estimation algorithm and also the problem itself. Depending on the type of the spacecraft (e.g. spinning spacecraft, 3-axis stabilized spacecraft), mission of the spacecraft (e.g. Earth observation, communication) and also the selection of the reference frame that we define the attitude with respect to (e.g. orbit frame, inertial frame – see Figure 1.1), one of these techniques may be more convenient than others. Table 1.1 presents a brief comparison of four attitude representation methods: Euler angles, quaternions, Gibbs vector and MRPs [1, 2].

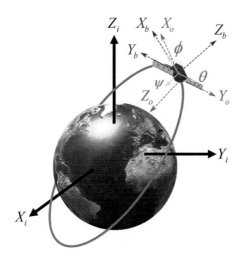

**FIGURE 1.1**   Coordinate frames and definition of Euler angles for a spacecraft.

**TABLE 1.1**

**Characteristics of Attitude Representations of Euler Angles and Quaternions**

| Representation | Number of Parameters | Advantages | Disadvantages |
|---|---|---|---|
| Euler Angles | 3 | – No redundant parameters.<br>– Clear physical interpretation.<br>– Minimal set for attitude representation. | – Trigonometric functions in both rotation matrix and kinematic relations.<br>– Singular for specific rotations.<br>– No convenient product rule. |
| Quaternions | 4 | – Convenient product rule.<br>– Simple kinematic relation.<br>– No trigonometric functions.<br>– No singularities. | – For paramaters do not have intuitive physical meaning.<br>– One redundant parameter. |
| Gibbs Vector | 3 | – No redundant parameters.<br>– There is no sign ambiguity in the definition.<br>– The components are independent parameters. | – It becomes infinite when the rotation angle is an odd multiple of 180°.<br>– No clear physical interpretation. |
| Modified Rodrigues Parameters | 3 | – No redundant parameters.<br>– The singularity is never encountered in practice. | – Singular at $2\pi$ |

## 1.1 EULER ANGLES

A transformation from one coordinate frame to another can be carried out by three consecutive rotations about different axes. While describing the rotation of an axis with respect to another, rotation matrixes formed by Euler angles are used. The direction cosine matrix of transformation in between the coordinate frames will be the product of these three matrices.

There are 12 possible Euler angle sequences and direction cosine matrixes that can be used for exactly the same transformation [1, 3]. They are categorized into two types:

*Type 1:* Case where three successive rotations take place around three different axes (asymmetric sequences).

*Type 2:* In this case first and third rotations are performed around same axis and the second one takes place about one of the other two axes (symmetric sequences).

### 1.1.1 EULER ANGLES FOR VECTOR TRANSFORMATION

Suppose that $\psi \rightarrow \theta \rightarrow \phi$ rotation order about $z$, $y$ and $x$, which may be also referred as 3-2-1 Euler angle rotation, is followed. This means a rotation $\psi$ about $z$ axis and a rotation matrix of

$$A_3 = \begin{bmatrix} \cos(\psi) & \sin(\psi) & 0 \\ -\sin(\psi) & \cos(\psi) & 0 \\ 0 & 0 & 1 \end{bmatrix}, \tag{1.1}$$

a rotation $\theta$ about axis $y$ and a rotation matrix of

$$A_2 = \begin{bmatrix} \cos(\theta) & 0 & -\sin(\theta) \\ 0 & 1 & 0 \\ \sin(\theta) & 0 & \cos(\theta) \end{bmatrix} \tag{1.2}$$

and a rotation $\phi$ about axis $x$ and a rotation matrix of

$$A_1 = \begin{bmatrix} 1 & 0 & 0 \\ 0 & \cos(\phi) & \sin(\phi) \\ 0 & -\sin(\phi) & \cos(\phi) \end{bmatrix}. \tag{1.3}$$

Then the direction cosine matrix that is defining the attitude of the body frame with respect to the chosen reference frame can be obtained as the product of these three matrices.

$$\begin{aligned} A_{321} &= A_3 A_2 A_1 \\ &= \begin{bmatrix} c(\theta)c(\psi) & c(\theta)s(\psi) & -s(\theta) \\ -c(\phi)s(\psi)+s(\phi)s(\theta)c(\psi) & c(\phi)c(\psi)+s(\phi)s(\theta)s(\psi) & c(\theta)s(\phi) \\ s(\phi)s(\psi)+c(\phi)s(\theta)c(\psi) & -s(\phi)c(\psi)+c(\phi)s(\theta)s(\psi) & c(\theta)c(\phi) \end{bmatrix} \end{aligned} \tag{1.4}$$

Here $c(\cdot)$ and $s(\cdot)$ represent the cosine and sine functions. The rotation matrix, which transforms a vector from body to reference frame, is simply the transpose of this matrix as $A_{321}^T = A_3^T A_2^T A_1^T$

Besides, for the small angle rotations, the sine functions can be approximated as, $\sin(\theta) \approx \theta$, $\sin(\phi) \approx \phi$, $\sin(\psi) \approx \psi$ as well as the cosine functions approaches to the unity. When these approximations are used and the products of angles are ignored, the skew symmetric direction cosine matrix for small angles can be obtained:

$$A_{321} \approx \begin{bmatrix} 1 & \psi & -\theta \\ -\psi & 1 & \phi \\ \theta & -\phi & 1 \end{bmatrix}. \tag{1.5}$$

## 1.1.2 PROPAGATION OF EULER ANGLES BY TIME

In order to find the kinematic equations, which relate the Euler angles with the angular velocities in body frame with respect to the reference frame ($\omega_{br} = [\omega_1 \ \omega_2 \ \omega_3]$), first, derivatives of the Euler angles must be transformed to the body angular rates.

$$\begin{bmatrix} \omega_1 \\ \omega_2 \\ \omega_3 \end{bmatrix} = A_1 A_2 A_3 \begin{bmatrix} 0 \\ 0 \\ \dot{\psi} \end{bmatrix} + A_1 A_2 \begin{bmatrix} 0 \\ \dot{\theta} \\ 0 \end{bmatrix} + A_1 \begin{bmatrix} \dot{\phi} \\ 0 \\ 0 \end{bmatrix}. \tag{1.6}$$

After the matrix multiplications:

$$\omega_1 = \dot{\phi} - \dot{\psi} \sin(\theta) \tag{1.7a}$$

$$\omega_2 = \dot{\theta} \cos(\phi) + \dot{\psi} \cos(\theta) \sin(\phi) \tag{1.7b}$$

$$\omega_3 = \dot{\psi} \cos(\theta) \cos(\phi) - \dot{\theta} \sin(\phi) \tag{1.7c}$$

If these equations are solved for $\dot{\phi}$, $\dot{\psi}$ and $\dot{\theta}$ the kinematic equations in Euler angles can be found.

$$\dot{\phi} = \omega_1 + \sin(\phi) \tan(\theta) \omega_2 + \cos(\phi) \tan(\theta) \omega_3 \tag{1.8a}$$

$$\dot{\theta} = \cos(\phi) \omega_2 - \sin(\phi) \omega_3 \tag{1.8b}$$

$$\dot{\psi} = \sin(\phi) \omega_2 / \cos(\theta) + \cos(\phi) \omega_3 / \cos(\theta) \tag{1.8c}$$

## 1.2 QUATERNIONS

The quaternion attitude representation is a technique based on the idea that a transformation from one coordinate frame to another may be performed by a single rotation about a vector $e$ defined with respect to the reference frame. The quaternion, denoted here by the symbol $q$, is a four element vector, the elements of which are functions of the vector $e$ and the angle of rotation, $\Phi$:

$$q_1 = e_1 \sin \frac{\Phi}{2} \tag{1.9a}$$

$$q_2 = e_2 \sin \frac{\Phi}{2} \tag{1.9b}$$

$$q_3 = e_3 \sin \frac{\Phi}{2} \tag{1.9c}$$

$$q_4 = \cos \frac{\Phi}{2} \tag{1.9d}$$

Here $e_1$, $e_2$, $e_3$ are the components of the vector $e$. The vector which shall be transformed is rotated around $e$ with an angle of $\Phi$. As a result, using the quaternions a transfer from reference frame to body frame can be denoted by a single rotation around a vector defined in the reference frame (Figure 1.2).

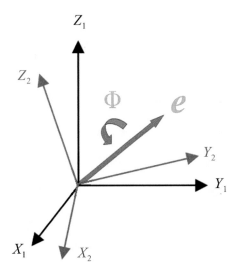

**FIGURE 1.2** Quaternion for defining transformation in between different frames.

A quaternion with components $q_1$, $q_2$, $q_3$ and $q_4$ may also be expressed as a four parameter complex number with a real component $q_4$ and three imaginary components, $q_1$, $q_2$ and $q_3$ as follows:

$$q = q_4 + iq_1 + jq_2 + kq_3, \tag{1.10}$$

where $i, j, k$ are hyper-imaginary numbers with the characteristics of

$$i^2 = j^2 = k^2 = -1 \tag{1.11}$$

$$ij = -ji = -k \tag{1.12a}$$

$$jk = -kj = -i \tag{1.12b}$$

$$ki = -ik = -j \tag{1.12c}$$

Note that this definition of quaternions does not correspond with the Hamilton's standard notation. Here we use rather the notation defined in [4] and most commonly used in attitude analysis.

As stated in Table 1.1, quaternion is not the minimal set for attitude representation and having four parameters to represent three rotations brings about 1 constraint on quaternion parameters. This redundancy of the quaternions must be noted as:

$$q_1^2 + q_2^2 + q_3^2 + q_4^2 = 1. \tag{1.13}$$

### 1.2.1 Vector Transformation by Quaternions

A vector quantity defined in the body axes, $r_B$, may be expressed in the reference axes as $r_R$ using the quaternions directly. First define a quaternion, $r_B^q$, in which the complex components are set equal to the components of $r_B$, and with a zero scalar component, that is, if

$$r_B = ix + jy + kz, \tag{1.14a}$$

$$r_B^q = 0 + ix + jy + kz. \tag{1.14b}$$

This is expressed in the reference axes as $r_R^q$ using:

$$r_R^q = q\, r_B^q\, q^*, \tag{1.15}$$

where $q^* = (q_4 - iq_1 - jq_2 - kq_3)$ is the complex conjugate of $q$.

Hence

$$
\begin{aligned}
r_R^q =& (q_4 + iq_1 + jq_2 + kq_3)(0 + ix + jy + kz)(q_4 - iq_1 - jq_2 - kq_3) \\
=& 0 + \left\{ \left(q_4^2 + q_1^2 - q_2^2 - q_3^2\right)x + 2(q_1q_2 - q_4q_3)y + 2(q_1q_3 + q_4q_2)z \right\} i \\
&+ \left\{ 2(q_1q_2 + q_4q_3)x + \left(q_4^2 - q_1^2 + q_2^2 - q_3^2\right)y + 2(q_2q_3 - q_4q_1)z \right\} j \\
&+ \left\{ 2(q_1q_3 - q_4q_2)x + 2(q_2q_3 + q_4q_1)y + \left(q_4^2 - q_1^2 + q_2^2 - q_3^2\right)z \right\} k
\end{aligned}
\tag{1.16}
$$

Alternatively, $r_R^q$ may be expressed in matrix form as follows:

$$r_R^q = A' r_B^q, \tag{1.17}$$

where $A' = \begin{bmatrix} 0 & 0 \\ 0 & A^T \end{bmatrix}$, $r_R^q = \begin{bmatrix} 0 & r_B \end{bmatrix}$ and

$$
A = \begin{bmatrix}
q_1^2 - q_2^2 - q_3^2 + q_4^2 & 2(q_1q_2 + q_3q_4) & 2(q_1q_3 - q_2q_4) \\
2(q_1q_2 - q_3q_4) & -q_1^2 + q_2^2 - q_3^2 + q_4^2 & 2(q_2q_3 + q_1q_4) \\
2(q_1q_3 + q_2q_4) & 2(q_2q_3 - q_1q_4) & -q_1^2 - q_2^2 + q_3^2 + q_4^2
\end{bmatrix}, \tag{1.18}
$$

which is equivalent to writing:

$$r_B = A r_R. \tag{1.19}$$

Here $A$ is the same direction cosine matrix that is used for transformation from body to reference frame.

## 1.2.2 Propagation of Quaternions by Time

While defining the kinematic equations of motion in quaternions, time dependence of them must be used. This time dependence can be derived from the product relation [1].

Multiplication of quaternion is performed in a way not too different from complex number multiplications. However, the order of the process is important. By using the characteristic of hyper-imaginary numbers:

$$\boldsymbol{q}'' = \boldsymbol{q}\boldsymbol{q}' = (q_4 + iq_1 + jq_2 + kq_3)(q_4' + iq_1' + jq_2' + kq_3'), \qquad (1.20)$$

$$\boldsymbol{q}'' = \left(-q_1q_1' - q_2q_2' - q_3q_3' + q_4q_4'\right)i\left(q_1q_4' + q_2q_3' - q_3q_2' + q_4q_1'\right)$$
$$+ j\left(-q_1q_3' + q_2q_4' + q_3q_1' + q_4q_2'\right) + k\left(q_1q_2' - q_2q_1' + q_3q_4' + q_4q_3'\right). \qquad (1.21)$$

If it is written in matrix form,

$$\begin{bmatrix} q_1'' \\ q_2'' \\ q_3'' \\ q_4'' \end{bmatrix} = \begin{bmatrix} q_4' & q_3' & -q_2' & q_1' \\ -q_3' & q_4' & q_1' & q_2' \\ q_2' & -q_1' & q_4' & q_3' \\ -q_1' & -q_2' & -q_3' & q_4' \end{bmatrix} \begin{bmatrix} q_1 \\ q_2 \\ q_3 \\ q_4 \end{bmatrix}. \qquad (1.22)$$

Now assume that, $\boldsymbol{q}$ and $\boldsymbol{q}''$ correspond to the orientation of the body at $t$ and $t + \Delta t$, respectively. Also $\boldsymbol{q}'$ is for the representation of position at $t + \Delta t$ in a relative way to the position that has been occupied at $t$,

$$q_1' \equiv e_1 \sin \frac{\Delta\Phi}{2}, \qquad (1.23a)$$

$$q_2' \equiv e_2 \sin \frac{\Delta\Phi}{2}, \qquad (1.23b)$$

$$q_3' \equiv e_3 \sin \frac{\Delta\Phi}{2}, \qquad (1.23c)$$

$$q_4' \equiv \cos \frac{\Delta\Phi}{2}. \qquad (1.23d)$$

When the necessary multiplication is done it is obvious that

$$\boldsymbol{q}(t + \Delta t) \approx \left\{ \cos \frac{\Delta\Phi}{2} I + \sin \frac{\Delta\Phi}{2} \begin{bmatrix} 0 & e_3 & -e_2 & e_1 \\ -e_3 & 0 & e_1 & e_2 \\ e_2 & -e_1 & 0 & e_3 \\ -e_1 & -e_2 & -e_3 & 0 \end{bmatrix} \right\} \boldsymbol{q}(t), \qquad (1.24)$$

where $e_1$, $e_2$, $e_3$ are the components of rotation axis unit vector and $I$ is the $4 \times 4$ identity matrix. After that by small angle approximation,

$$\cos \frac{\Delta\Phi}{2} \approx 1, \qquad (1.25a)$$

$$\sin \frac{\Delta\Phi}{2} \approx \frac{1}{2} \omega_{br} \Delta t. \qquad (1.25b)$$

It is possible to show that

$$q(t+\Delta t) \approx \left\{ I + \frac{1}{2} \begin{bmatrix} 0 & \omega_3 & -\omega_2 & \omega_1 \\ -\omega_3 & 0 & \omega_1 & \omega_2 \\ \omega_2 & -\omega_1 & 0 & \omega_3 \\ -\omega_1 & -\omega_2 & -\omega_3 & 0 \end{bmatrix} \Delta t \right\} q(t). \tag{1.26}$$

Hence if a skew-symmetric matrix is defined as

$$\Omega(\omega_{br}) = \begin{bmatrix} 0 & \omega_3 & -\omega_2 & \omega_1 \\ -\omega_3 & 0 & \omega_1 & \omega_2 \\ \omega_2 & -\omega_1 & 0 & \omega_3 \\ -\omega_1 & -\omega_2 & -\omega_3 & 0 \end{bmatrix}, \tag{1.27}$$

the equation becomes

$$q(t+\Delta t) \approx \left\{ I + \frac{1}{2}\Omega(\omega_{br})\Delta t \right\} q(t). \tag{1.28}$$

Finally, it is known that

$$\frac{dq(t)}{dt} \cong \frac{q(t+\Delta t)-q(t)}{\Delta t} = \frac{1}{2}\Omega(\omega_{br})q(t). \tag{1.29}$$

### 1.2.3 EULER ANGLES–QUATERNIONS RELATIONSHIP

In order to understand the physical meaning of the quaternions more easily, they can be related with the Euler angles. The formula for obtaining quaternion vector by the use of the Euler angles is given as

$$\begin{bmatrix} q_1 \\ q_2 \\ q_3 \\ q_4 \end{bmatrix} = \begin{bmatrix} \sin\left(\frac{\phi}{2}\right)\cos\left(\frac{\theta}{2}\right)\cos\left(\frac{\psi}{2}\right) - \cos\left(\frac{\phi}{2}\right)\sin\left(\frac{\theta}{2}\right)\sin\left(\frac{\psi}{2}\right) \\ \cos\left(\frac{\phi}{2}\right)\sin\left(\frac{\theta}{2}\right)\cos\left(\frac{\psi}{2}\right) + \sin\left(\frac{\phi}{2}\right)\cos\left(\frac{\theta}{2}\right)\sin\left(\frac{\psi}{2}\right) \\ \cos\left(\frac{\phi}{2}\right)\cos\left(\frac{\theta}{2}\right)\sin\left(\frac{\psi}{2}\right) - \sin\left(\frac{\phi}{2}\right)\sin\left(\frac{\theta}{2}\right)\cos\left(\frac{\psi}{2}\right) \\ \cos\left(\frac{\phi}{2}\right)\cos\left(\frac{\theta}{2}\right)\cos\left(\frac{\psi}{2}\right) + \sin\left(\frac{\phi}{2}\right)\sin\left(\frac{\theta}{2}\right)\sin\left(\frac{\psi}{2}\right) \end{bmatrix}. \tag{1.30}$$

Note that 3-2-1 sequence is used for representing the attitude in Euler angles here. On the other hand, the equations for finding the Euler angles using the quaternions are

$$\varphi = \text{atan2}\left[2\left(q_4 q_1 + q_2 q_3\right), 1 - 2\left(q_1^2 + q_2^2\right)\right], \tag{1.31a}$$

$$\theta = \text{asin}\left[2\left(q_4 q_2 - q_3 q_1\right)\right], \tag{1.31b}$$

$$\psi = \text{atan2}\left[2\left(q_4 q_3 + q_1 q_2\right), 1 - 2\left(q_2^2 + q_3^2\right)\right], \tag{1.31c}$$

where asin($\cdot$) is the arcsine function and atan2($\cdot$) is the arctangent function with two arguments, which is used to generate all the rotations (not just the rotations between $-\pi/2$ and $\pi/2$) and defined as

$$\text{atan2}(y,x) = \begin{cases} \arctan\left(\dfrac{y}{x}\right) & x > 0 \\[2mm] \arctan\left(\dfrac{y}{x}\right) + \pi & y \geq 0, x < 0 \\[2mm] \arctan\left(\dfrac{y}{x}\right) - \pi & y < 0, x < 0 \\[2mm] +\dfrac{\pi}{2} & y > 0, x = 0 \\[2mm] -\dfrac{\pi}{2} & y < 0, x = 0 \\[2mm] undefined & y = 0, x = 0 \end{cases} \tag{1.32}$$

## 1.3 GIBBS VECTOR

For finite rotations, the Gibbs vector, with the components of $\boldsymbol{g} = \begin{bmatrix} g_1 & g_2 & g_3 \end{bmatrix}^T$, can be defined as [4]

$$g_1 = q_1 / q_4 = e_1 \tan\left(\Phi/2\right)$$
$$g_2 = q_2 / q_4 = e_2 \tan\left(\Phi/2\right) \tag{1.33}$$
$$g_3 = q_3 / q_4 = e_3 \tan\left(\Phi/2\right)$$

on the basis of the quaternion vector.

The transformations matrix $A$ in terms of the Gibbs vector representation is as follows:

$$A = \frac{1}{1 + g_1^2 + g_2^2 + g_3^2} \begin{bmatrix} 1 + g_1^2 - g_2^2 - g_3^2 & 2\left(g_1 g_2 + g_3\right) & 2\left(g_1 g_3 - g_2\right) \\ 2\left(g_1 g_2 - g_3\right) & 1 - g_1^2 + g_2^2 - g_3^2 & 2\left(g_2 g_3 + g_1\right) \\ 2\left(g_1 g_3 + g_2\right) & 2\left(g_2 g_3 - g_1\right) & 1 - g_1^2 - g_2^2 + g_3^2 \end{bmatrix}. \tag{1.34}$$

Since the attitude matrix $A$ is an orthogonal matrix, it must satisfy $A^{-1} = A^T$ equality. Thus $A^T$ can be obtained by replacing $g_i$ by $-g_i$ in Eq. (1.34) for $i = 1, 2, 3$.

Expressions for the Gibbs vector components in terms of the direction cosine matrix elements can be as

$$g_1 = \frac{A_{23} - A_{32}}{1 + A_{11} + A_{22} + A_{33}}, \tag{1.35a}$$

$$g_2 = \frac{A_{31} - A_{13}}{1 + A_{11} + A_{22} + A_{33}}, \tag{1.35b}$$

$$g_3 = \frac{A_{12} - A_{21}}{1 + A_{11} + A_{22} + A_{33}}. \tag{1.35c}$$

Although the Gibbs vector itself is not widely used in aerospace applications, it may be preferred as an attitude error representation, especially when an attitude filter is being designed. As it has minimal set for attitude representation, attitude can be represented without any constraint in this case in contrast to the quaternions. Moreover, since one is working in the domain of attitude errors, singularity for Gibbs vector, which appears when the rotation angle is an odd multiple of 180°, is never experienced in practice.

## 1.4  MODIFIED RODRIGUES PARAMETERS

Another set of attitude parameters, which is especially preferred for attitude error representation in attitude filter design is the MRPs. They are related to the quaternion by [5],

$$\boldsymbol{p} = \frac{\boldsymbol{q}}{1 + q_4} = \boldsymbol{e} \tan\left(\frac{\Phi}{4}\right), \tag{1.36}$$

where $\boldsymbol{q} = \begin{bmatrix} q_1 & q_2 & q_3 \end{bmatrix}$ is the vector and $q_4$ is the scalar component of the quaternion vector, defined as (1.9).

The terms of the MRPs are as follows:

$$p_1 = \frac{q_1}{1 + q_4} = e_1 \tan\left(\frac{\Phi}{4}\right), \tag{1.37a}$$

$$p_2 = \frac{q_2}{1 + q_4} = e_2 \tan\left(\frac{\Phi}{4}\right), \tag{1.37b}$$

$$p_3 = \frac{q_3}{1 + q_4} = e_3 \tan\left(\frac{\Phi}{4}\right). \tag{1.37c}$$

The magnitude of $\boldsymbol{p}$ is given by $|\boldsymbol{p}|$. Then, $|\boldsymbol{p}| \to \infty$ when $\mu \to 2\pi$, meaning that MRPs are singular at $2\pi$. Since we only use this three-component representation for the attitude errors, the singularity is never encountered in practice.

## 1.5 SUMMARY

In this chapter, we presented a brief overview for different parameters that can be used to represent the attitude of a spacecraft. For this purpose, four commonly used techniques, which are Euler angles, quaternions (or Euler symmetric parameters), Gibbs vector and MRPs, are presented. The characteristics of attitude representations of these techniques are compared and advantages and disadvantages of all methods are discussed. Propagation of Euler angles and quaternions by time, which are the basis of attitude kinematics, are given.

## REFERENCES

1. J. R. Wertz, Ed., *Spacecraft Attitude Determination and Control*, vol. 73. Dordrecht: Springer Netherlands, 1978.
2. T. Bak, "Spacecraft Attitude Determination – a magnetometer approach," PhD Thesis, Aalborg University, 1999.
3. J. Diebel, Representing Attitude: Euler Angles, Unit Quaternions, and Rotation Vectors. Stanford, CA: Stanford University, 2006.
4. M. D. Shuster, "A survey of attitude representations," *J. Astronaut. Sci.*, vol. 41, no. 4. pp. 439–517, 1993.
5. F. L. Markley and J. L. Crassidis, *Fundamentals of Spacecraft Attitude Determination and Control*. New York, NY: Springer New York, 2014.

# 2 Mathematical Models for Small Satellite Attitude Dynamics and Kinematics

## 2.1 COORDINATE FRAMES

There are several reference frames which may be of interest when studying attitude dynamics for a spacecraft. In this book three of the most important ones are presented. These are satellite body frame, orbit reference frame and Earth-centered inertial reference frame. The definitions of these coordinate systems are given below.

*Earth-Centered Inertial (ECI) Frame:* The ECI frame is non-accelerated reference frame in which Newton's laws are valid. The origin of the frame is located at the center of the Earth. The $z$ axis shows the geographic North Pole while the $x$ axis is directed toward the Vernal Equinox ($\gamma$ – the point where the Sun crosses the celestial Equator in March on its way from south to north). The $y$ axis completes the coordinate system as the cross product of $z$ and $x$ axes (Figure 2.1a). This reference frame, which is very nearly non-rotating, is a suitable approximation to an inertial frame for the analysis of near-Earth space vehicle trajectories and for their attitude control.

Considering the precession and nutation of the Earth, the vernal equinox, the equatorial plane of the Earth and the ecliptic may vary according to time. Thus, depending on the definition of these parameters, whether we are using the mean or true values at a specific epoch, there are different ECI frames. Most commonly used one for attitude analysis is J2000, which is defined with the Earth's Mean Equator and Equinox at 12:00 on 1 January 2000.

*Orbit Reference Frame:* Specifically, for Earth-pointing spacecraft, it is convenient to define an orbit-centered frame as the reference frame. Even though the attitude of the spacecraft is estimated with respect to the ECI frame, the attitude may be controlled with respect to the orbit reference frame, which makes visualization of the attitude easier.

The origin of the orbit frame is at the mass center of the spacecraft. The $z$ axis is in nadir direction (toward the center of the Earth), the $y$ axis is normal to the orbit plane (in opposite direction to the orbital angular momentum vector) and the $x$ axis completes to the orthogonal right-hand system (aligns with velocity

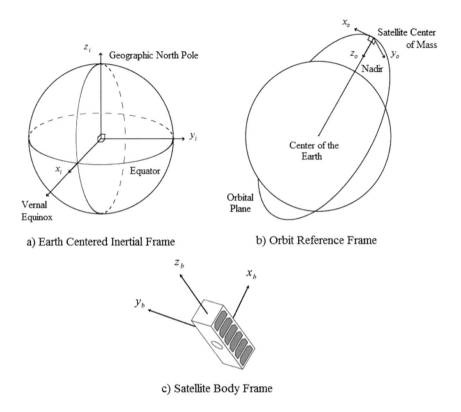

a) Earth Centered Inertial Frame       b) Orbit Reference Frame

c) Satellite Body Frame

**FIGURE 2.1**   Coordinate frames.

vector of the spacecraft in case of circular orbit) (Figure 2.1b). The body frame is aligned with the orbit frame when the satellite has an attitude of 0° in roll, pitch and yaw.

The transformation matrix from the ECI frame to the orbit frame can be defined using the Keplerian orbital elements for the spacecraft (Figure 2.2):

$$
C_{oi} = \begin{bmatrix} -s(u)c(\Omega)-c(u)c(i)s(\Omega) & -s(u)s(\Omega)+c(u)c(i)c(\Omega) & c(u)s(i) \\ -s(i)s(\Omega) & s(i)c(\Omega) & -c(i) \\ c(u)c(\Omega)+s(u)c(i)s(\Omega) & -c(u)s(\Omega)-s(u)c(i)c(\Omega) & -s(u)s(i) \end{bmatrix}.
$$

$$(2.1)$$

Here $u$ is the argument of latitude, which is simply sum of true anomaly ($v$) and argument of perigee ($\omega$) as $u = \omega + v$, $i$ is the inclination angle and $\Omega$ is the right ascension of ascending node.

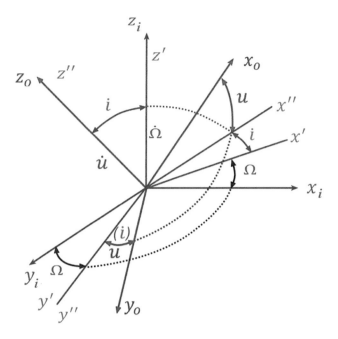

**FIGURE 2.2**   Coordinate transformation from ECI to orbit reference frame.

Another method of transformation is using the position and velocity vectors for the spacecraft which are defined in ECI frame:

$$C_{oi} = \begin{bmatrix} \mathbf{o}_1 & \mathbf{o}_2 & \mathbf{o}_3 \end{bmatrix}^T, \tag{2.2}$$

where the unit vector $\mathbf{o}_1, \mathbf{o}_2, \mathbf{o}_3$ are constituted using the $r$ position vector and $v$ velocity vector in inertial frame:

$$\mathbf{o}_3 = -r / \|r\|, \tag{2.3a}$$

$$\mathbf{o}_2 = -(r \times v) / \|r \times v\|, \tag{2.3b}$$

$$\mathbf{o}_1 = \mathbf{o}_2 \times \mathbf{o}_3. \tag{2.3c}$$

*Satellite Body Frame:* The origin of the frame is located usually at the center of mass (CM) of the satellite and the frame itself is defined referencing the rigid structure of the satellite (Figure 2.1c). All the internal components including the attitude sensors and actuators take this frame as the basis. During the assembly, a special interest is paid to align the axes of the body frame with the principal inertial axes of the spacecraft. Although this is usually difficult for larger spacecraft, for small spacecraft such as nanosatellites alignment can be successful to a great extent as the satellite is mostly a rigid structure and misalignments introduce almost negligible products of inertias.

## 2.2 SATELLITE ATTITUDE DYNAMICS

The fundamental equation of the satellite attitude dynamics relates the time derivative of the angular momentum vector with the overall torque affecting the satellite [1],

$$\frac{d\boldsymbol{L}}{dt} = \boldsymbol{N} - \boldsymbol{\omega}_{bi} \times \boldsymbol{L} = \boldsymbol{J}\frac{d\boldsymbol{\omega}_{bi}}{dt}, \tag{2.4}$$

$$\boldsymbol{L} = \boldsymbol{J}\boldsymbol{\omega}_{bi}, \tag{2.5}$$

where $\boldsymbol{L}$ is the angular momentum vector, $\boldsymbol{N}$ is the torque vector, $\boldsymbol{\omega}_{bi}$ is the angular velocity vector of the body frame with respect to the inertial frame and $\boldsymbol{J}$ is the moment of inertia matrix.

In general, without any necessity to have a diagonal inertia matrix, the dynamic equation of the satellite attitude can be given as

$$\frac{d\boldsymbol{\omega}_{bi}}{dt} = \boldsymbol{J}^{-1}\left[\boldsymbol{N} - \boldsymbol{\omega}_{bi} \times \left(\boldsymbol{J}\boldsymbol{\omega}_{bi}\right)\right]. \tag{2.6}$$

The torque vector $\boldsymbol{N}$ can be defined as the sum of the disturbance torques and control torque acting on the satellites as

$$\boldsymbol{N} = \boldsymbol{N}_d + \boldsymbol{N}_c. \tag{2.7}$$

Here $\boldsymbol{N}_c$ is the control torque and $\boldsymbol{N}_d$ is the vector of disturbance torques affecting the satellite which can be given as a sum of

$$\boldsymbol{N}_d = \boldsymbol{N}_{gg} + \boldsymbol{N}_{ad} + \boldsymbol{N}_{sp} + \boldsymbol{N}_{md}. \tag{2.8}$$

Here $\boldsymbol{N}_{gg}$ is the gravity gradient torque, $\boldsymbol{N}_{ad}$ is the aerodynamic disturbance torque, $\boldsymbol{N}_{sp}$ is the solar pressure disturbance torque and $\boldsymbol{N}_{md}$ is the residual magnetic torque which is caused by the interaction of the satellite's residual dipole and the Earth's magnetic field.

## 2.3 DISTURBANCE TORQUES FOR A SMALL SATELLITE

Different than a large spacecraft, small spacecraft and especially nanosatellites do not have large appendages that introduce uneven inertia distribution and large solar panels that can cause large aerodynamic and solar pressure torques in certain attitudes. In this sense, although the disturbance torque modeling, which will be briefly reviewed in this section, is similar to that for a large spacecraft, the attitude specialist may not need to care for these torques in the design process, except the residual magnetic moment (RMM). The RMM is caused by the interaction of the residual dipole and the Earth's magnetic field and is, generally, the main attitude disturbance source for the low Earth orbit (LEO) small satellites [2, 3]. Other disturbances such as the gravity gradient, sun pressure, aerodynamic drag have relatively less effect because of the small size of the satellite and can be minimized during the design process [4]. Eventually, the magnitude of each disturbance torque depends on the mission orbit and the satellite specifications but experiences show that the RMM can be 2–3 order larger than the other disturbance torques for a small satellite.

## 2.3.1 GRAVITY GRADIENT TORQUE

Gravity gradient torque acting on a satellite can be calculated departing from the Newton's laws of gravitation and taking the satellite as a whole volume orbiting the Earth, rather than a point mass. Within this volume, small differences of distance to the main body of gravitation, which is Earth in our case, cause an acting torque on the satellite. The resulting torque around the CM of the satellite is

$$N_{gg} = \int_m r_m \times a_g dm,$$

(2.9)

where $r_m$ is the position vector of the mass element and $a_g$ is the acceleration acting on the element as

$$a_g = -\frac{\mu}{|r_m|^3} r_m.$$

(2.10)

Here $\mu$ is the gravitational parameter for Earth. If the position of the mass element is defined with respect to the CM of the satellite as $r_m = r + \delta r$, we can solve (2.9) to get the gravity gradient torque equation as

$$N_{gg} = \frac{3\mu}{|r|^3} n \times (Jn).$$

(2.11)

Here, $n$ is the nadir-pointing vector in the satellite body frame. Obviously, we need to know the attitude of the spacecraft to calculate the corresponding terms of the gravity gradient torque. A simple derivation can be done for an Earth-pointing spacecraft if the reference frame that we define the attitude with respect to is selected as the orbit frame. In this case

$$n = A \begin{bmatrix} 0 \\ 0 \\ 1 \end{bmatrix}.$$

(2.12)

Then assuming the body and the principal axes frames are aligned for the space-craft $N_{gg}$ can be given as,

$$N_{gg} = \frac{3\mu}{|r|^3} \begin{bmatrix} (J_3 - J_2) A_{23} A_{33} \\ (J_1 - J_3) A_{13} A_{33} \\ (J_2 - J_1) A_{13} A_{23} \end{bmatrix}. \tag{2.13}$$

$A_{ij}$ represents the corresponding element of the attitude matrix.

It is trivial that for a cube-satellite, with an evenly distributed mass the gravity gradient torque is negligible. For a 3U nanosatellite, the magnitude of the torque may be on the order of $10^{-7}$ Nm [5].

### 2.3.2 AERODYNAMIC (ATMOSPHERIC) TORQUE

At LEOs, where small satellites operate, especially at altitudes of 500 km and below, total atmospheric density is not totally negligible. According to the US Standard Atmosphere 1976 model, it is about $2.8 \times 10^{-12}$ kg/m³ at the altitude of International Space Station, which is approximately 400 km [6]. Together with the altitude it is a strong function of solar activity and whether the Sun is visible or not (orbit day or night) [7]. Furthermore, at these orbits the satellite's velocity is very high. Thus, the fact that the aerodynamic pressure is directly proportional to the air density and the square of the relative air velocity shows that the aerodynamic torque for a small satel-lite may reach to a considerable amount depending on the area of the exposed sur-face. This torque can be even used as a passive attitude stabilization method for a small satellite together with an active magnetic controller [7, 8].

The aerodynamic disturbance torque vector on a spacecraft structure can be obtained by taking the cross product of the aerodynamic pressure vector on the total projected area, and the vector from the CM to the center of pressure (CP) of the total structure,

$$N_{ad} = \frac{1}{2} \rho_a V^2 C_d A_p \left[ \bar{V} \times c_p \right], \tag{2.14}$$

where, $\rho_a$ is the atmospheric density, $V$ is the magnitude of satellite's velocity vector, $\bar{V}$ is the unit velocity vector, $A_p$ is the total projected area of satellite, $c_p$ is the vector between CM and CP and $C_d$ is the drag coefficient.

### 2.3.3 SOLAR RADIATION TORQUE

Solar radiation pressure generally has a major effect at high altitudes of 1000 km and above and may be at a negligible level for most of the LEO small satellites. Yet it should be accounted for especially satellites with a large surface exposed to the solar radiation and a small mass (such as a solar sail nanosatellite [9], which is designed to make use of this pressure itself for orbit and attitude control).

The reasons of solar radiation are photons over a wide spectrum and energetic particles of solar wind which are emitted by the Sun. These particles travel at the speed of light and create a radiation pressure on the surfaces of the satellite that they meet. The solar radiation pressure can be calculated as,

$$P_s = \frac{S}{c} \tag{2.15}$$

where $S = 1367$ W/m$^2$ is the radiation intensity at the Earth's orbit, which is known as solar constant and $c = 2.998 \times 10^8$ m/s is the speed of light. Then the force that is acting on this surface due to the solar radiation will be

$$F_{sr} = -\nu P_s C_r A_p \frac{r_{s/s}}{|r_{s/s}|}. \tag{2.16}$$

Here, $\nu$ is the shadow function which is simply 0 if the satellite is in eclipse and $\nu = 1$ if the satellite is in light. Moreover, $A_p$ is the projected area, $r_{s/s}$ is the vector pointing from satellite toward the Sun and $C_r$ is the radiation pressure coefficient. If the surface is a blackbody $C_r = 1$. When $C_r = 2$ it means that the surface is a mirror-like surface and all the radiation is reflected. A typical value of $C_r$ used for analyses is $C_r \simeq 1.2$.

If the sum of the solar radiation forces does not pass through the CM of satellite, a torque will be produced. For design purposes, the resulting solar radiation forces are assumed to pass through a CP. The distance between the points CP and CM, which is the moment arm for satellite, allows the designers to measure the total radiative torque. Then the solar radiation torque is the sum of all radiation forces acting on each individual part of the satellite:

$$N_{sp} = \sum_k r_{cp,k} \times F_{sr,k}. \tag{2.17}$$

The shadow function in (2.16) can be calculated using the Earth to Sun ($r_s$), Earth to satellite ($r$) vectors in ECI frame and the radius of the Earth ($R_e$). For

$$\theta = \text{acos}\left(\frac{r \cdot r_s}{|r||r_s|}\right); \quad \theta_1 = \text{acos}\left(\frac{R_e}{|r|}\right) \quad \text{and} \quad \theta_2 = \text{acos}\left(\frac{R_e}{|r_s|}\right), \tag{2.18}$$

$$\nu = \begin{cases} 0 & \theta_1 + \theta_2 \leq \theta \\ 1 & \theta_1 + \theta_2 > \theta \end{cases}. \tag{2.19}$$

### 2.3.4 Magnetic Torque

Generally, the main attitude disturbance source for the LEO small satellites is the RMM [2, 10]. Other disturbances such as the gravity gradient, sun pressure and aerodynamic drag have relatively less effect because of the small size of the satellite and can be further minimized during the design process. However, the magnetic disturbance is mainly caused by the onboard electric current loop, small permanent magnet in some devices or some special material on the satellite, and does not strongly depend on the satellite size.

The RMM causes the magnetic disturbance torque by interacting with the surrounding magnetic field. The torque in body frame can be obtained by

$$N_{md} = M \times B_b, \qquad (2.20)$$

Here, $M = \begin{bmatrix} M_x & M_y & M_z \end{bmatrix}^T$ is the RMM vector and $B_b$ is the Earth's magnetic field vector in the body frame, which is measured by the magnetometers.

As mentioned, the RMM does not strongly depend on the satellite size. In-orbit experiences for various small satellite missions prove this fact. For 6.5 SNAP-1 nanosatellite, the estimated RMM was $M = \begin{bmatrix} 0.034 & 0.036 & -0.12 \end{bmatrix}^T$ Am$^2$ [11]. As another example, in-orbit estimated RMM for 70 kg REIMEI satellite was $M = \begin{bmatrix} -0.51 & 0.042 & 0.11 \end{bmatrix}^T$ Am$^2$ [2]. For a LEO satellite at an approximate altitude of 700 km, this level of RMM may cause disturbance torques as large as $2 \times 10^{-5}$ Nm in magnitude (Figure 2.3). Of course, this value increases as the altitude gets lower.

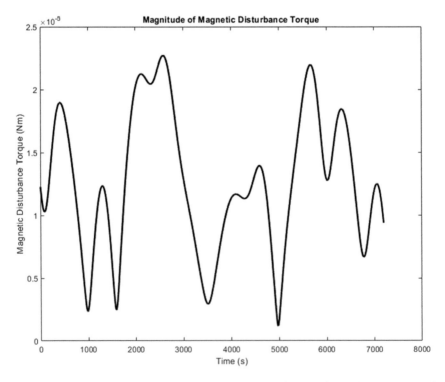

**FIGURE 2.3** Magnitude of magnetic disturbance torque for a small satellite at an approximate altitude of 700km. RMM values for REIMEI satellite are assumed.

Last, for 1U cubesat ESTCube-1 (1 kg) the calculations for the in-orbit values showed that the RMM was as high as $M = \begin{bmatrix} -0.0986 & -0.001 & 0.0953 \end{bmatrix}^T$ Am$^2$ [12]. This value is higher than the one for SNAP-1 in magnitude, although the satellite is approximately 20 times smaller in volume and was also higher than the maximum torque capacity of the onboard magnetic torquers for attitude control. Such large RMM, which could not be compensated, eventually altered the inertial attitude of the satellite.

Usually RMM components are considered to be constant in time. However, in practice, these parameters may change with sudden shifts because of the instantaneous variations in the onboard electrical current. Such instantaneous variations in the current may be caused by switching on/off the onboard electronic devices or going into/out of eclipse. In this case, considering the shadow function introduced in Eqs. (2.18) and (2.19), a straightforward model for RMM may be

$$M = \begin{cases} M_e & v = 0 \\ M_l & v = 1 \end{cases}, \tag{2.21}$$

where $M_e$ and $M_l$ represent two different levels of RMM for eclipse and light conditions, respectively.

The effects of the RMM on the attitude determination and control accuracy and the necessity for the RMM compensation are well discussed in references [2, 4, 12]. The orthodox way to cancel out the disturbance caused by the RMM is to use a feedback controller but the efficiency of the method depends on several conditions such as the sensor noise, computational performance or plant model accuracy. Another method is to use a feedforward cancellation technique and when this technique is used the performance depends on the accuracy of the RMM estimation.

The RMM can be estimated on-ground during the spacecraft design phase. This is also when the engineers try to minimize the RMM [13]. Yet, in any case, as the several mission experiences validate, the RMM must be in-orbit estimated for small satellites. An in-orbit RMM estimation method will be presented in Chapter 9.

## 2.4 SATELLITE KINEMATICS

The kinematics equation of motion of the satellite via the quaternion attitude representation can be given as in Eq. (1.29):

$$\dot{q}(t) = \frac{1}{2}\Omega(\omega_{br})q(t). \tag{2.22}$$

Here, $q$ is the quaternion formed of four attitude parameters, $q = \begin{bmatrix} q_1 & q_2 & q_3 & q_4 \end{bmatrix}^T$. First three of the terms represent the vector part and the last one is the scalar term.

An important factor that is affecting the kinematics equation is the selection of the reference frame. If it is selected as the ECI than $\omega_{br} = \omega_{bi} = \begin{bmatrix} \omega_x & \omega_y & \omega_z \end{bmatrix}^T$ and we can directly use the angular velocity terms that are measured by the onboard gyros (or predicted using the satellite dynamics) in the kinematics equation. On the other hand, if the reference frame is selected as the orbit frame, then $\omega_{br}$ should be related to $\omega_{bi}$ via the following equation:

$$\omega_{bo} = \omega_{bi} - A\begin{bmatrix} 0 & -\omega_o & 0 \end{bmatrix}^T. \tag{2.23}$$

Here, $\omega_o$ denotes the orbital angular velocity of the satellite, found as $\omega_o = (\mu/r^3)^{1/2}$ for a circular orbit using $\mu$, the gravitational constant and $r$, the distance between the CM of the satellite and the Earth.

When the Euler angles are used for representing the attitude, usually the reference frame is the orbit frame. Then as in Eq. (1.8), the kinematics equation is

$$\begin{bmatrix} \dot{\phi} \\ \dot{\theta} \\ \dot{\psi} \end{bmatrix} = \begin{bmatrix} 1 & \sin(\phi)\tan(\theta) & \cos(\phi)\tan(\theta) \\ 0 & \cos(\phi) & -\sin(\phi) \\ 0 & \dfrac{\sin(\phi)}{\cos(\theta)} & \dfrac{\cos(\phi)}{\cos(\theta)} \end{bmatrix} \omega_{bo}. \tag{2.24}$$

## 2.5 SUMMARY

In this chapter, the mathematical models for small satellite attitude dynamics and kinematics are presented. First the coordinate frames, which may be of interest when studying attitude dynamics for a spacecraft, are introduced. Then the attitude dynamics equations are given in detail. As a part of the attitude dynamics, the disturbance torques are discussed by considering their dominance on small satellite motion and presenting the necessary equations to model the torques. Last, the satellite kinematics equations are given both in terms of attitude quaternions and Euler angles.

## REFERENCES

1. J. R. Wertz, Ed., *Spacecraft Attitude Determination and Control*, vol. 73. Dordrecht: Springer Netherlands, 1978.
2. S. I. Sakai, Y. Fukushima, and H. Saito, "Design and on-orbit evaluation of magnetic attitude control system for the REIMEI microsatellite," in *Proceedings of the 10th IEEE International Workshop on Advanced Motion Control*, Trento, Italy, 2008, pp. 584–589.
3. H. E. Soken, S. I. Sakai, and R. Wisniewski, "In-orbit estimation of time-varying residual magnetic moment," *IEEE Trans. Aerosp. Electron. Syst.*, vol. 50, no. 4, pp. 3126–3136, 2014.
4. T. Inamori, N. Sako, and S. Nakasuka, "Magnetic dipole moment estimation and compensation for an accurate attitude control in nano-satellite missions," *Acta Astronaut.*, vol. 68, no. 11–12, pp. 2038–2046, 2011.
5. S. A. Rawashdeh, "Attitude analysis of small satellites using model-based simulation," *Int. J. Aerosp. Eng.*, vol. 2019, 1–11, 2019, Article ID 3020581.
6. H. D. Curtis, "Chapter 10 - Introduction to orbital perturbations," in *Orbital Mechanics for Engineering Students* (Fourth Edition), H. D. Curtis, Ed. Butterworth-Heinemann, 2020, pp. 479–542.
7. W. H. Steyn and M. Kearney, *An Attitude Control System for ZA-AeroSat subject to significant Aerodynamic Disturbances*, vol. 47, no. 3. IFAC, 2014.

8. M. L. Psiaki, "Nanosatellite Attitude Stabilization Using Passive Aerodynamics and Active Magnetic Torquing," *J. Guid. Control. Dyn.*, vol. 27, no. 3, pp. 347–355, May 2004.

9. V. Lappas et al., "CubeSail: A low cost CubeSat based solar sail demonstration mission," *Adv. Sp. Res.*, vol. 48, no. 11, pp. 1890–1901, 2011.

10. S. Busch, P. Bangert, S. Dombrovski, and K. Schilling, "UWE-3, in-orbit performance and lessons learned of a modular and flexible satellite bus for future pico-satellite formations," *Acta Astronaut.*, vol. 117, pp. 73–89, 2015.

11. W. H. Steyn and Y. Hashida, "In-orbit attitude performance of the 3-axis stabilised SNAP-1 nanosatellite," in *Proceedings of the 15th AIAA/USU Conference on Small Satellites*, Utah, US, 2001, pp. 1–10.

12. H. Ehrpais, J. Kütt, I. Sünter, E. Kulu, A. Slavinskis, and M. Noorma, "Nanosatellite spin-up using magnetic actuators: ESTCube-1 flight results," *Acta Astronaut.*, vol. 128, pp. 210–216, 2016.

13. A. Lassakeur and C. Underwood, "Magnetic cleanliness program on cubesats for improved attitude stability," *J. Aeronaut. Sp. Technol.*, vol. 13, no. 1, pp. 25–41, 2020.

# 3 Attitude Sensors

The attitude determination and control system of a satellite requires sensors for determining the attitude and attitude rate. The most common sensors used for this purpose are magnetometer, Sun sensor, horizon (Earth) sensor, star tracker and gyros. They provide various types of information from attitude rate with respect to the inertial frame to direction vector to some reference, which will be eventually used in the attitude estimation algorithms.

In the early years of their first appearance – a few decades ago – small satellites (especially nanosatellites) could carry only small and light attitude sensors such as magnetometers and Sun sensors, and the total sensor number should be optimized. Star trackers, like more sophisticated equipment, were too bulky to carry onboard the small satellites. However, these days, as the technology has advanced, even a few kilograms of nanosatellite can carry a set of these sensors, including the star trackers.

This section gives a brief overview of the small satellite attitude sensors, emphasizing their functioning principles. We also present some commercially available examples for each sensor.

## 3.1 MAGNETOMETERS

Three-axis magnetometers (TAMs) are part of the attitude sensor package for almost all of the low Earth orbit (LEO) small satellites [1–4]. Being lightweight, small, reliable and having low power requirements make them ideal for small satellite applications. These sensors provide both the direction of magnetic field and its magnitude; therefore, they are useful for attitude determination and control. Operating magnetometers as primary sensor in small satellite missions is a common method for achieving attitude information. Moreover, satellite detumbling algorithms and attitude rate estimators, which are especially useful at the early stages of the mission after the launch, rely on magnetometer measurements.

There are a number of magnetometers available today. They can be classified in accordance with their sensitivities as well as their physical size and power consumption. Some applications requiring magnetic field information can afford size, power and cost in order to get the desired sensitivities. The following are the most commonly used magnetometers as a part of the satellite attitude determination and control systems:

- Search-coil magnetometers
- Fluxgate magnetometers
- MEMS (Micro-Electro-Mechanical Systems) magnetometers

Among them, search-coil magnetometers are used on relatively larger small satellites (e.g. microsatellites), and especially for phase angle estimation on spinning

**TABLE 3.1**
**Examples for Small Satellite Magnetometers**

| Sensor | Manufacturer | Mass (g) | Size (mm³) |
|---|---|---|---|
| MAG-3 Three-Axis (Fluxgate) Magnetometer | SpaceQuest | 100 | 35.1 × 32.3 × 82.6 |
| Magneto-Resistive Magnetometer | NewSpace | 85 | 99 × 43 ×17 |
| MicroMag3 Magneto-Inductive Magnetometer | PNI | 200 | 25.4 × 25.4 × 19 |

spacecraft. On the other hand, fluxgate magnetometers and MEMS magnetometers are for general use, including the 3-axis stabilized spacecraft and ideal even for nano-satellites. Some commercially available small satellite magnetometers are given in Table 3.1.

## 3.1.1 SEARCH-COIL MAGNETOMETER

Search-coil magnetometer works based on the Faraday's law of magnetic induction. It comprises copper coils that are wrapped around a magnetic core. The core gets magnetized by the magnetic field lines produced inside the coils. The fluctuations in the magnetic field results with the flow of electrical current and induces a time-varying voltage. This voltage is measured and recorded by the magnetometer.

The induced voltage for a search-coil magnetometer is proportional to the time-rate change of the magnetic field and may be due to the temporal or spatial variations [5]:

$$V = -c\left[ \boldsymbol{B} \cdot \frac{d\boldsymbol{n}}{dt} + \boldsymbol{n} \cdot \frac{d\boldsymbol{B}}{dt} + \boldsymbol{u} \cdot \nabla\left(\boldsymbol{B} \cdot \boldsymbol{n}\right)\right]. \tag{3.1}$$

Here, $c$ is a constant that depends on the area of the loop and on the properties of the core material, $\boldsymbol{B}$ is the magnetic field vector, $\boldsymbol{n}$ is the unit vector along the coil axis and $\boldsymbol{u}$ is the velocity vector relative to the magnetic field. Eq. (3.1) shows that the measured voltage may be due to angular motion of the satellite (e.g. when the satellite is spinning), time-dependent variations in the magnetic field and the orbital motion of the spacecraft.

Now let us consider the search-coil magnetometer is placed on a spinning spacecraft such that the coil axis is perpendicular to the spin axis and makes an angle $\theta$ with the local magnetic field vector. For a typical search-coil magnetometer that has a coil of $N$ turns on a material with a relative magnetic permeability $\mu_t$ and cross-sectional area $A$, the measured voltage will be

$$V = -NA\mu_t \frac{dB_p}{dt}. \tag{3.2}$$

$B_p$ is simply the magnetic field parallel to the magnetometer's coil axis (Figure 3.1). For a spacecraft with a spin rate of $\omega_z$ this is

$$B_p = B\cos\theta \sin\left(\omega_z t\right). \tag{3.3}$$

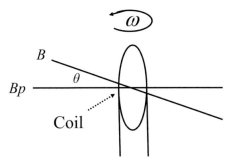

**FIGURE 3.1**   Search-coil magnetometer.

Hence the measured voltage due to the spin motion will be

$$V = -AN\mu_r B\omega_z \cos\theta \cos(\omega_z t). \tag{3.4}$$

Thus, depending on the accuracy of our knowledge for the $B$ and the spin rate of the spacecraft, the angle $\theta$ can be determined from the amplitude of $V$. The phase angle of the spacecraft can be also estimated by measuring the current voltage relative to its peak value.

### 3.1.2 FLUXGATE MAGNETOMETER

The fluxgate magnetometers are used for the high sensitive applications. A fluxgate sensor drive has an alternating drive current that runs a permeable core material. It consists of a magnetically susceptible core wound by two coils of wire. One coil is excited by the AC supply and a constantly changing field induces an electrical current in the second coil. This current change is based on the background field. Hence the alternating magnetic field, and the induced output current, will be out of step with the input current. The extent to which this is the case will depend on the strength of the background magnetic field.

Fluxgate magnetometers are vector magnetometers that provide both the strength and direction information for the magnetic field. They are the most capable magnetometers for a small satellite application. They give the best performance in terms of noise and also the thermal robustness. However, the performance for a fluxgate magnetometer reduces as the sensor itself gets smaller. Thus a high-grade fluxgate magnetometer can be only used on rather larger small satellites such as microsatellites. They are generally too heavy and demand high power for use on nanosatellites, although there are some examples for such implementation [6].

The fluxgate magnetometer consists of a magnetic core which is driven into saturation by a triangular current waveform applied to the primary coil in Figure 3.2.

If the ambient magnetic field is zero, the secondary coil produces a symmetrical waveform. However, if there is an ambient magnetic field the core will be more easily saturated in one direction than the other, causing a mismatch in the input and output current. So the intensity in the core becomes unsymmetrical and this changes the

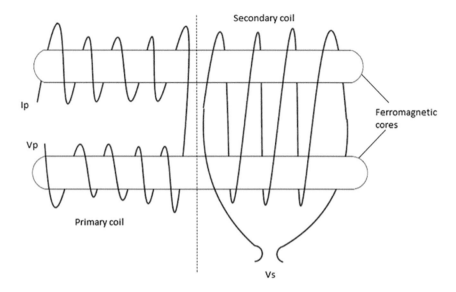

**FIGURE 3.2**  Fluxgate magnetometer.

phasing of the secondary coil's output. In the end the voltage output of the secondary coil will show an impulsive behavior (Figure 3.3). By measuring the phase difference between two impulses the component of the external magnetic field in the direction of coil axis can be calculated. By the use of three orthogonal coils, a three-axis magnetic field measurement can be obtained.

The measurement accuracy of the fluxgate magnetometer is limited by the ability of producing the required symmetrical current waveform with no offset errors using the core driving circuitry. This is a hard-to-satisfy requirement since the strength of the magnetic field to saturate the core is orders of magnitude higher than that of the Earth's magnetic field [7]. Moreover, the current to drive and saturate the core may require high power. This is why, implementation of this type of magnetometer on nanosatellite is limited. Last but not least, the current also causes a Residual Magnetic Moment (RMM), which eventually may lead to significant amount of magnetic disturbance torque.

### 3.1.3 MEMS Magnetometer

MEMS-type magneto-resistive and magneto-inductive magnetometers are especially favorable for nanosatellite missions. Due to their small size, very low power demand, low cost and commercial availability, a vast majority of LEO small satellites use MEMS magnetometers as the primary attitude sensor.

A typical example of magneto-resistive MEMS magnetometer is Anisotropic Magneto-Resistive (AMR) sensor. The AMR elements, which are used as the four primary components of a Wheatstone bridge, change their effective resistance when they pass through a magnetic field (Figure 3.4). In the absence of a magnetic field, the voltage out of the Wheatstone bridge will be half of the applied voltage [8].

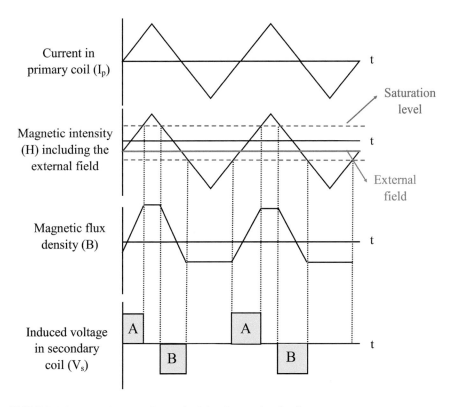

**FIGURE 3.3**   Waveforms and the principle of operation for fluxgate magnetometer.

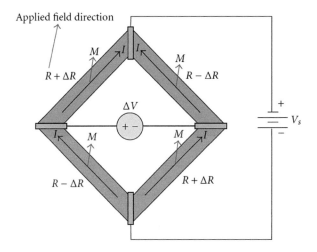

**FIGURE 3.4**   Wheatstone bridge for AMR sensor (courtesy of Honeywell).

In contrast, in the presence of magnetic field, all four AMR elements will have different resistances changing the output voltage form the nominal value. A simple relationship in between the voltage deviation from the nominal and the applied magnetic field can be used to measure the magnetic field strength. Such three Wheatstone bridges in a single sensor module are needed to measure the magnetic field in three dimensions.

The main problem of AMR magnetometer is being highly prone to errors [9]. First of all, it is hard to fabricate identical AMR elements. Slight errors, which appear as different resistance values, cause an offset in the voltage. This is read as a bias in the sensor measurements. Besides, temperature variation may also affect the resistance values and cause a temperature-dependent bias error [10]. Last, the AMR elements are nonlinear by nature. This means that the sensitivity of the AMR sensors varies with the variations in the magnitude of the sensed magnetic field. This nonlinearity results in scale factor error. To conclude, the AMR magnetometers onboard a nano-satellite must be in-orbit calibrated.

Another type of MEMS magnetometer uses the magneto-inductive technology. The sensor, which is a simple resistor–inductor circuit, measures the time it takes to charge and discharge an inductor between an upper and lower threshold by means of a Schmitt trigger oscillator. This time is proportional to the applied field strength [11].

## 3.2 SUN SENSORS

Sun sensors are devices used for detecting the direction of the Sun with respect to the satellite's body frame. They are widely used in space applications because of several factors like being light, cheap and small. Also Sun-dependent working principle gives them several advantages. Unlike the Earth the angular radius of the Sun is not related to the position of the satellite and is sufficiently small (0.267° at 1 AU). Thus for most applications a point-source approximation is valid. At the same time the Sun is bright enough and it can be detected with considerably high reliability even with simple, low-power equipment.

Today many space missions have solar experiments, most have the Sun-related thermal limitations and almost all need the Sun as a power source. Consequently, missions are concerned with the orientation and time evolution of the Sun vector in body coordinates. Sun sensors are also needed to turn the solar panels through the Sun and to protect the sensitive equipment such as star trackers (by preventing their long exposure to the sunlight). Hence, for safe-mode of the spacecraft they are the primary attitude sensor. They are even ideal for planetary rover missions because of the negligible magnetic field in other planets like Mars.

Because of all these listed reasons and especially because of being light, small, low-power sensor, together with the magnetometers, Sun sensors are one of the most favorable couple of attitude sensors for LEO small satellites. Few examples of Sun sensors that can be used on small satellites are given in Table 3.2.

The wide range of Sun sensor applications caused the development of numerous sensor types. Simply the Sun sensors can be examined in three classes: Analog sensors which are monotonic and have an output signal of continuous function of the

## TABLE 3.2
### Examples for Small Satellite Sun Sensors

| Sensor | Manufacturer | Mass (g) | Size (mm³) | Accuracy (°) |
|---|---|---|---|---|
| Digital Fine Sun Sensor | NewSpace | 35 | 34 × 32 × 20 | 0.1 |
| Analog Cubesat Sun Sensor | NewSpace | 5 | 33 × 11 × 6 | 0.5 |
| Digital Two-axis Sun Sensor | SolarMEMS | 35.5 | 50 × 30 × 12 | 0.1 |

Sun angle, Sun presence sensors which send out a constant signal only when the Sun is in the field of view (FOV) and digital sensors whose output signal is encoded and discontinuous. Sun sensors can be also categorized in terms of their accuracy (e.g. coarse vs fine) and the number of angular information that they provide (e.g. 1D vs 2D). Usually digital sensors provide better accuracy than analog sensors but with a narrower FOV. Note that in contrast to usual understanding, fine Sun sensors can be both analog and digital, there is no requirement such that a fine Sun sensor must be digital. MEMS fabrication process is commonly used for small satellite Sun sensors.

### 3.2.1 ANALOG SUN SENSORS

Two different types of analog Sun sensors are introduced here. One of them is cosine Sun sensor which provides a coarse Sun direction in one axis. This coarse direction is usually used to maneuver the spacecraft and put the Sun in the FOV of fine Sun sensors.

Cosine sun sensors are based on the sinusoidal variation of the output current of a silicon solar cell, which Sun rays come through with an angle, as shown in Figure 3.5. The energy flux, $E$, through a surface of area $dA$ with unit normal $\bar{n}$ is [12]

$$E = \boldsymbol{P} \cdot \boldsymbol{n} dA. \tag{3.5}$$

Here $\bar{P}$ is Poynting vector that gives the direction and the magnitude of energy flow for electromagnetic radiation. Thus the energy stored in the photocell and, as a

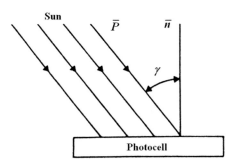

**FIGURE 3.5**   Cosine detector sun sensor.

**FIGURE 3.6**   Cosine Sun Sensor by Bradford Space (courtesy of Bradford Space).

result, the output current, $I$, are proportional to the cosine of the angle of incidence of the solar radiation:

$$I(\gamma) = I(0)\cos\gamma. \tag{3.6}$$

$I(0)$ is the current in the photocell when $\gamma = 0°$ ; $\gamma$ is the angle between the Poynting vector $P$ and unit normal $n$.

Model given by Eq. (3.6) is simple and transmission loses due to Fresnel reflection, the effective photocell area and angle-dependent reflection at the air-cell interface are neglected.

Large FOV and low power consumption characteristics of Cosine Sun sensor have made them suitable for sun-tracking and positioning systems. Figure 3.6 gives an example for such Sun sensor, which has a FOV as large as 160° and approximate accuracy of 3°. Because the cosine Sun sensor is so simple, the difference between the direct sunlight and the albedo cannot be distinguished. The error due to the albedo can be as high as 30°. Thus it is reasonable to use cosine Sun sensors when the accuracy requirement is low such as pointing control during safe-mode, attitude sanity checking, etc.

Another type of analog Sun sensor, which is favorable also for small satellite missions, is quadrant Sun sensor. This sensor is formed of a cover (mask) with a pinhole and a quadrant photodiode placed beneath the cover. Figure 3.7 presents a quadrant

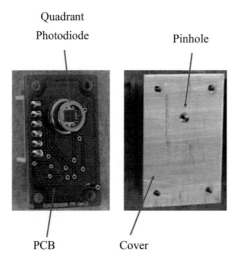

**FIGURE 3.7**   Quadrant Sun Sensor for HORYU-IV Nanosatellite [13].

Sun sensor designed and manufactured for HORYU-IV nanosatellite. The cover, which is functioning as an aperture mask, is accurately positioned above the photodiode. Sunlight reaches the photodiode through the pinhole. The distribution of the incident sunlight across the four quadrants and the induced photocurrents, which are proportional to the illuminated quadrant areas, depend on the solar angles of incidence, $\alpha$ and $\beta$, the angles about x and y axes, respectively (Figure 3.8). Following simple formulas are used to define the angles in terms of the signals from each quadrant:

$$S_x = \frac{A+B-C-D}{A+B+C+D} = \frac{\tan(\alpha)}{\tan(\alpha_{max})} \tag{3.7a}$$

$$S_y = \frac{B+C-A-D}{A+B+C+D} = \frac{\tan(\beta)}{\tan(\beta_{max})} \tag{3.7b}$$

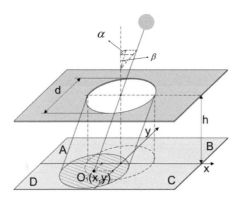

**FIGURE 3.8**   Schematic for a quadrant Sun sensor (modified from [13]).

$\alpha_{max}$ and $\beta_{max}$ are maximum solar angles of incidence which are defined by the sensor geometry as

$$\alpha_{max} = \beta_{max} = \frac{d}{2h}. \tag{3.8}$$

Quadrant Sun sensors are fine sensors, which are capable of reaching an accuracy as high as 0.1°. They often use look-up tables to describe relations between output signals from photodiodes and the Sun vector [13]. These look-up tables are prepared based on Eqs. (3.7–3.8) and after several calibration processes. If the primary concern is the accuracy, look-up tables must be used. However, when we use these tables a large amount of data has to be saved and for especially nanosatellites storing such huge amount of data may be a problem. Instead fitting equations with few coefficients may be considered. They give less accuracy than look-up tables, but they are more convenient for especially nanosatellite missions.

### 3.2.2 SUN PRESENCE SENSORS

Sun presence sensors are simple equipment used for not obtaining the Sun's direction but rather getting a binary output indicating the Sun's presence in the FOV of the sensor. On 3-axis stabilized spacecraft they are mostly carried for protecting instrumentation (e.g. camera), activating hardware and coarse positioning of the spacecraft.

Sun presence sensors may have different geometries and numbers of photocells [14]. A common example is binary type presence sensor with a single photocell. The sensor consists of two slits and a photocell. When the Sun is in the plane formed by the entrance and reticle slits and the angle with normal to the sensor face is in the limitations (typically 32° or 64°), the photocell indicates the Sun's presence (Figure 3.9).

Two of binary type Sun presence sensors can be used to build a spin-type Sun aspect sensor which is capable of providing Sun direction vector on a spinning spacecraft. Two sensors should be placed in a V-configuration such that they make an angle of $v$ to each other (Figure 3.10). The spin rate for the spacecraft ($\omega_z$) can be calculated

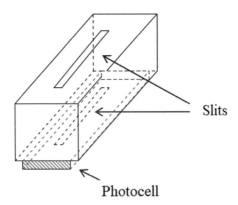

Slits

Photocell

**FIGURE 3.9**   Binary type Sun presence sensor (modified from [14]).

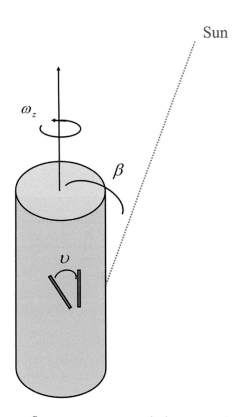

**FIGURE 3.10**  Spin-type Sun aspect sensor on a spinning spacecraft.

(measured) based on successive detections of Sun by one of these slits within one spin period. Furthermore, $\tau$, which is the interval in between Sun crossings for each slit, depends on the spin rate, the angle $v$ and the Sun aspect angle $\beta$. Using the cosine theorem for spherical triangles, the Sun aspect angle can be calculated as

$$\cot(\beta) = \frac{\sin(\tau\omega_z)}{\tan(v)}. \tag{3.9}$$

Once the Sun aspect angle $\beta$ is calculated, the unit Sun direction vector in spacecraft body frame can be obtained considering the placement of sensors.

### 3.2.3 DIGITAL SUN SENSORS

Digital Sun sensors provide more accurate measurements compared to other Sun sensor types. They are also mostly insensitive to the albedo, which is a large advantage over analog sensors [15]. However, because of their complicated structure and precision engineering that is required during the manufacturing, they are usually more expansive as well.

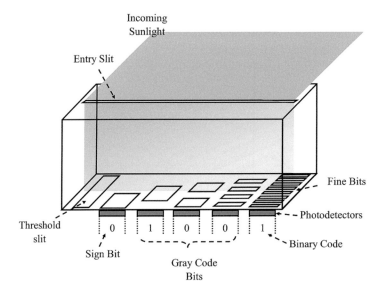

**FIGURE 3.11**    Measurement unit example for a digital Sun sensor.

Digital fine sun sensors historically relied on arrays of slits (or reticles) with linear photosensitive surfaces behind them. The first row of these slits (the leftmost slit in Figure 3.11) is used as a threshold setting when the data from other slits are used. Specifically, measurement slits (and the bits they represent in the Sun angle measurement) are deemed to be "on" if their intensity is greater than the threshold slit output [16]. This is also how the Earth albedo can be eliminated. Notice that the threshold slit is half the width of other slits. So, a measurement slit's bit is "on" if more than half of it is illuminated by the Sun. The slit right next to the threshold slit is sign bit. Depending on this bit's value, the sensor detects which side the sunlight is coming from. The bits defining the magnitude of the Sun angle then are derived from. The rest of the bits are for decoding the Sun aspect angle. The gray code bits provide the Sun angle up to mid-level of accuracy and the fine bits further improve the accuracy. This measurement unit of digital Sun sensor measures the Sun direction in one axis. Thus for two-axis information another unit, whose slit is orthogonally placed to that of current unit, is needed. As to the authors' knowledge there is no example for recent small satellite mission with this type of digital Sun sensor.

Newer digital sun sensors rely on the MEMS manufacturing and in principle they are not too different from the analog quadrant Sun sensor presented in Figure 3.8. There is also a pinhole for this type of digital Sun sensor. However, the photodiode is replaced with an active-pixel sensor (APS) such as CMOS (complementary metal–oxide–semiconductor) sensor and the detected sunlight by the sensor is digitally processed. In fact, considering that most of the star trackers for especially nanosatellites are also based on APS, this type of digital Sun sensors has similarities to star trackers as well. However, as will be introduced in the rest of this chapter, star trackers are more sophisticated equipment with light-gathering optics, baffle-mechanism

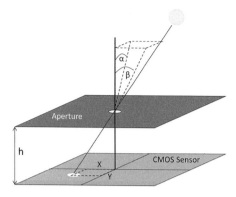

**FIGURE 3.12**   Schematic for CMOS sensor based digital Sun sensor.

to prevent stray lights and an integrated higher capacity processor for operating the sensor.

A schematic for a CMOS sensor based digital Sun sensor is given in Figure 3.12. Ignoring the pixel size, following simple relations for Sun aspect angles can be obtained:

$$\alpha = \text{atan}\left(\frac{Y}{h}\right), \quad \beta = \text{atan}\left(\frac{X}{h}\right). \tag{3.10}$$

This is a digital Sun sensor type frequently used on small satellites [15, 17]. Nature of the technology allows manufacturing sensors small and light enough for using on nanosatellites.

## 3.3  EARTH HORIZON SENSORS

For an Earth orbiting spacecraft, it is natural to think of the Earth itself as a reference for measuring the attitude. Especially for an Earth imaging satellite, which should be nadir pointing during most of the mission, having a sensor which can detect the deviations from nadir can be useful. This is what an Earth horizon sensor does. Simply by detecting the Earth's horizon, it gets the nadir vector in the body frame and provides two-axis attitude information (roll and pitch angles). However, although the concept can be considered rather easy in theory, there are several difficulties in practice.

First of all, the nadir vector measurement is not straightforward as in case for the Sun. Compared to the Sun, which can be easily assumed as a point light source, the Earth itself has a large apparent radius. So, first the horizon should be detected to get the nadir vector. If the horizon detection is done in visible light, there are many disturbances affecting the accuracy of the detection such as atmospheric conditions (clouds) and variations due to the illumination. Thus, most of the Earth sensors view Earth in infrared and utilize $CO_2$ band. As a result, most of the stray lights can be eliminated (including those of the albedo) and the Earth can be viewed more uniformly. Yet the accuracy depends on the following:

- how well the radiation distribution can be modeled.
- how accurately the angle between the Earth's horizon and the Earth's center, which is changing due to the oblateness, is known.
- and, of course, the actual measurement accuracy, which strongly depends on the temperature.

An Earth horizon sensor has detector and optics components for all types and an additional scanner for scanner types for 3-axis stabilized spacecraft (for spinning spacecraft the scan naturally occurs). The detector is usually a bolometer, photodiode, pyroelectric device or a thermopile. Among those, thermopile is especially preferred as detector for Earth horizon sensors that are suitable to be used on nanosatellites [18–20]. Optics is used to limit the spectral band (e.g. to infrared) and also to reduce the effective FOV of the sensor, especially for lowering the probability of interference by the Sun and Moon.

There are two general types of Earth horizon sensors: Scanning Earth horizon sensors and static Earth horizon sensors. Due to the necessity for moving parts, the scanning Earth horizon sensors are usually not used on nanosatellites. Yet if the satellite is a spin-stabilized one, as the scan naturally occurs, they can be still used without necessity to the moving parts. Table 3.3 presents few examples for Earth sensors that can be used on small satellites.

### 3.3.1 SCANNING EARTH HORIZON SENSORS

As the name suggests, a scanning Earth horizon sensor detects the horizon by scanning. On a 3-axis stabilized spacecraft the sensor is mounted on a (wheel-like) moving mechanism to scan for the horizon. This happens naturally due to the spin motion on a spinning spacecraft. Unless the spacecraft is not a spin-stabilized one, this type of Earth horizon sensor is not widely used especially on nanosatellites since the moving part requires additional power and complexity. Thus the functional principle of the sensor will be described based on measurements on a spinning small spacecraft.

The scanning Earth sensor has two pencil-beams oriented at angles $\mu_i$ ($i = 1, 2$) with respect to the spin axis. Each pencil-beam measures the crossing times for

---

**TABLE 3.3**
**Examples for Small Satellite Earth Horizon Sensors**

| Sensor | Manufacturer | Mass (g) | Size (mm³) | Accuracy (°) |
|---|---|---|---|---|
| COTS Based Thermopile Earth Sensor | Meisei | <250 | 40 × 40 × 55 | 1.5 |
| Horizon Sensor for Nano & Micro Satellites | Solar MEMS | 120 | 90 × 90 × 50 | <0.33 |
| MAI-SES | Maryland Aerospace | 33 | 43 × 30 × 30 | 0.25 |

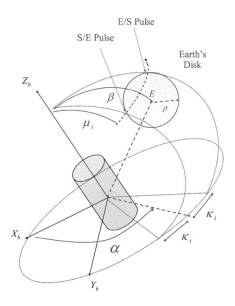

**FIGURE 3.13** Measurement geometry for a horizon scanning Earth sensor.

Space/Earth (S/E) and Earth/Space (E/S) infrared boundaries. With the knowledge of the spin rate, these crossing pulses can be transformed into the half-chord angles $\kappa_i$ ($i = 1, 2$), which are the fundamental measurements produced by the Earth sensor. By using these measurements and as well the apparent Earth radius angle, $\rho$, we find the relationship for the Earth aspect angle (EAA) $\beta$, from the spherical geometry (Figure 3.13) [21]:

$$\cos \mu_i \cos \beta + \sin \mu_i \cos \kappa_i \sin \beta = \cos \rho \quad i = 1, 2. \tag{3.11}$$

Here, the instantaneous value for the apparent Earth radius angle, $\rho$, can be calculated as $\rho(t) = \sin^{-1}(R_{IR}/r(t))$, where $r$ is the orbital radius and $R_{IR}$ is the infrared (IR) Earth radius. Note that the nominal value of $R_{IR}$ is ~40 km above the Earth radius. The actual value of $R_{IR}$ is unknown and varies over time and location.

In fact, the scan paths of each of the two IR pencil-beams over the Earth are different due to different mounting angles ($\mu_1$ and $\mu_2$). As a result, their S/E and E/S crossings are at *different* locations on the Earth's IR rim. Therefore, the two $\rho$ values observed by the IR sensors will in general differ under seasonal, diurnal and local variations in the IR radiation intensities. Yet it is very hard to model all of these effects realistically so we assume, a priori, that the Earth's IR radius is perfectly uniform so we may use the same $\rho$ value for both IR beams.

We may use different methods to derive the EAA when there are two individual pencil-beams. To start with, we may rewrite the Eq. (3.11) in the form [22]:

$$b_i \cos \chi_i \cos \beta + b_i \sin \chi_i \sin \beta = \cos \rho \quad i = 1, 2. \tag{3.12}$$

where the auxiliary functions are defined as:

$$b_i = \sqrt{1-\left(\sin\mu_i\sin\kappa_i\right)^2}\;;\quad \chi_i = \arctan\left(\tan\mu_i\cos\kappa_i\right)\quad i=1,2 \tag{3.13}$$

Thus, each pencil-beam gives its own EAA solution, $\beta_i$:

$$\beta_i^\pm = \chi_i \pm \arccos\left\{\cos\left(\rho\right)/b_i\right\}\quad i=1,2 \tag{3.14}$$

One method to construct a single EAA value to use in the estimator is to take the average of the two different $\beta_i$ values produced by each of the two pencil-beams as $\beta_{ave} = \left(\beta_1^\pm + \beta_2^\pm\right)/2$, after the sign ambiguity is resolved. The second method is to calculate an optimal (in a minimum variance sense) $\beta_{opt}$ value as a weighted combination of the two individual $\beta_i$ angles. This method is based on the sensitivity analysis for each of the two $\beta_i$ solutions with respect to the half-chord angle measurements, $\kappa_i$ [22].

Once the EAA is known the nadir vector can be easily calculated using the rotation angle shift of the sensor's vertical slit plane relative to the body $X_b$ axis, i.e. angle $\alpha$. It is common that this angle is defined with respect to the Sun crossing when the Earth sensor is coupled with a Sun sensor onboard the spacecraft.

### 3.3.2 STATIC EARTH HORIZON SENSORS

Static Earth sensors use multiple detectors that are aligned with the nadir pointing axis of the spacecraft (usually $Z_b$ axis) such that the Earth's disc is continuously detected. Assuming that, for detectors, the measurement principle of a typical static Earth sensor is straightforward. In case of zero pitch and roll angles (with respect to the orbit frame for an orbit with low eccentricity), the detectors will generate equal signals. Any change in the pitch and roll angles will be sensed by detector couples as difference in the generated signal (see Figure 3.14).

Figure 3.15 gives an example for static Earth sensor, which is also usable on nanosatellite missions. Since these sensors have a limited field of view (FOV) and are

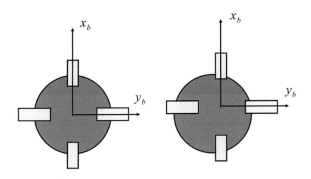

**FIGURE 3.14** Static Earth sensor. Left: Earth is centered (spacecraft is nadir pointing with no attitude error); Right: Earth is not centered (roll and pitch angles are non-zero). Body z axis is nadir pointing for the spacecraft.

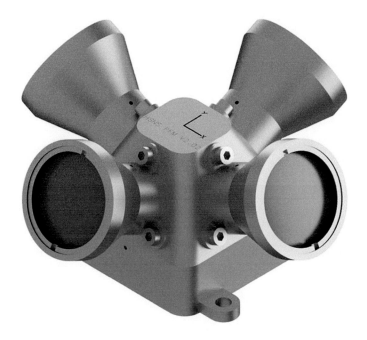

**FIGURE 3.15**    A static Earth sensor suitable for nanosatellite missions (courtesy of Solar MEMS).

rigidly mounted on the satellite, they are suitable for a limited range of missions in terms of attitude, altitude and orbit type.

## 3.4  STAR TRACKERS

Star trackers are the most sophisticated attitude sensors with the highest accuracy. In the early days of small satellite applications, they could not be easily used on these satellites because of being rather bulky and large. However, today, thanks to the developments in technology and miniaturization of many components, there are star trackers even suitable to be used on nanosatellites. Table 3.4 presents examples for small satellite star trackers and Figure 3.16 shows one of them.

A star tracker is different from other sensors since it is capable of providing the 3-axis attitude information alone. To do that it must obtain an image of multiple stars,

**TABLE 3.4**
**Examples for Small Satellite Star Trackers**

| Sensor | Manufacturer | Mass (g) | Size (mm³) | Accuracy (arcsec) |
|---|---|---|---|---|
| Auriga Star Tracker | Sodern | 210 | 56 × 66 × 94 | 6 (cross boresight) 40 (boresight) |
| Star Tracker | KU Leuven | 250 | 45 × 40 × 95 | 2 (cross boresight) 10 (boresight) |

**FIGURE 3.16**   Auriga star tracker (courtesy of Sodern).

measure their apparent position in the reference frame of the spacecraft and identify the stars, so their directions can be compared with their known absolute directions from a star catalog. After that the inertial attitude of the spacecraft is provided as quaternion vector with a sampling rate up to few hertz.

In terms of structure a star tracker has generally a baffle, light-gathering optics, a sensor for imaging and a processor for operating the sensor. The camera part for imaging and the processor might be both integrated and separated. It is also possible to use multiple cameras with a single processor for increased redundancy and better accuracy in all axes on rather larger small satellites. The baffle of the sensor is used to prevent the stray lights coming from the Sun, the Earth's albedo and the spacecraft's structure itself. The sensor part is usually a CCD (charge-coupled device) or a CMOS sensor. The processor runs for several tasks including the star identification by comparing the pattern of observed stars with the known pattern of stars in the catalog and attitude estimation.

Regarding the functioning principle of a star tracker, first the light from the stars is collected by optics and focused on the detector. The detector will convert the

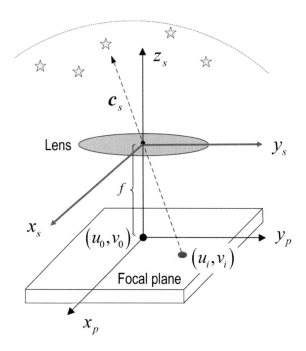

**FIGURE 3.17**    Star tracker measurement geometry.

received luminous flux into electrons. The image is slightly defocused so that the star covers a certain number of pixels. The position of the star on the matrix is then obtained by calculating the barycenter of the energy collected per useful pixel. This position for $i$th star is used to calculate the star direction vector ($c_s$) in the sensor frame in the end as given in Eq. (3.15) (see also Figure 3.17).

$$c_s = \frac{1}{\sqrt{f^2 + (u_i - u_0)^2 + (v_i - v_0)^2}} \begin{bmatrix} -(u_i - u_0) \\ -(v_i - v_0) \\ f \end{bmatrix}. \tag{3.15}$$

When the unwanted electrons from sources like other celestial objects or a reflection from satellite's solar panels fall on the imager, they create noise in the images. Noise filtering removes these noises and prepares the image for further processing. Then the thresholding module separates the pixels that belong to stars from those that are part of the background. With a good threshold, centroid locations of the stars on the image can be calculated more accurately and processing times to find stars will be reduced. Since noise level changes with every image, threshold level also has to change. After a threshold level is set, stars on the image can be detected. Algorithm first checks every pixel and compares them with the threshold level and considers all the pixels that have brightness more than the threshold level as possible star. Both noise filtering and thresholding are also necessary for preventing faulty detections.

As mentioned, electrons coming from a star do not fall on a single pixel but a group of pixels, so it is necessary to find the pixel that represents the star best.

Centroiding algorithm is used for that purpose and a good optimized threshold level would make this algorithm work more efficiently. Once the stars on the image are found they should be matched with the catalog stars by identification. Identification algorithms typically use the star-to-star angular distances and the star magnitudes (brightness of the stars) to distinguish the star. Once an initial attitude is acquired from the lost-in-space condition, the identification algorithm may change to a tracking one. In this mode, which works well at slow angular velocities, the star tracker just tracks the once identified stars by comparing successive images. Refs. [23, 24] provide a comprehensive survey of star identification algorithms.

Identification returns with both the sensor frame vectors for the captured stars ($c_s$) and their corresponding inertial vectors ($c_i$). The attitude estimation algorithm estimates the inertial to sensor frame quaternion attitude vector $\hat{q}_{si}$, which can be then easily used to estimate the inertial attitude of the spacecraft for known sensor alignment. The algorithm, which is used for the estimation may vary, but one of the single-frame algorithms (these algorithms will be introduced in Chapter 6) such as QUEST is usually preferred since they are robust and not computationally demanding. Functioning flow for a star tracker is summarized in Figure 3.18.

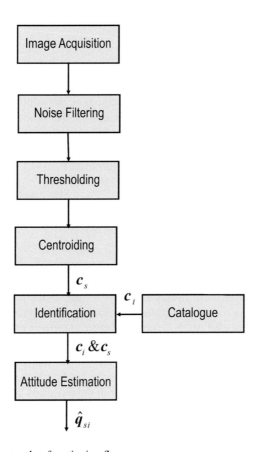

**FIGURE 3.18** Star tracker functioning flow.

## 3.5 GYROSCOPES

All of the four attitude sensors presented so far measure the direction with respect to a reference (and its magnitude for magnetometers) and their measurements provide absolute attitude information. In contrast, the gyroscopes (or gyros in short) measure the dynamic motion itself for the spacecraft without needing any external reference and give relative attitude information. They are ideal for coupling with one of other sensors (e.g. star tracker) since, by using their measurements, which are usually obtained with a higher frequency, the attitude can be easily propagated in the absence of absolute attitude measurements by other sensors. This is a concept that we will introduce as a part of attitude filtering in Chapter 9 of this book.

Gyros have several types. On a spacecraft they are almost exclusively used in strap-down mode, which means that the gyro is solidly attached to the spacecraft rather than being used to control a separate gimbaled platform [25]. Depending on the nature of the measurement that they provide, they can be categorized into two broad classes: rate gyros that measure the angular rate of the spacecraft and rate-integrating gyros (RIGs) that measure the integrated rates or angular displacements. Moreover, regarding the functioning principle and how they measure the angular rate, there are three general gyro types: Mechanical gyros that works based on the principle of angular momentum conversation, optical gyros that make use of the Sagnac effect and the MEMS gyros whose measurements are based on the Coriolis force (Figure 3.19).

Due to size and power constraints and the available budget, usually mechanical gyros are not used in small spacecraft. Fiber Optic Gyros (FOG) and MEMS gyros are those which are mostly preferred for such missions. FOGs perform better in terms of accuracy and bias stability but at a cost and mass penalty [26]. In contrast, MEMS gyros are ideal even for low-cost nanosatellite missions. Table 3.5 presents examples for gyros that can be used in small satellites.

### 3.5.1 FIBER OPTIC GYROS

All the optical gyros measure the angular rate or displacement based on the Sagnac effect. To describe the measurement principle, let us assume a circular optical path

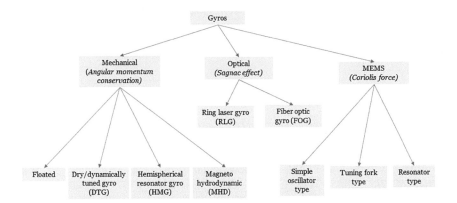

**FIGURE 3.19** Gyro types.

**TABLE 3.5**
**Examples for Small Satellite Gyros**

| Sensor | Manufacturer | Mass (g) | Size (mm³) | Accuracy |
|--------|-------------|----------|-----------|----------|
| FOG 1-axis µFORS-3 U | Northrop Grumman | 150 | 21 × 65 × 88 | 0.08° / $\sqrt{h}$ (random walk) 3 ° /h (bias drift) |
| ADIS16405 | Analog Devices | 16 | 32 × 23 × 23 | 2° / $\sqrt{h}$ (random walk) 25 ° /h (bias drift) |

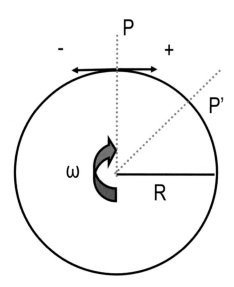

**FIGURE 3.20**   A conceptual scheme for optical gyro.

(Figure 3.20). The transmitter at P produces two signals (+ ***and*** −) in two opposite directions of the optical path. Since the gyroscope is rotating, the receiver, collocated with the transmitter, will receive the two signals at different time instants since the two signals will have travelled for different lengths.

$$ct^- = \left(2\pi - \omega t^-\right)R \tag{3.16a}$$

$$ct^+ = \left(2\pi + \omega t^+\right)R \tag{3.16b}$$

Here $c$ is the speed of light, $ct^-$ is the length of path in the "−" direction, $ct^+$ is the length of path in the "+" direction, $\omega$ is the angular velocity to be measured, $R$ is the radius of the ring, $t^-$ is the instant in which the "−" direction signal is received at $P'$, $t^+$ is the instant in which the "+" direction signal is received at $P'$.

The instants, in which the signals are received, are then:

$$t^- = \frac{2\pi R}{c + \omega R},$$                                          (3.17a)

$$t^+ = \frac{2\pi R}{c - \omega R}.$$                                          (3.17b)

Measuring the time difference, the angular velocity can be computed as:

$$\Delta t = \left(t^+ - t^-\right) = \frac{4\pi R^2 \omega}{c^2 - \omega^2 R^2} \cong \frac{4\pi R^2 \omega}{c^2} = \frac{4A\omega}{c^2},$$       (3.18a)

$$\omega = \frac{\Delta t c^2}{4A}.$$                                           (3.18b)

If the optical path is made of a fiber optic, it is possible to wind the fiber optic for more than one winding, increasing $A$ and improving the instrument sensitivity. This instrument is known as FOG. These gyroscopes are not subject to friction and motion problems and can be used for long durations with high reliability. Compared to the Ring Laser Gyros (RLGs), which functions based on the same principle, but is different in terms of the instruments used for measuring the angular velocity, FOGs are usually cheaper, yet less accurate. FOGs suffer from lock-in effect, which prevents them from measuring very low angular velocities.

## 3.5.2 MEMS GYROS

MEMS gyros are ideal angular rate sensors for especially nanosatellite applications. They are cheap, commercially available and having sufficient theoretical and technical experience, they can be even designed and developed at home [27]. At the moment, they are usually less accurate compared to other types of gyros including the FOGs, but the as the technology advances, the performance gap is closing day by day.

Although they may differ structurally, all MEMS gyros function based on the principle of measuring the Coriolis force due to the rotation. A MEMS gyro can be dynamically modeled using a spring-damper-mass system (Figure 3.21) to describe the measurement principle. In order to obtain the dynamics of the gyroscope attached to a rotating object, this analysis should be applied to the position vector of a vibratory gyroscope proof mass. Assuming that the gyro is measuring the angular rate about z axis, $\omega_z$, and its driving and sensing axes are x and y axes respectively, the simplified 2° of freedom equations of motion of a MEMS gyro can be written as

$$m\ddot{x} + c_x \dot{x} + k_x x = F_d + 2m\omega_z \dot{y},$$                  (3.19a)

$$m\ddot{y} + c_y \dot{y} + k_y y = -2m\omega_z \dot{x}.$$                       (3.19b)

Here, $m$ is the mass for the vibrating proof mass, $k_x$ and $k_y$ are spring coefficients, $c_x$ and $c_y$ are damping coefficients and $F_d$ is the vibratory driving force. $2m\omega_z \dot{y}$ and $2m\omega_z \dot{x}$ are Coriolis force terms that appear due to the rotation.

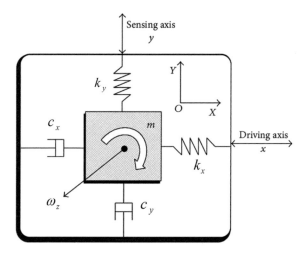

**FIGURE 3.21**   Schematic illustration of MEMS vibratory gyroscope.

Naturally, when the gyro is at rest (no rotation to be sensed), the proof mass works under simple harmonic vibration. The source for the force that is causing the vibration may differ for types of MEMS gyros but it is usually an electrostatic, piezoelectric, electromagnetic or electrothermal force [27]. When the rotation starts, Coriolis terms cause dynamic coupling between the drive and sense modes of vibration. As a result of the coupling, the suspended proof mass starts to vibrate in the sense direction as well and by measuring this effect the rotation rate about z axis ($\omega_z$), which is the measurement output of the gyro, can be obtained.

## 3.6  AUXILIARY ATTITUDE SENSORS FOR SMALL SATELLITES

So far we have presented the attitude sensors that are common for most spacecraft missions. They are "standard" sensors that the attitude system designers and engineers are used to. However, size, mass and power budgets limit the possibilities for using any sensor on small satellites, especially on very small ones such a cubesats. Consequently, to save space and mass, using any of the mission payloads or other equipment onboard as an attitude sensor has been one of the research areas for small satellite attitude estimation.

Certainly, one of the most prominent payloads that can be used as attitude sensors are cameras. The recent advancement in technology has enabled smaller cameras with high resolution. Moreover, the obtained images can be processed in a short time and processes such as limb detection and image recognition can be done in real time. As a result, there is extensive research on methods to use these cameras for attitude determination on small satellites [28–32]. In case the camera, which is already on the satellite for other tasks such as ground observation, is also used for attitude determination, a vast space can be saved since there is no more need for rather big sensors (e.g. star tracker) for precise attitude determination. Besides, unlike traditional attitude sensors, the reference for attitude estimation (e.g. Sun, stars) can be different at different stages of the mission, when the camera is implemented as an attitude sensor. Thus far, there are studies to use the camera to estimate the full attitude using the Earth images [29, 33], star images (like a star tracker) [31] or to provide a direction vector to the Earth, Moon and Sun [34]. They can be used to estimate the attitude rate as well based on the Earth or star images [30, 35].

Another useful equipment onboard the satellite in term of attitude estimation is solar panels. Depending on the voltage information read by the solar cells a coarse direction to the Sun can be estimated. Nevertheless, considering the very low accuracy of such attitude measurement and also how tiny the Sun sensors are today, this is not a preferred method for recent missions. Yet solar panel data may be used to verify the attitude estimation results [36].

Last, although it is traditionally a positioning sensor, the Global Navigation Satellite System (GNSS) measurements can be also used for attitude estimation when there are more than two receivers onboard the small satellite [37]. A more detailed discussion for this concept will be provided in Chapter 6.

## 3.7 SUMMARY

This chapter presents the functioning principles for small satellite attitude sensors. All of the standard attitude sensors, which are magnetometers, Sun sensors, Earth horizon sensors, star trackers and gyroscopes are discussed in turn enveloping their different types. Specifically, the newer types of these sensors, mostly based on MEMS technologies and COTS (commercial off-the-shelf) equipment are introduced. In this scope, several recent publications on small satellite attitude determination were surveyed. Schematics and figures for measurement geometries are included to help the reader when investigating how the sensor works.

Sections for each sensor also include a discussion regarding popularity and suitability of the sensor type for small satellite missions. Example sensors, suitable to be used on small satellites, are presented in associated tables. Last section of the chapter discusses the auxiliary attitude sensors, which can be used instead or in addition to the standard sensors, especially for nanosatellite missions.

## REFERENCES

1. J. C. Springmann and J. W. Cutler, "Flight results of a low-cost attitude determination system," *Acta Astronaut.*, vol. 99, no. 1, pp. 201–214, 2014.
2. D. Ivanov, M. Ovchinnikov, N. Ivlev, and S. Karpenko, "Analytical study of microsatellite attitude determination algorithms," *Acta Astronaut.*, vol. 116, pp. 339–348, 2015.
3. S. Nakasuka, K. Miyata, Y. Tsuruda, Y. Aoyanagi, and T. Matsumoto, "Discussions on attitude determination and control system for micro/nano/pico-satellites considering survivability based on Hodoyoshi-3 and 4 experiences," *Acta Astronaut.*, vol. 145, no. (December 2017), pp. 515–527, 2018.
4. A. Slavinskis et al., "Flight results of ESTCube-1 attitude determination system," *J. Aerosp. Eng.*, vol. 29, no. 1, pp. 4015014–4015017, 2015.
5. G. B. Hospodarsky, "Spaced-based search coil magnetometers," *J. Geophys. Res. Sp. Phys.*, vol. 121, no. 12, pp. 12068-12079, 2016.

6.  E. Matandirotya, R. R. Van Zyl, D. J. Gouws, and E. F. Saunderson, "Evaluation of a commercial-off-the-shelf fluxgate magnetometer for CubeSat Space magnetometry," *J. Small Satell.*, vol. 2, no. 1, pp. 133–146, 2013.

7.  A. M. Cruise, J. A. Bowles, T. J. Patrick, and C. V. Goodall, *Principles of Space Instrument Design*. Cambridge: Cambridge University Press, 1998.

8.  V. Renaudin, M. H. Afzal, and G. Lachapelle, "Complete triaxis magnetometer calibration in the magnetic domain," *J. Sensors*, vol. 2010, 1–10, 2010, Article ID 967245.

9.  L. Y. Liu et al., "The magneto-resistive magnetometer of BCU on the tatiana-2 satellite," *Terr. Atmos. Ocean. Sci.*, vol. 23, no. 3, pp. 317–326, 2012.

10. M. Hong, "SSC18-PII-25 In-Flight Magnetometer Calibration with Temperature Compensation for PHOENIX CubeSat."

11. L. H. Regoli et al., "Investigation of a low-cost magneto-inductive magnetometer for space science applications," *Geosci. Instrumentation, Methods Data Syst.*, vol. 7, no. 1, pp. 129–142, 2018.

12. J. R. Wertz, Ed., *Spacecraft Attitude Determination and Control*, vol. 73. Dordrecht: Springer Netherlands, 1978.

13. D. Faizullin, Improvement of Analog Sun Sensor Accuracy and Data Processing for Sun Vector Determination. Kita-kyushu, Japan: Kyushu Institute of Technology, 2018.

14. F. B. Zazzera, "Course Notes Spacecraft Attitude Dynamics and Control Part 2 : Attitude determination and control systems," 2011.

15. B. M. de Boer et al., "MiniDSS: a low-power and high-precision miniaturized digital sun sensor," in *International Conference on Space Optics — ICSO 2012*, Corsica, France, October 2012, p. 88.

16. H. L. Hallock, G. Welter, D. G. Simpson, and C. Rouff, *ACS Without an Attitude*. London: Springer London, 2017.

17. A. Antonello, L. Olivieri, and A. Francesconi, "Development of a low-cost sun sensor for nanosatellites," *Acta Astronaut.*, vol. 144, no. February 2017, pp. 429–436, 2018.

18. A. G. Sáez, J. M. Quero, and M. A. Jerez, "Earth sensor based on thermopile detectors for satellite attitude determination," *IEEE Sens. J.*, vol. 16, no. 8, pp. 2260–2271, 2016.

19. S. W. Janson, B. S. Hardy, A. Y. Chin, D. L. Rumsey, D. A. Ehrlich, and D. A. Hinkley, "Attitude control on the Pico Satellite Solar Cell Testbed-2," *Proceedings of the 26th Annual AIAA/USU Conference Small Satell.* Utah, USA, vol. SSC12, pp. SSC12-II–1, 2012.

20. T. Nguyen, K. Cahoy, and A. Marinan, "Attitude determination for small satellites with infrared earth horizon sensors," *J. Spacecr. Rockets*, vol. 55, no. 6, pp. 1466–1475, 2018.

21. H. E. Soken, J. C. Van Der Ha, and S. Sakai, "Spin-axis tilt estimation algorithm with validation by real data," *Adv. Astronaut. Sci.*, vol. 160, pp. 2631–2646, 2017.

22. J. C. Van Der Ha, "Spin-axis attitude determination using in-flight data," *J. Guid. Control. Dyn.*, vol. 33, no. 3, pp. 768–781, 2010.

23. K. Ho, "A survey of algorithms for star identification with low-cost star trackers," *Acta Astronaut.*, vol. 73, pp. 156–163, 2012.

24. B. B. Spratling and D. Mortari, "A survey on star identification algorithms," *Algorithms*, vol. 2, no. 1, pp. 93–107, 2009.

25. F. L. Markley and J. L. Crassidis, *Fundamentals of Spacecraft Attitude Determination and Control*. New York, NY: Springer New York, 2014.

26. NASA, "State of the Art Small Spacecraft Technology.," no. December, p. 207, 2018.

27. D. Xia, C. Yu, and L. Kong, "The development of micromachined gyroscope structure and circuitry technology," *Sensors (Switzerland)*, vol. 14, no. 1, pp. 1394–1473, 2014.

28. Y. Iwasaki et al., "Development and initial on-orbit performance of multi-functional attitude sensor using image recognition," in *AIAA/USU Conference on Small Satellites*, Utah, USA, 2019.

29. D. Modenini and M. Zannoni, "A high accuracy horizon sensor for small satellites," in *2019 IEEE International Work. Metrol. AeroSpace, Metroaerosp. 2019 - Proc.*, Torino, Italy, pp. 451–456, 2019.

30. S. A. Rawashdeh and J. E. Lumpp, "Image-based attitude propagation for small satellites using RANSAC," *IEEE Trans. Aerosp. Electron. Syst.*, vol. 50, no. 3, pp. 1864–1875, 2014.

31. N. Korn, F. Baumann, R. Wolf, and K. Brieß, "Multifunctional optical attitude determination sensor for Picosatellites," in *68th International Astronautical Congress*, Adelaide, Australia, no. September, pp. 25–29, 2017.

32. D. Amartuvshin and K. Asami, "Vision-based attitude determination system for small satellites using unscented Kalman filter," in *68th International Astronautical Congress*, Adelaide, Australia, no. September, pp. 25–29, 2017.

33. S. Koizumi, Y. Kikuya, K. Sasaki, Y. Masuda, Y. Iwasaki, and K. Watanabe, "Development of attitude sensor using deep learning," in *AIAA/USU Conference on Small Satellites*, Utah, USA, 2018.

34. A. Dagvasumberel, "A visual-inertial attitude propagation for resource-constrained small satellites," *J. Aeronaut. Sp. Technol.*, vol. 12, no. 1, pp. 65–74, 2019.

35. B. Y. K. Ikuya and S. M. Atunaga, "On-board relative attitude determination and propagation using earth sensor," in *32nd International Symposium on Space Technology and Science*, Fukui, Japan, 2019.

36. O. J. Kim et al., "In-orbit results and attitude analysis of the snuglite cube-satellite," *Appl. Sci.*, 10, no. 7, 2020.

37. A. Hauschild, U. Mohr, M. Markgraf, and O. Montenbruck, "Flight results of GPS-based attitude determination for the microsatellite Flying Laptop," *Navig. J. Inst. Navig.*, vol. 66, no. 2, pp. 277–287, 2019.

# 4 Attitude Sensor Measurement Models

To estimate the attitude of the satellite, we need the unit vectors measured by the sensors in the satellite body frame[1] and the corresponding unit vectors calculated in the reference frame, with respect to which the attitude is estimated. The measurement models simply show us how these measured and calculated unit vectors can be related to estimate the attitude.

In this section, sensor measurement models for attitude estimation are presented along with the procedures to calculate the reference directions for each sensor. These procedures are discussed in terms of their accuracy, computational load and suitability to be run in real-time on a small satellite microprocessor. Various advantages and drawbacks of using different sensors and so having different references (e.g. Sun, stars) for attitude estimation are listed.

## 4.1 MAGNETOMETER MODELS

### 4.1.1 MAGNETOMETER MEASUREMENT MODEL

The three-axis magnetometer (TAM) onboard the satellite measures the components of the magnetic field vector in the body frame. Different than other sensors, it provides both the direction and magnitude. Assuming a full model for the magnetometers including various errors that may be effecting a small satellite magnetometer, the measurement equation can be given as

$$B_b = \left(I_{3\times3} + D\right)^{-1}\left(AB_r + b_m + \eta_b\right),\tag{4.1}$$

where, $B_b$ is the magnetometer measurements in the body frame, $B_r$ is the reference magnetic field vector in the reference frame (ECI or orbit), $b_m = [b_{mx} \ b_{my} \ b_{mz}]^T$ is the bias vector, $D$ is the scaling matrix which reflects the scaling, symmetrical soft iron effects and nonorthogonality $\eta_b$ is the zero mean Gaussian white noise with the characteristic of

$$E\left[\eta_{bk}\eta_{bj}^T\right] = I_{3\times3}\sigma_m^2\delta_{kj}\tag{4.2}$$

and $\sigma_m$ is the standard deviation of each magnetometer error.

Note that here we assume a symmetrical $D$ matrix with six independent terms as

$$D = \begin{bmatrix} D_{11} & D_{12} & D_{13} \\ D_{12} & D_{22} & D_{23} \\ D_{13} & D_{23} & D_{33} \end{bmatrix}\tag{4.3}$$

and neglect any non-symmetrical soft-iron errors and sensor misalignment. $D$ matrix with nine independent components, accounting also for these errors, can be estimated without ambiguity only if we have an independent knowledge about attitude, such as the one from a star-tracker, or if we can apply a specific calibration maneuver [1].

Magnetometers are sensors that are highly prone to errors. Their measurements are affected by the magnetic disturbances generated by the onboard electronics etc. On large spacecraft, these errors can be minimized at a degree by locating the sensor as far as possible from the disturbance sources. However, especially for nanosatellite missions, this is not an option as the size of the satellite should be kept to a minimum. In this case, the magnetometers must be in-flight calibrated. For a fully calibrated magnetometer, the model reduces to

$$\tilde{\boldsymbol{B}}_b = A\boldsymbol{B}_r + \boldsymbol{\eta}_b, \tag{4.4}$$

where $\tilde{\boldsymbol{B}}_b$ denotes the calibrated measurements. Magnetometer calibration algorithms will be discussed in detail in Chapters 15 and 16.

Table 4.1 presents the advantages and disadvantages of using the magnetometers as attitude sensor[2].

### 4.1.2 MODELS FOR THE EARTH'S MAGNETIC FIELD IN THE REFERENCE FRAME

One of the well-known models for obtaining the Earth magnetic field vector components in the reference frame is the International Geomagnetic Reference Field (IGRF) model. Inputs to the IGRF model are the radial distance from the center of the Earth $(r)$, geocentric colatitude $(\theta$ – in degree), the east longitude $(\phi$ – in degree) and time $t$. Eq. (4.5) returns the magnetic field vector in spherical polar coordinates using Gauss coefficients $(g_n^m, h_n^m)$, which are functions of time and given in units of nT.

$$B(r,\theta,\phi,t) = -\nabla \left\{ a \sum_{n=1}^{N} \sum_{m=0}^{n} \left(\frac{a}{r}\right)^{n+1} \left[ g_n^m(t)\cos(m\phi) + h_n^m(t)\sin(m\phi) \right] P_n^m(\cos\theta) \right\}. \tag{4.5}$$

---

**TABLE 4.1**

**Advantages and Disadvantages of Using Magnetometer as Attitude Sensor**

| Advantages | Disadvantages |
|---|---|
| - Sensor itself is cheap, lightweight and small. It also consumes low power.<br>- Sensor can operate over a wide range of temperatures and robust against radiation.<br>- Sensor can operate stably for extended durations.<br>- Magnetic field is a continuous reference for LEO satellites without any occultation.<br>- Measuring both the direction and magnitude of the magnetic field, a magnetometer alone can provide 3-axis attitude estimate as the satellite travels on its orbit. | - Available only at lower altitudes.<br>- Limited by modeling accuracy due to inaccuracies in the knowledge of the Earth's magnetic field.<br>- Sensor must be in-orbit calibrated for small satellites. |

---

Here $a$ is magnetic reference spherical radius as $a = 6371.2$ km, and $P_n^m(\cos\theta)$ are the Schmidt quasi-normalized associated Legendre functions of degree $n$ and order $m$ [2]. Coefficients of the IGRF model are updated every 5 years and published by NOAA (National Oceanic and Atmospheric Administration). The time dependence of the Gauss coefficients can be denoted as:

$$g_n^m(t) = g_n^m(T_0) + \dot{g}_n^m(T_0)(t - T_0), \tag{4.6a}$$

$$h_n^m(t) = h_n^m(T_0) + \dot{h}_n^m(T_0)(t - T_0). \tag{4.6b}$$

$t$ is time of interest in the units of years and $T_0$ is the epoch time that is preceding $t$ and an exact multiple of 5 year such that $T_0 \le t < (T_0 + 5.0)$. The most recent IGRF model is the IGRF-13 as of the preparation date of this book and it is valid until 2025.

Inputs to the IGRF model should be provided by the orbit estimation/propagation algorithm or directly by a GNSS (Global Navigation Satellite System) receiver onboard the satellite. Outputs, which are in spherical polar coordinates, must be transformed to the reference frame of interest before being used in Eq. (4.1). Although the IGRF model allows calculations up to the maximum truncation degree of $N = 13$, such high degree of model is almost never used for attitude estimation purposes. Because after a certain degree, the magnetometer sensor noise, especially for small satellites, is the dominating error source in the estimation process. Calculating the model at a higher degree does not improve the accuracy at all and such computationally demanding calculation is not reasonable for running on-orbit. Subject to changes depending on the processor capacity and the magnetometer quality, a model at a degree of $N = 4$ or $N = 5$ is usually preferred. Figure 4.1 presents the components of the Earth's magnetic field vector in orbit frame for a LEO small satellite for 2 days, when the IGRF model (at $N = 5°$) is used for calculations. The satellite is at Sun-synchronous orbit at an approximate altitude of 680 km. The orbit is circular one with an inclination of 98.1°. The unit for components is nanoTesla (nT).

Especially for nanosatellites, the IGRF model, even at degrees such as $N = 4$, may be computationally demanding and a dipole approximation to the model with $N = 1$ may be sufficient regarding the desired attitude estimation accuracy. This is a useful approximation especially for satellites at a higher altitude (above 7000 km from ground) and on a circular orbit. The dipole model in this case is as given below [3, 4]:

$$B_{ox}(t) = \frac{M_e}{r_0^3}\left\{\cos(\omega_0 t)\left[\cos(\varepsilon)\sin(i) - \sin(\varepsilon)\cos(i)\cos(\omega_e t)\right]\right.$$
$$\left. - \sin(\omega_0 t)\sin(\varepsilon)\sin(\omega_e t)\right\},$$

$$B_{oy}(t) = -\frac{M_e}{r_0^3}\left[\cos(\varepsilon)\cos(i) + \sin(\varepsilon)\sin(i)\cos(\omega_e t)\right] \tag{4.7}$$

$$B_{oz}(t) = \frac{2M_e}{r_0^3}\left\{\sin(\omega_0 t)\left[\cos(\varepsilon)\sin(i) - \sin(\varepsilon)\cos(i)\cos(\omega_e t)\right]\right.$$
$$\left. - 2\sin(\omega_0 t)\sin(\varepsilon)\sin(\omega_e t)\right\},$$

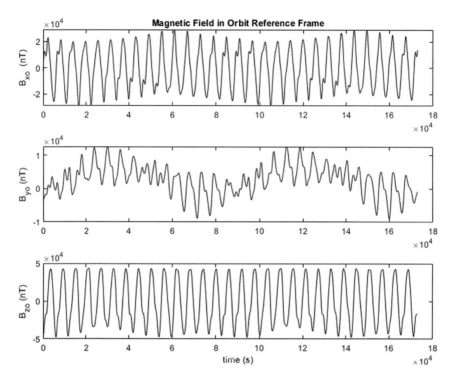

**FIGURE 4.1** Magnetic field components in orbit frame for a Sun-synchronous LEO satellite. IGRF model is used for calculations.

Here $M_e = 7.71 \times 10^{15}$ Wb. m is the magnetic dipole moment of the Earth[3], $i$ is the orbit inclination, $\omega_e = 7.29 \times 10^{-5}$ rad/s is the spin rate of the Earth, $\varepsilon = 9.3°$ is the magnetic dipole tilt angle, $r_0$ is the distance between the center of mass of the satellite and the Earth, $\omega_0$ is the orbital angular velocity of the spacecraft found as $\omega_0 = \sqrt{\mu / r_0^3}$ and $\mu = 3.9860044 \times 10^{14}$ m³/s² is the Earth's gravitational constant.

Note that, the outputs of the model, $B_{ox}$, $B_{oy}$ and $B_{oz}$, are the components of the magnetic field vector in the orbit frame. It is possible to further simplify these equations assuming a non-tilted model ($\varepsilon = 0°$) as given in reference [5]. Figure 4.2 gives the simulated results for the dipole model calculated magnetic field components in orbit frame. Calculations are for the same spacecraft, for which we presented the IGRF model calculated magnetic field components in Figure 4.1. When two figures are compared, it can be easily noticed that on average the calculations for two models match. Yet it should be noted that using the dipole model degrades the accuracy of reference magnetic field direction more than 10° on average.

## 4.2 SUN SENSOR MODELS

### 4.2.1 SUN SENSOR MEASUREMENT MODEL

The Sun sensor measures the Sun direction vector in the body axes of the spacecraft. Unless a coarse Sun sensor is used, errors other than the noise may be assumed

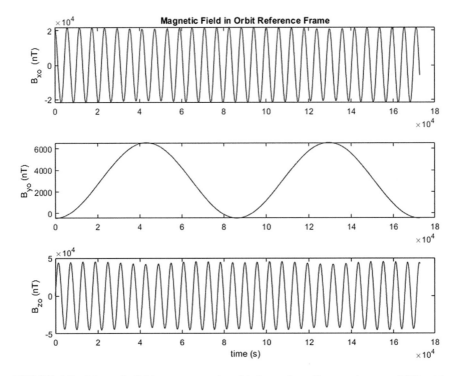

**FIGURE 4.2** Magnetic field components in orbit frame for a Sun-synchronous LEO satellite. Dipole model is used for calculations.

negligible, when building the measurement model. In this case the Sun sensor measurements can be given as

$$S_b = AS_r + \eta_s. \tag{4.8}$$

where $S_r$ is the Sun direction vector in reference frame and $S_b$ is the Sun sensor measurements in body frame which are corrupted with, $\eta_s$, the zero mean Gaussian white noise with the characteristic of

$$E\left[\eta_{sk}\eta_{sj}^T\right] = I_{3\times3}\sigma_s^2\delta_{kj}. \tag{4.9}$$

$\sigma_s$ is the standard deviation of Sun sensor noise. In contrast to the assumption for presented model, if a coarse Sun sensor is used an in-orbit calibration of the sensor for compensating various errors including the Earth's albedo effects may be necessary before the measurements are used for attitude estimation. See references [6–8] for detailed discussion on calibration of Sun sensors used on small satellites.

Table 4.2 presents using the advantages and disadvantages of using the Sun sensors as attitude sensor.

**TABLE 4.2**

**Advantages and Disadvantages of Using Sun Sensor as Attitude Sensor**

| Advantages | Disadvantages |
|---|---|
| - Sensor itself is cheap, lightweight and small.<br>- It consumes very low power.<br>- As a reference source Sun is bright enough, so the detection procedure is not complicated.<br>- Sun is an unambiguous orientation source.<br>- Sun direction is vital information for most missions. Sun sensor alone can provide this information. | - Sun is not visible during eclipse.<br>- For analog sensors albedo effect, which is a function of the attitude, may cause significant errors.<br>- Sun's angular diameter of 0.53° (viewed from Earth) limits the accuracy to 1 arc minute. |

### 4.2.2 MODELS FOR THE SUN DIRECTION VECTOR IN THE REFERENCE FRAME

The Sun model calculates the Sun direction vector in inertial frame and determines whether the Sun is visible or not to the spacecraft. As the satellite travels along its orbit around the Earth, the Sun direction vector in the inertial frame (centered to the spacecraft) changes. However, for a LEO spacecraft, the Earth center to spacecraft radial distance is negligible compared to 1AU (astronomical unit) distance from the Earth to the Sun. These cyclic changes, in the order of few arcseconds, are well below the accuracy of the Sun sensors and the required accuracy of the reference model [9].Thus we can assume the Sun direction vector in an inertial frame centered to the spacecraft is same as the direction vector in ECI:

$$S_i^{s/c \to Sun} \approx S_i^{Earth \to Sun}. \tag{4.10}$$

In result the Sun direction vector can be calculated with a time-dependent model. The following standard algorithm describes the true motion of the Earth around the inertially fixed Sun in the form of the Sun orbiting the Earth being fixed in inertial space. The right ascension of the ascending node of this virtual orbit of the Sun around the Earth is 0° by definition; the argument of perigee ($\omega$) changes negligibly and is assumed constant as 282.94°. The reference epoch of this first-order model is the 1st of January 2000, 12:00:00 pm. This model is accurate up to an accuracy of 0.01° and valid from 1950 to 2050 [10]. We first start by calculating the Julian centuries from the epoch to the exact date:

$$T_{UT1} = \frac{JD_{UT1} - 2451545.0}{36525}, \tag{4.11}$$

where

$$JD_{UT1} = 367 \text{ y} - INT \left\{ \frac{7 \left\{ y + INT \left[ (m+9)/12 \right] \right\}}{4} \right\} + INT \left\{ \frac{275 \text{ m}}{9} \right\}$$
$$+ \frac{h + min/60 + s/3600}{24} + d + 1721013.5 \tag{4.12}$$

is the Julian date. Here y, m, d, h, min and s denote the year, month, day, hour, minute and second, respectively, for the exact date and time. INT means only the integer part of the variable should be taken.

Then using this time information, we calculate parameters that are defining the Sun position in ECI. The mean longitude of the Sun is

$$\lambda_{M_{Sun}} = 280.460° + 36000.770 T_{UT1}. \tag{4.13}$$

The mean anomaly of the sun is

$$M_{Sun} = 357.5277233° + 35999.05034 T_{UT1}. \tag{4.14}$$

Here the first term on the right is the mean anomaly of the Sun at epoch and the second term is the change of the mean anomaly during a Julian Day ($0.9856474° \cdot 365.25 = 360°$).

The ecliptic longitude of the Sun introduces the correction of the Sun "traveling" on an eccentric orbit.

$$\lambda_{ecliptic} = \lambda_{M_{Sun}} + 1.914666471° \sin(M_{Sun}) + 0.019994643 \sin(2M_{Sun}) \tag{4.15}$$

The linear model of the ecliptic of the Sun is

$$\varepsilon = 23.439291° - 0.0130042 T_{UT1} \tag{4.16}$$

which is close to constant.

Finally, the unit sun vector is given in the inertial frame as

$$S_i = \begin{bmatrix} \cos \lambda_{ecliptic} \\ \sin \lambda_{ecliptic} \cos \varepsilon \\ \sin \lambda_{ecliptic} \sin \varepsilon \end{bmatrix}. \tag{4.17}$$

If the reference frame is selected as the orbit frame instead, the calculated Sun direction vector, $S_i$ must be transformed to the orbit frame using the transformation in Eq. 2.1 (or Eqs. 2.2 and 2.3) whose inputs are inertial position and velocity of the satellite. Moreover, the shadow function given in Eqs. (2.18) and (2.19) can be used to define the eclipse conditions, when the Sun is not visible to the Sun sensor. Figure 4.3 presents the modeled Sun direction vector components in orbit reference frame for the same Sun-synchronous spacecraft for which we presented the magnetic field models in the previous section. The shadowed areas represent the eclipse that is defined using the shadow function. It is clearly seen that whenever the $S_{o,z}$ components get closer to 1, the spacecraft goes into eclipse.

## 4.3 EARTH HORIZON SENSOR MODELS

### 4.3.1 EARTH HORIZON SENSOR MEASUREMENT MODEL

Assuming that the Earth sensor is providing the direction vector to the Earth's center (nadir vector) as measurements, a standard vector measurement model can be used

$$E_b = AE_r + \eta_e. \tag{4.18}$$

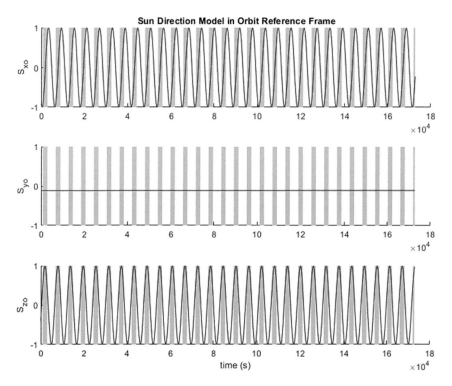

**FIGURE 4.3** Sun direction vector in orbit frame for a Sun-synchronous LEO satellite. Shadowed areas represent the eclipse.

Here $E_r$ is the nadir vector in reference frame and $E_b$ is the Earth sensor measurements in body frame which are corrupted with, $\eta_e$, the zero mean Gaussian white noise with the characteristic of

$$E\left[\eta_{ek}\eta_{ej}^T\right] = I_{3x3}\sigma_e^2\delta_{kj}. \tag{4.19}$$

$\sigma_e$ is the standard deviation of Earth sensor noise.

It should be noted here the model in Eq. (4.18) assumes a sensor that is calibrated against the biases and systematic errors. Bias is inherent to sensors and arises due to the electronics and mechanics (if the sensor is a scanning type). Systematic errors, which are similar to bias in nature, may be due to the Earth's oblateness and variation of Earth's radiation. A large portion of these systematic errors can be compensated with appropriate modeling of the Earth's shape and modeling of sensor optics and Earth's radiation [11]. Moreover, seasonal and latitudinal variation of Earth's radiation contributes also to the stochastic error. This error term may be incorporated into the sensor noise when modeling the measurements and should be accounted for when designing the attitude estimation algorithm.

The advantages and disadvantages of using the Earth horizon sensor for attitude estimation are listed in Table 4.3.

**TABLE 4.3**
**Advantages and Disadvantages of Using Earth Horizon Sensor as Attitude Sensor**

| Advantages | Disadvantages |
|---|---|
| - Earth is always visible to the LEO satellites. <br> - Earth is largely an unambiguous source (Moon interference is possible). <br> - A static Earth sensor provides direct roll and pitch angle measurements with respect to the orbit frame. | - Limited accuracy not just because of sensor errors but also because of the modeling errors (e.g. for Earth's limb). <br> - Sensors must be protected from the Sun. <br> - Scanning-type Earth horizon sensors require moving parts (unless the satellite itself is spinning). <br> - Sensors are affected by Earth's oblateness and variation of Earth radiation. <br> - Especially static Earth sensors can be used for a limited range of missions in terms of attitude, altitude and orbit type. |

### 4.3.2 Models for the Earth Direction Vector in the Reference Frame

Modeling the nadir vector (direction vector to the Earth's center) is a straightforward routine. If the attitude is defined with respect to the orbit frame, then $E_o = \begin{bmatrix} 0 & 0 & 1 \end{bmatrix}^T$. If the ECI frame is selected as the reference frame, then the Earth direction vector in ECI frame is defined using the spacecraft's position information ($r_i$) as:

$$E_i = -\frac{r_i}{|r_i|}. \tag{4.20}$$

## 4.4 STAR TRACKER MEASUREMENT MODEL

A star tracker provides directly the attitude quaternion as the output. Accounting for only the noise and neglecting any bias or systematic errors for the measurements, the measurement model for the star tracker can be given as

$$q_{str} = q_{noise}^{-1} \otimes q_{true}, \tag{4.21}$$

where, $q_{str}$ is the measured quaternion vector by the star tracker, $q_{true}$ is the true attitude quaternion for the sensor and $q_{noise}$ is the sensor noise quaternion vector, which can be modeled as $I = \pi \left( d_o^4 - d_i^4 \right)/64$ as a result of small angle approximation. Here $\eta_{str,\,c}$ and $\eta_{str,\,b}$ are random Gaussian white noise for star tracker measurements in cross-boresight ($x$, $y$) and boresight axes ($z$), respectively. In Eq. (4.21) $\otimes$ defines the quaternion multiplication. If we rewrite the quaternion vector by $q = \begin{bmatrix} g^T & q_4 \end{bmatrix}^T$, so $g = \begin{bmatrix} q_1 & q_2 & q_3 \end{bmatrix}^T$, then for $\Xi(q)$ matrix shown as

$$\Xi(q) = \begin{bmatrix} q_4 I_{3\times3} + [g\times] \\ -g^T \end{bmatrix}, \tag{4.22}$$

the quaternion multiplication can be denoted with

$$q' \otimes q = \left[ \Xi(q) \quad \vdots \quad q \right] q'. \tag{4.23}$$

Here $[g\times]$ is the cross-product matrix as,

$$\left[ g\times \right] = \begin{bmatrix} 0 & -g_3 & g_2 \\ g_3 & 0 & -g_1 \\ -g_2 & g_1 & 0 \end{bmatrix}. \tag{4.24}$$

One should be careful here regarding the definition of quaternion multiplication. We use $q' \otimes q$ product, rather than Hamilton's $q' \odot q$ product. Last, the quaternion inverse, which also appears in Eq. (4.21) is defined as:

$$q^{-1} = \left[ -g^T \quad q_4 \right]^T / \|q\|^2. \tag{4.25}$$

Note that we assumed a multiplicative error model for the star trackers, which appears to be useful when designing an attitude filter and for attitude error analyses compared to an additive error model [12].

In Eq. (4.21) the star tracker gives the measurements in its sensor frame. Thus the output is the attitude quaternion of sensor frame with respect to the inertial frame. To have the measurements in the spacecraft body frame they must be multiplied with the alignment quaternion as $q_{str}^{s/c \to i} = q_{align} \otimes q_{str}^{str \to i}$.

Table 4.4 provides the advantages and disadvantages of using star tracker as attitude sensor.

Please note that, almost all of the recent star trackers provide directly the attitude quaternion as output. However, if we have the star direction vector in body frame as measurements than the measurement model takes form of the standard vector measurement equation:

$$C_b = AC_i + \eta_c. \tag{4.26}$$

Here $\eta_c$ is the zero mean Gaussian white noise with the standard deviation of $\sigma_c$. Direction vectors for the stars in ECI, $C_i$, are all obtained from star catalogs.

---

**TABLE 4.4**

**Advantages and Disadvantages of Using Star Tracker as Attitude Sensor**

| Advantages | Disadvantages |
|---|---|
| - Directly provides the attitude in terms of quaternions. | - Requires Sun protection. |
| - Essentially orbit independent (except for velocity aberration). | - Sensor itself is expensive and rather large compared to other sensors. |
| - Can provide measurements without interruption. | - Processing of the measurements (e.g. identification) causes time delay. |
| - Attitude sensor with highest accuracy. | |

---

## 4.5 GYROSCOPE MEASUREMENT MODEL

Gyros onboard the satellite measure the body angular rate vector with respect to the ECI frame. The measurement model for gyros is

$$\tilde{\omega}_{bi} = \omega_{bi} + b_g + \eta_{gn}, \tag{4.27}$$

where, $\tilde{\omega}_{bi}$ is the measured angular rates of the satellite, $b_g$ is the gyro bias vector as $b_g = \begin{bmatrix} b_{gx} & b_{gy} & b_{gz} \end{bmatrix}^T$ and $\eta_{gn}$ is the zero mean Gaussian white noise with the characteristic of

$$E\left[\eta_{gnk}\eta_{gnj}^T\right] = I_{3x3}\left(\frac{\sigma_v^2}{\Delta t}\right)\delta_{kj}. \tag{4.28}$$

Here, $\sigma_v$ is the gyro angular random walk value and $\Delta t$ is the gyro sampling period. Gyro bias is modeled as a random walk process,

$$\dot{b}_g = \eta_{gb}, \tag{4.29}$$

where $\eta_{gb}$ is also zero mean Gaussian white noise with the characteristic of

$$E\left[\eta_{gbk}\eta_{gbj}^T\right] = I_{3x3}\left(\frac{\sigma_u^2}{\Delta t}\right)\delta_{kj}. \tag{4.30}$$

Here, $\sigma_u$ is the gyro rate random walk value. Variation in the gyro bias may be due to various factors depending on the used gyro type but one main factor is the temperature.

A more general gyro model includes scale factors and misalignments, as

$$\tilde{\omega}_{bi} = \left(I_{3\times3} + S\right)\omega_{bi} + b_g + \eta_{gn}, \tag{4.31}$$

where $S$ is the matrix of scale factor and misalignment. These terms are usually modeled as constant and can be estimated and compensated in real-time in orbit.

Table 4.5 gives the advantages and disadvantages of using the gyro as attitude sensor.[4]

---

**TABLE 4.5**

**Advantages and Disadvantages of Using Gyro as Attitude Sensor**

| Advantages | Disadvantages |
|---|---|
| - Sensor itself is cheap, lightweight and small. | - Provides only relative attitude information. |
| - Self-dependent (does not require external reference for measurements). | - Bias is subject to drift in time and must be estimated and compensated in real-time. |
| - Provides measurements with high sampling frequency. | |
| - Having gyros provides advantages in the attitude filter design. | |

---

## 4.6 SUMMARY

In this chapter, the measurement models for the attitude sensors are presented. For each attitude sensor, first the measurement model is given. Then the procedure to compute the reference direction model is described. These algorithms for reference direction calculation are presented considering they are to be used onboard a small satellite and concerning about their computational load as much as the accuracy. In the sensor measurement models, only for the magnetometer and gyro the full-model accounting for different sensor errors is given. As will be discussed in Chapters 15 and 16, a real-time calibration for small satellite magnetometers is usually inevitable. As to the gyros, their calibration (estimation of gyro bias) is a part of the usual attitude filtering algorithm. For other sensors we give the common models neglecting the sensor errors other than Gaussian noise. Including the other errors into the model may be needed considering the quality of the sensor (i.e. if a coarse Sun sensor is used).

In this chapter, we also discussed the advantages and disadvantages of using each attitude sensor and the associated models. We provided this information considering the general specifications of each sensor. The reader shall keep in mind that using a specific type of each sensor may bring about additional advantages/disadvantages.

## NOTES

1 Although most attitude sensors provide the measurements in its own frame, this measurement can be easily transformed to the satellite body frame using the known sensor alignment.
2 There are necessary conditions for such estimation. Besides, although there are different methods to estimate the attitude using only magnetometer measurements, they all also require the modeled magnetic field information in the reference frame. See for example [13–16]. Attitude filtering methods, which can provide attitude estimate using only the magnetometer measurements, will be discussed in Chapter 9.
3 Reader shall note that this value together with the magnetic dipole tilt angle is showing yearly variations. Please see the website by Kyoto University's Data Analysis Center for Geomagnetism and Space Magnetism http://wdc.kugi.kyoto-u.ac.jp/poles/polesexp.html
4 When there are gyros onboard the spacecraft, there is no need to use dynamics information in an attitude filter. This provides certain advantages especially if the dynamic model has several uncertainties. Details will be discussed in Chapter 9.

## REFERENCES

1. B. A. Riwanto, T. Tikka, A. Kestila, and J. Praks, "Particle swarm optimization with rotation axis fitting for magnetometer calibration," *IEEE Trans. Aerosp. Electron. Syst.*, vol. 9251, no. c, pp. 1–1, 2017.
2. E. Thébault et al., "International geomagnetic reference field: The 12th generation international geomagnetic reference field - The twelfth generation," *Earth, Planets Sp.*, vol. 67, no. 1, 2015.
3. H. E. Soken and C. Hajiyev, "UKF-based reconfigurable attitude parameters estimation and magnetometer calibration," *IEEE Trans. Aerosp. Electron. Syst.*, vol. 48, no. 3, pp. 2614–2627, July 2012.
4. P. Sekhavat, Q. G. Q. Gong, and I. M. Ross, "NPSAT1 parameter estimation using unscented Kalman filtering," *2007 American Control Conference.*, New York, USA, pp. 4445–4451, 2007.

5. D. S. Ivanov, M. Y. Ovchinnikov, V. I. Penkov, D. S. Roldugin, D. M. Doronin, and A. V. Ovchinnikov, "Advanced numerical study of the three-axis magnetic attitude control and determination with uncertainties," *Acta Astronaut.*, vol. 132, no. October 2016, pp. 103–110, 2017.
6. S. A. O'Keefe and H. Schaub, "Consider-filter-based on-orbit coarse sun sensor calibration sensitivity," *J. Guid. Control. Dyn.*, vol. 40, no. 5, pp. 1296–1299, 2017.
7. J. C. Springmann and J. W. Cutler, "On-orbit calibration of photodiodes for attitude determination," *J. Guid. Control. Dyn.*, vol. 37, no. 6, pp. 1808–1823, 2014.
8. D. D. V. Bhanderi, Spacecraft Attitude Determination with Earth Albedo Corrected Sun Sensor Measurements. Aalborg, Denmark: Aalborg University, 2005.
9. J. Gießelmann, Development of an Active Magnetic Attitude Determination and Control System for Picosatellites on Highly Inclined Circular Low Earth Orbits. Melbourne, Australia: RMIT University, 2006.
10. D. Vallado, *Fundamentals of Astrodynamics and Applications*, 4th ed. Hawthorne: Microcosm Press, 2013.
11. V. V. Unhelkar and H. B. Hablani, "Spacecraft attitude determination with sun sensors, horizon sensors and Gyros: Comparison of steady-state Kalman filter and extended Kalman filter," in *Advances in Estimation, Navigation, and Spacecraft Control*, D. Choukroun, Y. Oshman, J. Thienel and M. Idan (Eds.) Berlin, Heidelberg: Springer Berlin Heidelberg, 2015, pp. 413–437.
12. F. L. Markley and J. L. Crassidis, *Fundamentals of Spacecraft Attitude Determination and Control*. New York, NY: Springer New York, 2014.
13. J. D. Searcy and H. J. Pernicka, "Magnetometer-only attitude determination using novel two-step Kalman filter approach," *J. Guid. Control. Dyn.*, vol. 35, no. 6, pp. 1693–1701, 2012.
14. H. Ma and S. Xu, "Magnetometer-only attitude and angular velocity filtering estimation for attitude changing spacecraft," *Acta Astronaut.*, vol. 102, pp. 89–102, 2014.
15. H. E. Söken and S.-I. Sakai, "Magnetometer only attitude estimation for spinning small satellites," in *Proceedings of 8th International Conference on Recent Advances in Space Technologies, RAST 2017*, Istanbul, Turkey, 2017.
16. G. A. Natanson, S. F. Mclaughlin, and R. C. Nicklas, "A method of determining attitude from magnetometer data only," *Flight Mech. Theory Symp.*, pp. 359–378, 1990.

# 5 Attitude Determination Using Two Vector Measurements – TRIAD Method

Attitude can be determined by using at least two vectors measured by the attitude sensors (Sun sensor, Earth horizon sensor, magnetometer, etc.) in the spacecraft body frame and the models describing the corresponding directions (to the Sun, nadir and magnetic field) in the reference frame[1]. The most straightforward and the earliest algorithms that estimate the attitude using these measured and calculated vectors depend on the measurements taken at the same time (or close enough to ignore the dynamic motion of the spacecraft) and provide a single-frame attitude estimate. This class of methods is called "single-frame (point) methods" or "attitude determination methods based on vector measurements" since the attitude is estimated at a single point in time when more than two vector measurements are available. These methods require enough observations at each time to fully compute the attitude. Otherwise the attitude cannot be estimated. This is in contrast to the attitude filtering algorithms (will be presented in Chapter 9) that can estimate the attitude by incorporating the spacecraft mathematical models even if there is no available measurement at exactly same time of the estimation.

Single-frame attitude estimation methods are further classified as TRIAD method and statistical methods. This chapter discusses the TRIAD method. In this method, attitude matrix is determined directly from two vector observations. The name "TRIAD" can be considered either as the word "triad" or as an acronym for TRIaxial Attitude Determination [1]. TRIAD, the earliest published algorithm for satellite attitude determination from two vector measurements, has been widely used in both ground-based and onboard attitude determination. Another common name of TRIAD algorithm is "two-vector algorithm." We will use both names in this book.

## 5.1 TRIAD METHOD

The TRIAD method is based on the rotation matrix representation of the attitude. Any two vectors, $u$ and $v$, define an orthogonal coordinate system with the basis vectors, $q, r$ and $s$ as given below [2, 3]:

$$q = u, \tag{5.1a}$$

$$r = u \times v / |u \times v|, \tag{5.1b}$$

$$s = q \times r \tag{5.1c}$$

provided that $u$ and $v$ are not parallel:

$$|u \cdot v| < 1 \tag{5.2}$$

At a given time, two measured vectors in the spacecraft body coordinates (with the index $B$), $u_b$ and $v_b$, can be used to form the body matrix $M_b$:

$$M_b = \begin{bmatrix} q_b & r_b & s_b \end{bmatrix} \tag{5.3}$$

These measured vectors may be the Sun direction vector from a Sun sensor, position vector of an identified star from a star tracker, the nadir vector from an Earth horizon sensor or the Earth's magnetic field vector from a magnetometer. The associated reference directions for these measured vectors (denoted by the subscript $r$) can be calculated using the satellite ephemeris, star catalogs, and magnetic field model as discussed in Chapter 4. The reference matrix, $M_r$, is constructed from these two reference directions $u_r$ and $v_r$ again using Eq. (5.1) as:

$$M_r = \begin{bmatrix} q_r & r_r & s_r \end{bmatrix}. \tag{5.4}$$

The attitude matrix, or direction cosine matrix, $A$, is given by the coordinate transformation,

$$AM_r = M_b, \tag{5.5}$$

as it transfers any vector given in the reference frame to the body frame. This equation may be solved for $A$ to give

$$A = M_b M_r^{-1}. \tag{5.6}$$

Since $M_r$ is orthogonal, $M_r^{-1} = M_r^T$ and

$$A = M_b M_r^T. \tag{5.7}$$

Nothing in the development thus far has limited the choice of the reference frame or the form of the attitude matrix. The only requirement is that $M_r$ possess an inverse, which follows because the vectors $q$, $r$ and $s$ are linearly independent provided that Eq. (5.2) holds. The simplicity of Eq. (5.7) makes it particularly attractive for onboard processing. Inverse trigonometric functions are not required; a unique, unambiguous attitude is obtained; and computational requirements are minimal.

At this point we shall note that the preferential treatment of the vector $u$ over $v$ in Eq. (5.1) suggests that $\hat{u}$ should be the more accurate measurement. In solution given with Eq. (5.7) $u$ is emphasized more. The attitude matrix exactly transforms $u$ from the reference frame to the body frame and $v$ is used only to determine the phase angle about $u$. On the contrary, if we selected $q$ as $q = v$ at the beginning in Eq. (5.1), $v$ would be emphasized more in the solution. In either case the parallel component of two measured vectors, i.e. $u_b \cdot v_b$, is implicitly discarded by the

TRIAD algorithm. All of the error in $u_b \cdot v_b$ is assigned to the less accurate measurement, which accounts for the lost information [4]. In ideal case, where the measurements are error free $u_b \cdot v_b = u_r \cdot v_r$, and all the solutions by the TRIAD method are identical.

Any convenient representation may be used to parameterize the attitude matrix. Quaternions and Euler angles are commonly used in this matter [5–7].

### 5.1.1 TRIAD Algorithm Using Magnetometer and Sun Sensor Measurements

As discussed, the TRIAD algorithm requires two vectors measured in the body frame and modeled in the reference frame to determine the attitude. Let these vectors be the unit vectors to the Sun and for the Earth's magnetic field. In orbital frame, which is selected as the reference frame for estimating the attitude with respect to, these vectors are indicated by $S_o$ and $B_o$, respectively. In the body frame, these vectors are measured by the Sun sensors and magnetometers. The measured vectors are denoted with $S_b$ and $B_b$. We would like to estimate the attitude matrix between the mentioned coordinate systems. If the attitude matrix between these frames is $A$, and if $n_o = S_o \times B_o$, $n_b = S_b \times B_b$, then the following equalities can be written:

$$S_b = AS_o, \quad B_b = AB_o, \quad n_b = An_o. \tag{5.8}$$

Let us form the matrices $M_o$ and $M_r$ whose columns are made up of the above vectors:

$$M_o = \begin{bmatrix} S_o & B_o & n_o \end{bmatrix} \text{ and } M_b = \begin{bmatrix} S_b & B_b & n_b \end{bmatrix} \tag{5.9}$$

Then Eq. (5.6) can be used to determine the attitude matrix $A$. The attitude is represented in terms of Euler angles as given below in Eq. (5.10) (in $A$ $c(\cdot)$ and $s(\cdot)$ represent the cosine and sine functions).

$$A = \begin{bmatrix} c(\theta)c(\psi) & c(\theta)s(\psi) & -s(\theta) \\ -c(\phi)s(\psi)+s(\phi)s(\theta)c(\psi) & c(\phi)c(\psi)+s(\phi)s(\theta)s(\psi) & c(\theta)s(\phi) \\ s(\phi)s(\psi)+c(\phi)s(\theta)c(\psi) & -s(\phi)c(\psi)+c(\phi)s(\theta)s(\psi) & c(\theta)c(\phi) \end{bmatrix}$$
$$\tag{5.10}$$

It is obvious that the Euler angles defining the attitude of the spacecraft can be found as a function of $S_o$, $B_o$, $S_b$ and $B_b$ vectors.

$$\theta = f_\theta\left(S_o,B_o,S_b,B_b\right), \quad \psi = f_\psi\left(S_o,B_o,S_b,B_b\right), \quad \phi = f_\phi\left(S_o,B_o,S_b,B_b\right). \tag{5.11}$$

To extract the Euler angles from the attitude matrix, following equations can be used:

$$\theta = a\sin\left[-A(1,3)\right] \tag{5.12a}$$

$$\psi = \text{atan2}\left[A(1,2), A(1,1)\right], \tag{5.12b}$$

$$\phi = \text{atan2}\left[A(2,3), A(3,3)\right]. \tag{5.12c}$$

Instead of Eq. (5.12a)

$$\theta = \text{atan2}\left[-A(1,3), \sqrt{A(2,3)^2 + A(3,3)^2}\right] \tag{5.12d}$$

can be also used.

Here atan2($\cdot$) is the arctangent function with two arguments, which is used to generate all the rotations and introduced in Eq. (1.32).

These expressions in Eq. (5.11) will be used in computing the accuracy of the satellite's attitude angles.

## 5.1.2 Quaternion Estimates from the Attitude Matrix

When we would like to estimate the attitude in term of quaternions, we need to use Eq. (1.18) to extract the quaternion vector from the TRIAD estimated attitude matrix. For ease of reading, we present the attitude matrix here again:

$$A = \begin{bmatrix} q_1^2 - q_2^2 - q_3^2 + q_4^2 & 2(q_1 q_2 + q_3 q_4) & 2(q_1 q_3 - q_2 q_4) \\ 2(q_1 q_2 - q_3 q_4) & -q_1^2 + q_2^2 - q_3^2 + q_4^2 & 2(q_2 q_3 + q_1 q_4) \\ 2(q_1 q_3 + q_2 q_4) & 2(q_2 q_3 - q_1 q_4) & -q_1^2 - q_2^2 + q_3^2 + q_4^2 \end{bmatrix}. \tag{5.13}$$

We first build the following four four-component vectors using the $A$ matrix components:

$$\begin{bmatrix} 1 + 2A(1,1) - \text{tr}(A) \\ A(1,2) + A(2,1) \\ A(1,3) + A(3,1) \\ A(2,3) - A(3,2) \end{bmatrix} = 4q_1\boldsymbol{q}, \quad \begin{bmatrix} A(2,1) + A(1,2) \\ 1 + 2A(2,2) - \text{tr}(A) \\ A(2,3) + A(3,2) \\ A(3,1) - A(1,3) \end{bmatrix} = 4q_2\boldsymbol{q}. \tag{5.14a}$$

$$\begin{bmatrix} A(1,3) + A(3,1) \\ A(2,3) + A(3,2) \\ 1 + 2A(3,3) - \text{tr}(A) \\ A(1,2) - A(2,1) \end{bmatrix} = 4q_3\boldsymbol{q}, \quad \begin{bmatrix} A(2,3) - A(3,2) \\ A(3,1) - A(1,3) \\ A(1,2) - A(2,1) \\ 1 + \text{tr}(A) \end{bmatrix} = 4q_4\boldsymbol{q}. \tag{5.14b}$$

The quaternion vector can be found by normalizing any of these four vectors [1]. However, to minimize the numerical errors the vector with the greatest norm so the one with the largest value of $|q_i|$ for $i = 1, 2, 3$ on the right side should be chosen. This is same as selecting the largest of $\text{tr}(A)$ or $A_{ii}$. If $\text{tr}(A)$ is largest then $|q_4|$ is largest. Otherwise one of $|q_i|$ with the same index of largest $A_{ii}$ is the largest one.

Once the quaternion vector is found, the twofold ambiguity for the sign of the quaternions still remains. Since $q = -q$ and they represent the same attitude, this is not a problem in theory. However, in practice it introduces difficulties especially when visualizing the quaternions and checking the errors during attitude analyses. To overrun this difficulty, the Nearest Neighboring Method [8] can be used.

The Nearest Neighboring Method suggests that the value of the quaternion in each step should be as close to the previous value as possible. By following the steps below, the nearest neighboring method can be applied:

- the quaternion values in steps $k$ and $k + 1$ are compared;
- $q_k$ and $q_{k+1}$ are dot multiplied.

If the multiplication result is negative, then negative (adjoin) of $q_{k+1}$ is taken. If the value is positive, $q_k$ is left unchanged.

Figure 5.1 gives errors for quaternions estimated by TRIAD algorithm for a small satellite using the magnetometer and Sun sensor measurements. Sun sensor

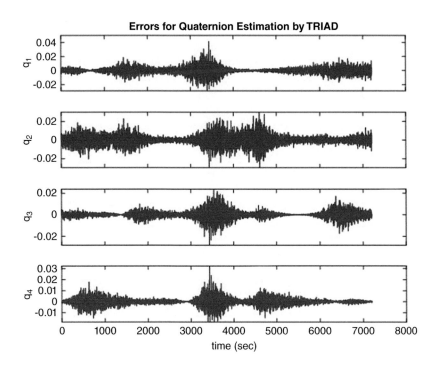

**FIGURE 5.1**   Errors for Quaternion Estimation by TRIAD.

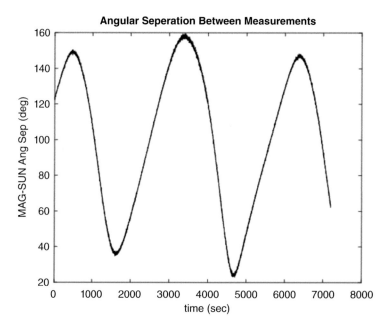

**FIGURE 5.2**    Angular separation in between the magnetometer and Sun sensor measurements.

and magnetometer are calibrated against systematic errors (bias, misalignment, etc.) and the standard deviation of their noises are $\sigma_s = 0.1°$ and $\sigma_b = 300nT$, respectively. No eclipse is assumed in this scenario. To better evaluate the results, Figure 5.2 presents the separation angle in between the magnetometer and Sun sensor measurements. As can be seen the main reason of increase in the error for quaternion estimation about 3400th s. is the measurement vectors getting close to be anti-parallel to each other. Likewise, the increase at about 4700th s. is due to the vectors getting close to be parallel. The magnitude of the error is observed different in each quaternion component due to the specific absolute attitude of the spacecraft.

### 5.1.3  USING TRIAD ALGORITHM FOR MORE THAN TWO VECTOR MEASUREMENTS

When there are more than two vector measurements, obviously a single TRIAD algorithm is not sufficient to implement all the measurements and estimate the attitude. In this case there is need for multiple TRIAD algorithms which are using two of these measurement vectors in different combinations:

    1st TRIAD algorithm: Magnetometer and Sun sensor measurements
    2nd TRIAD algorithm: Magnetometer and Earth sensor measurements
    3rd TRIAD algorithm: Sun sensor and Earth sensor measurements

Such method can be useful because of having multiple attitude solutions and enabling fault tolerant attitude determination (by easily detecting the faulty sensor by comparing multiple solutions) for especially cubesats. Separate estimates can be merged to find a single solution considering the covariances of each algorithm [9]. An example for such integration will be provided in next sections of this chapter.

## 5.2 ANALYSIS OF THE TRIAD METHOD ACCURACY

When we are estimating the attitude of the satellite using the TRIAD algorithm, there are several errors affecting the attitude determination accuracy. The most important of these errors are as follows [7]:

1. Errors due to the orbit measurements: These orbit measurements may be ground based or by using the Global Navigation Satellite System (GNSS) receivers onboard the satellite.
2. Errors due to the estimation of orbital parameters: Estimation may be a straightforward propagation algorithm (especially when the ground-based orbit measurements are used) or by using an estimator (e.g. Kalman filter).
3. Errors due to the models of the reference vectors in the orbital frame.
4. Errors due to the measurements in the body frame.
5. Errors due to the TRIAD algorithm itself.

The scheme of the Earth's magnetic field and Sun vector based two-vector algorithm, used for determination of satellite's attitude, is shown in Figure 5.3.

Determination of Euler attitude angles with the TRIAD algorithm has the following steps:

1. Measuring the orbital parameters (position and velocity) of the satellite.

$$\tilde{r} = \Phi_1(\rho),$$ (5.15a)

$$\tilde{v} = \Phi_1(\rho).$$ (5.15b)

Here $\Phi_1$ is the algorithm for relating the measurements ($\rho$) to the position and velocity of the satellite and $r$ is the position and $v$ is the velocity that we get using this algorithm.

2. Estimation (propagation) of the orbit

$$r = \Phi_2(\tilde{r}, \tilde{v}),$$ (5.16a)

$$v = \Phi_2(\tilde{r}, \tilde{v}).$$ (5.16b)

Here $\Phi_2$ is the algorithm for estimation (propagation) of the orbital parameters at any time other than the measurements.

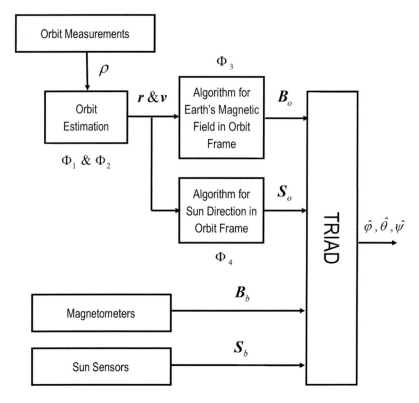

**FIGURE 5.3**   Magnetometer and Sun sensor measurements-based TRIAD algorithm.

3. Calculation of the Earth's magnetic field in the orbit frame

$$\boldsymbol{B}_o = \Phi_3(\boldsymbol{r},\boldsymbol{v}). \tag{5.17}$$

Here $\Phi_3$ is the algorithm for Earth's magnetic field calculation (e.g. IGRF model).

4. Calculation of Sun direction in orbit frame

$$\boldsymbol{S}_o = \Phi_4(\boldsymbol{r},\boldsymbol{v}). \tag{5.18}$$

Here $\Phi_4$ is the algorithm for Sun direction calculation.

5. Using TRIAD for the attitude estimation

$$\theta = f_\theta\left(\boldsymbol{S}_o,\boldsymbol{B}_o,\boldsymbol{S}_b,\boldsymbol{B}_b\right), \quad \psi = f_\psi\left(\boldsymbol{S}_o,\boldsymbol{B}_o,\boldsymbol{S}_b,\boldsymbol{B}_b\right), \quad \phi = f_\phi\left(\boldsymbol{S}_o,\boldsymbol{B}_o,\boldsymbol{S}_b,\boldsymbol{B}_b\right). \tag{5.19}$$

In general, the attitude determination algorithm is a nonlinear function of random variables. Thus, in order to find the error of the algorithm given, it was linearized by expanding to Taylor series. Correlation between different parameters was ignored. So, after linearization, accuracy (variance) of the algorithm is found as:

$$D_{i\theta} = \left(\frac{\partial f_\theta}{\partial \mathbf{B}_o}\right)^2_m D_{iB_o} + \left(\frac{\partial f_\theta}{\partial \mathbf{S}_o}\right)^2_m D_{iS_o} + \left(\frac{\partial f_\theta}{\partial \mathbf{B}_b}\right)^2_m D_{iB_b} + \left(\frac{\partial f_\theta}{\partial \mathbf{S}_b}\right)^2_m D_{iS_b}. \qquad (5.20)$$

Here $D_{i\theta}$ is error computing variance for angle $\theta$ at step $i$. The subscript $m$ for each partial differentiation in Eq. (5.20) means that mean values of the parameters must be used for calculation. In a similar way,

$$D_{i\psi} = \left(\frac{\partial f_\psi}{\partial \mathbf{B}_o}\right)^2_m D_{iB_o} + \left(\frac{\partial f_\psi}{\partial \mathbf{S}_o}\right)^2_m D_{iS_o} + \left(\frac{\partial f_\psi}{\partial \mathbf{B}_b}\right)^2_m D_{iB_b} + \left(\frac{\partial f_\psi}{\partial \mathbf{S}_b}\right)^2_m D_{iS_b}, \qquad (5.21)$$

$$D_{i\phi} = \left(\frac{\partial f_\phi}{\partial \mathbf{B}_o}\right)^2_m D_{iB_o} + \left(\frac{\partial f_\phi}{\partial \mathbf{S}_o}\right)^2_m D_{iS_o} + \left(\frac{\partial f_\phi}{\partial \mathbf{B}_b}\right)^2_m D_{iB_b} + \left(\frac{\partial f_\phi}{\partial \mathbf{S}_b}\right)^2_m D_{iS_b}. \qquad (5.22)$$

Now suppose that there is also Earth horizon sensor measurements and the attitude is estimated with multiple TRIAD algorithms as described in Section 5.1.3. The presented procedure for the accuracy analyses can be easily generalized for all the algorithms. In the rest of this section, we present numerical results for such analyses.

In the analyses, the satellite's orbital parameters are assumed as: inclination $i = 97°$; right ascension of the ascending node $\Omega = 15°$; eccentricity $e = 0$ and orbital altitude of $h = 550$ km. The Earth radius is $R = 6378.14$ km; the Earth angular velocity is; the Earth magnetic field moment is $M = 7.86 \times 10^{15}$ Wb $\cdot$ m; the magnetic tilt angle is $\varepsilon = 11.4°$. The accuracy of the orbital parameters $i$, $\Omega$ and $u$ are 5e-6 rad, 1e-5 rad and 1.5e-4 rad, respectively. The attitude sensors' accuracy is ~1° for magnetometer, 0.1° for sun sensor and 0.36° for horizon sensor (horizon sensor determines the roll and pitch angles). It is assumed that eccentric anomaly is equal to the mean anomaly. Only one orbital period was simulated. In Figure 5.4, the change of the satellite attitude accuracy throughout the orbit is shown when the first algorithm (SUN-MAG) is used (required accuracy is 1°). As accuracy characteristics pitch ($\theta$), yaw ($\psi$) and roll ($\varphi$) angles' variances are taken. The results for the second and third algorithms are given in Figures 5.5 and 5.6, respectively.

When the results were examined it is seen that, when the reference vectors become almost parallel or the value of the pitch angle $\theta$ approaches to (90°+$n\pi$) degrees, the accuracy of the outputs is below the requirement. Furthermore, depending on the orbit, the Sun may be out of sight for some periods of orbit (as it is for the most LEO satellites). It makes the first and third algorithm unusable.

Regarding the simulation results, the following statements can be made:

a. the accuracy of the spacecraft's attitude is changing in a wide range along the orbit; the accuracy is worst when the reference vectors are close to parallel or the value of the pitch angle $\theta$ approaches to 90°+$n\pi$ degrees;

b. the attitude determination accuracy is affected by different factors in a different manner; the most influent factors on the accuracy are the sensor errors;

c. to increase the accuracy of attitude determination, redundant data processing methods (statistical methods) can be used.

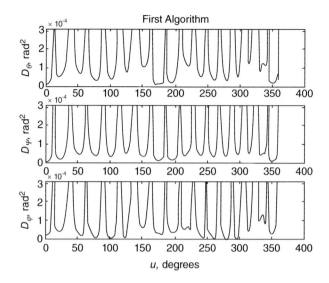

**FIGURE 5.4** Change of the variance of the attitude angles, obtained by the magnetometer-Sun sensor TRIAD algorithm, along one whole orbit.

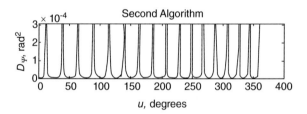

**FIGURE 5.5** Change of the yaw angle variance obtained by the Earth sensor-magnetometer TRIAD algorithm along one whole orbit.

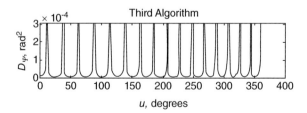

**FIGURE 5.6** Change of the yaw angle variance obtained by the Sun sensor-Earth sensor TRIAD algorithm along one whole orbit.

## 5.3  INCREASING ACCURACY OF THE TRIAD METHOD USING REDUNDANCY TECHNIQUES

In order to increase the attitude determination accuracy, the redundant data processing algorithm, based on the Maximum Likelihood Method (MLM), can be used to statistically process the measurements of the three algorithms mentioned in the previous section.

Let's assume that the output $x$ of a system is measured simultaneously with $n$ different measurement devices with different measuring principles. Then the measurement equation of the $i^{th}$ device will be

$$z_{x_i} = x + \delta_i, \quad i = \overline{1,n}. \tag{5.23}$$

Here $z_{x_i}$ is the measurement of the $i^{th}$ device; $\delta_i$ is the measurement error of the $i^{th}$ device. It is assumed that there is no correlation between the measurement errors of the measurement channels. Another assumption is that the measurement errors are subject to normal distribution with zero mean and finite $\sigma_i^2$ variance,

$$p(\delta_i) = \frac{1}{\sqrt{2\pi}\sigma_i} \exp\left\{-\frac{\delta_i^2}{2\sigma_i^2}\right\}, \quad i = \overline{1,n}. \tag{5.24}$$

Thus, the distribution density of measurement is known as a function of evaluated parameter:

$$p(z_{x_i} / x) = \frac{1}{\sqrt{2\pi}\sigma_i} \exp\left\{-\frac{\left(z_{x_i} - x\right)^2}{2\sigma_i^2}\right\}. \tag{5.25}$$

It is desired to find the optimal value of $x$ which maximizes the Likelihood function. With the assumption that the measurements $z_{x_i}(i = \overline{1,n})$ are independent, the Likelihood function $p(z_{x1}, z_{x2}, ..., z_{xn}/x)$ can be expressed as,

$$p\left(z_{x_1}, z_{x_2}, ..., z_{x_n} / x\right) = \frac{1}{\sqrt{(2\pi)^n} \prod_{i=1}^{n} \sigma_i} \exp\left\{-\sum_{i=1}^{n} \frac{\left(z_{x_i} - x\right)^2}{2\sigma_i^2}\right\} \tag{5.26}$$

After mathematical operations the expression for the estimated value that is searched is found as

$$\hat{x} = \left(\sum_{i=1}^{n} \frac{z_{x_i}}{\sigma_i^2}\right) / \left(\sum_{i=1}^{n} \frac{1}{\sigma_i^2}\right) \tag{5.27}$$

whose covariance can be shown as

$$D = \left\{-E\left[\frac{\partial^2 \ln p\left(z_{x_1}, z_{x_2}, ..., z_{x_n} / x\right)}{\partial x^2}\right]^{-1}\right\} = \left(\sum_{i=1}^{n} \frac{1}{\sigma_i^2}\right)^{-1}. \tag{5.28}$$

Here $E$ -denotes the operator for mathematical expectation.

**Theorem 5.1: The inequality** $D < \sigma_i^2, \forall i \in [1,n]$ **is true for the variance of the estimated value (5.27).**

Proof of the theorem is given in [10].

Pitch, roll and yaw angles that characterize the angular position of the satellite were found with three different algorithms in the previous section. It means that, there is more information than required. The scheme of the redundancy techniques for increasing accuracy of the satellite attitude determination is presented in Figure 5.7.

By adapting the redundant data processing method based on the MLM to the case examined, the satellite attitude accuracy results shown in Figure 5.8 are obtained.

When the results, given in Figures 5.4–5.6, are compared with results, given in Figure 5.8, it is obvious that the results obtained by redundant data processing algorithm are better than all other given results.

In order to get good results from the redundant data processing method, at least one of the other algorithms has to produce available data (equal or better than the required accuracy).

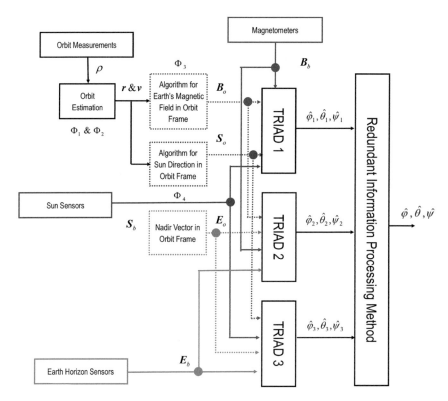

**FIGURE 5.7** Redundancy techniques for increasing accuracy of the satellite attitude determination.

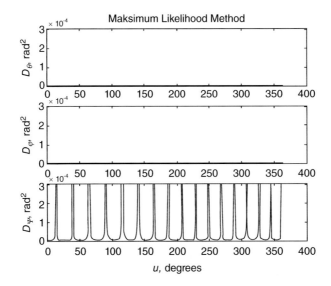

**FIGURE 5.8**   Change of the variance of the attitude angles obtained by redundant data processing algorithm through the whole orbit.

The "*bad areas*" (areas where the attitude accuracy is worse than the required one), formed due to the parallelism of the reference vectors, can be removed by using the redundant data processing algorithm. But it is impossible to remove the "*bad areas*", formed due to the pitch angle's value. A method to remove such ambiguity is to use quaternions instead of Euler angles as an attitude parameterization technique.                                                                                        ■

## 5.4  CONCLUSION AND DISCUSSION

In this chapter the TRIAD method (two-vector algorithm) is introduced. As the TRIAD method is one of the mostly used attitude determination methods, its error analysis is performed in detail.

In the conduced simulations for a small satellite for attitude estimation using the TRIAD algorithm we use Sun sensor, magnetometer and Earth horizon sensor measurements. Extraction of attitude in terms of both the Euler angles and the quaternions from the TRIAD estimated attitude matrix is shown.

According to the simulation results, the following conclusions are drawn:

a. the accuracy of the spacecraft's attitude is changing in a wide range along the orbit; the accuracy is worst when the reference vectors are close to parallel (colinearity) or the value of the pitch angle $\theta$ approaches to $90°+n\pi$ degrees when the estimated attitude is in terms of Euler angles;
b. the attitude determination accuracy is affected by different factors in a different manner; the most influent factors on the accuracy are the sensor errors.

In order to increase the attitude determination accuracy, the redundant data processing algorithm, based on the Maximum Likelihood Method, is used. The accuracy of the values found with redundant data processing is better than the values found with each other three algorithms. The effect of colinearity of the reference vectors can be removed by using redundant data processing method. But the ambiguity due to the pitch angle's value cannot be removed with this method one way to remove this effect is to use quaternions instead of Euler angles as an attitude parameterization technique.

In the second case (attitude determination in terms of quaternions), TRIAD uses one two-vector algorithm based on the Earth's magnetic field vector and Sun vector. The simulation results show that the TRIAD method allows to determine the satellite attitude in terms of quaternions with high accuracy. The quaternions, determined via TRIAD, have sufficiently small error variances. Using quaternions instead of Euler angles allows to remove the ambiguity formed due to the pitch angle's value (when the value of the pitch angle $\theta$ approaches to $(90° + n\pi)$ degrees).

In the TRIAD method the combination of the attitude sensor measurements is not optimal in any statistical sense. These methods are not easily applied to combinations of attitude sensors which provide many simultaneous vector measurements. This is the main disadvantage of the TRIAD method.

## NOTE

1 Magnetometer-only attitude determination methods that are based on filtering the magnetometer data are an exception to this.

## REFERENCES

1. F. L. Markley and J. L. Crassidis, *Fundamentals of Spacecraft Attitude Determination and Control*. New York, NY: Springer New York, 2014.
2. H. D. Black, "A passive system for determining the attitude of a satellite," *AIAA J.*, vol. 2, no. 7, pp. 1350–1351, July 1964.
3. F. L. Markley and D. Mortari, "Quaternion attitude estimation using vector observations," *J. Astronaut. Sci.*, vol. 48, no. 2–3, pp. 359–380, 2000.
4. F. L. Markley and D. Mortari, "How to estimate attitude from vector observations," *Adv. Astronaut. Sci.*, vol. 103, no. PART III, pp. 1979–1996, 2000.
5. D. Cilden Guler, E. S. Conguroglu, and C. Hajiyev, "Single-frame attitude determination methods for nanosatellites," *Metrol. Meas. Syst.*, vol. 24, no. 2, pp. 313–324, 2017.
6. D. C. Guler and C. Hajiyev, *Gyroless Attitude Determination of Nanosatellites: Single-Frame Methods and Kalman Filtering Approach*. Saarbrucken, Germany: LAP LAMBERT Academic Publishing, 2017.
7. C. Hajiyev and M. Bahar, "Increase of accuracy of the small satellite attitude determination using redundancy techniques," *Acta Astronaut.*, vol. 50, no. 11, pp. 673–679, 2002.
8. A. J. Hanson and S. Cunningham, *Visualizing Quaternions*. San Francisco, USA: Morgan Kaufmann, 2006.
9. M. D. Shuster, "The optimization of TRIAD," *J. Astronaut. Sci.*, vol. 55, no. 2, pp. 245–257, 2007.
10. C. Hajiyev, *Radio Navigation (in Turkish)*. Istanbul: Istanbul Technical University Press, 1999.

# 6 Statistical Methods for Three-Axis Attitude Determination

In the TRIAD method presented in Chapter 5, the two vector measurements are used to provide an attitude estimate, but the estimation is not optimal in any statistical sense. Furthermore, although using multiple TRIAD algorithms and a redundancy technique enables attitude estimate from more than two vector measurements, the overall process is computationally burdensome. It cannot be easily adopted for combinations of sensors which provide many simultaneous vector measurements.

Statistical methods for attitude estimation need at least two vector measurements and knowledge of the relative precision of the sensors. The aim of all these statistical attitude estimation methods is to minimize a well-known cost function that is proposed by Wahba [1]. They provide means for computing an optimal three-axis attitude from many vector observations.

In this chapter, we present Davenport's q-method, Quaternion Estimator (QUEST) algorithm and singular value decomposition (SVD) algorithm, which are alternative methods for solving the Wahba problem and estimating the attitude. Readers shall note that there are various other single-frame attitude estimators such as Estimator of the Optimal Quaternion (ESOQ), Second Estimator of the Optimal Quaternion (ESOQ2) and Fast Optimal Attitude Matrix (FOAM) [2]. Despite having different names, they slightly depart from the methods, which will be presented here. For example, FOAM method solves the attitude matrix iteratively in contrast to SVD method and tries to decrease the computational load. With all these new methods for single-frame attitude estimation, the search is always for a more robust, accurate and computationally lighter attitude estimator. However, in practice there is a little change in decades and QUEST can be accounted as the most widely used single-frame attitude estimation algorithm [3, 4].

The single-frame attitude estimation algorithms can be both used as the sole attitude estimator on a spacecraft or as one of the many algorithms. It is very common to use the attitude estimate by one of these algorithms for initializing an attitude filter. Moreover, today's star trackers may implement a single-frame attitude estimator to output the quaternion measurements (estimates) once the detected starts are identified. Last but not least, as will be presented in Chapter 10, single-frame estimator can be integrated with the attitude filter to provide directly the quaternion measurements to the filter and to form a more robust estimation algorithm.

## 6.1 WHAT IS WAHBA'S PROBLEM?

The major disadvantage of the TRIAD algorithm is not providing an optimal solution for the attitude. Furthermore, it is not easily usable in case we have more than two measurements a single frame. Given a set of $n \geq 2$ vector measurements, $b_i$ in the body system, one choice for an optimal attitude matrix, $A$, is to minimize the loss function given as

$$J(A) = \sum_{i=1}^{n} a_i \| b_i - Ar \|_i^2, \qquad (6.1)$$

where $a_i$ is the weight of the $i^{\text{th}}$ vector measurement and $r_i$ is the vector in the reference coordinate system.

This problem is defined as the Wahba's problem and was first proposed by Grace Wahba in 1965 [1]. It seeks the proper orthogonal attitude matrix, $A$, which minimizes the cost function (6.1). As apparent the loss function is the weighted sum squared of the difference between the measured and transformed vectors.

## 6.2 DAVENPORT'S Q-METHOD

The very first useful solution of Wahba's problem for spacecraft attitude determination was proposed by Paul Davenport [5]. To track the solution of the problem with the q-method, first let us rewrite the loss function as

$$J(A) = -2 \sum_{i=1}^{n} W_i A V_i + C, \qquad (6.2)$$

where $W_i$ and $V_i$ are unnormalized vectors and C symbolizes the constant terms. Vectors $W_i$ and $V_i$ can be defined as

$$W_i = \sqrt{a_i} b_i \quad V_i = \sqrt{a_i} r_i. \qquad (6.3)$$

It is obvious that minimum condition for the loss function $J(A)$ may be found as

$$J'(A) = \sum_{i=1}^{n} W_i A V_i \equiv \mathrm{tr}\left(W^T A V\right), \qquad (6.4)$$

where the $3 \times n$ matrices $W$ and $V$ are defined by

$$W \equiv \left[ W_1 \, W_2 \ldots W_n \right], \qquad (6.5a)$$

$$V \equiv \left[ V_1 \, V_2 \ldots V_n \right]. \qquad (6.5b)$$

We need to parameterize $A$ in terms of the quaternion, $q$ to find the attitude matrix $A$, which maximizes Eq. (6.2). Recall Eq. (1.18) which can be also written as

$$A = \left(q_4^2 - g^2\right)I_{3\times3} + 2gg^T - 2q_4\left[g\times\right], \tag{6.6}$$

for $q = [g^T \ q_4]^T$. Substituting Eq. (6.6) into Eq. (6.4) and necessary matrix algebra brings about the following convenient form for the modified loss function:

$$J'(q) = q^T Kq, \tag{6.7}$$

where $K$ is a $4 \times 4$ matrix and

$$K = \begin{pmatrix} S - I_{3\times3}\sigma & Z \\ Z^T & \sigma \end{pmatrix}. \tag{6.8}$$

Moreover, the intermediate $3 \times 3$ matrices $B$ and $S$, the vector $Z$ and the scalar $\sigma$ are given by

$$B \equiv WV^T = \sum_{i=1}^{n} a_i b_i r_i^T, \tag{6.9a}$$

$$S \equiv B^T + B, \tag{6.9b}$$

$$Z \equiv \begin{bmatrix} B_{23} - B_{32} \\ B_{31} - B_{13} \\ B_{12} - B_{21} \end{bmatrix} = \sum_{i=1}^{n} a_i \left(b_i \times r_i\right), \tag{6.9c}$$

$$\sigma \equiv \operatorname{tr}(B) \tag{6.9d}$$

Considering the normalization constraint $q^T q = 1$, the extrema of modified loss function $J'$ can be found by the Lagrange multipliers method. Hence we define a new function at this point:

$$g(q) = q^T Kq - \lambda q^T q, \tag{6.10}$$

where $\lambda$ is Lagrange multiplier that is chosen to satisfy the normalization constraint. Differentiating (6.10) with respect to $q^T$ shows that $g'(q)$ is stationary if

$$Kq = \lambda q. \tag{6.11}$$

Thus, the optimal attitude is an eigenvector of $K$ matrix. To find which eigenvalue corresponds to the optimal eigenvector that we are searching for, let us substitute (6.11) into (6.7):

$$J'(q) = q^T Kq = q^T \lambda q = \lambda. \tag{6.12}$$

The maximum value of the modified loss function is then associated with the maximum eigenvalue of the Davenport's $K$ matrix. As a consequence, the optimal attitude vector, from a statistical point of view, is the eigenvector corresponding to the maximum eigenvalue of matrix $K$. Therefore, it can be stated that, the optimum quaternion is the eigenvector of Davenport's $K$ matrix (6.8) corresponding to its largest eigenvalue.

It can be shown that if at least two of the vectors $W_i$ are not collinear, the eigenvalues of $K$ are distinct and this procedure yields an unambiguous quaternion or, equivalently, three-axis attitude. If $K$ has two maximum eigenvalues, the q-method cannot provide attitude estimate. This is not a failure of the algorithm but rather shows that there are no sufficient data to estimate the attitude (as in parallel measurement vectors when we use Sun sensor and magnetometer measurements).

There are many methods to calculate the eigenvalues or eigenvectors of a matrix and solve the attitude by q-method. QUEST algorithm that we present next alternatively approximates the largest eigenvalue for the optimal attitude rather than directly solving for.

## 6.3 QUEST METHOD

The key for the q-method is to solve for the eigenvalues and eigenvectors of the $K$ matrix. Although this is an easy problem using the modern computational tools, the solution is numerically intensive for applying onboard the spacecraft. Especially for nanosatellites, the computational load of an algorithm is one of the main concerns for the designer.

The QUEST algorithm that is discussed in this section provides a computationally lighter solution for solving the eigenvalue/eigenvector problem. Since it was first applied in 1979, it has been the most widely used algorithm for solving Wahba's problem. Let us first rewrite (6.1) in a more convenient form as

$$J(A) = \lambda_0 - \text{tr}(AB^T),\tag{6.13}$$

where

$$\lambda_0 = \sum_{i=1}^{n} a_i,\tag{6.14}$$

and $B$ is as defined in Eq. (6.9a). As stated, the eigenvector of $K$ with the largest eigenvalue, i.e. the solution of $Kq_{opt} = \lambda_{max}q_{opt}$ gives the optimal attitude. In this case the optimized loss function is

$$L(A_{opt}) = \lambda_0 - \lambda_{max}.\tag{6.15}$$

As clearly discussed in [4], if this optimized loss function is small then $\lambda_{max} \approx \lambda_0$. In this case, we can apply the Newton–Raphson method to the characteristic equation

$\det(K - \lambda_{max}I) = 0$, which is a fourth order equation in terms of $\lambda$, by choosing the $\lambda_0$ as the initial estimate. Explicit form of the characteristic equation is[1]

$$f(\lambda) = \lambda^4 - (a+b)\lambda^2 - c\lambda + (ab + c\sigma - d) = 0. \tag{6.16}$$

Here, $\sigma$ is as given in Eq. (6.9d). Other parameters are

$$a = \sigma^2 - \mathrm{tr}\big(\mathrm{adj}(S)\big), \tag{6.17a}$$

$$b = \sigma^2 + Z^T Z, \tag{6.17b}$$

$$c = \det(S) + Z^T S Z, \tag{6.17c}$$

$$d = (Z^T S)(SZ). \tag{6.17d}$$

$\mathrm{adj}(S)$ denotes the adjoint of matrix $S$.

Once the characteristic equation in Eq. (6.16) is solved for the $\lambda_{max}$ using the Newton–Raphson method, the optimal quaternion estimate for QUEST algorithm can be found using below-given three equations:

$$\gamma \equiv \det\Big[\big(\lambda_{max} + \mathrm{tr}(B)\big)I_{3\times3} - S\Big] = \alpha\big(\lambda_{max} + \mathrm{tr}(B)\big) - \det(S), \tag{6.18a}$$

$$\mathbf{x} \equiv \big(\alpha I + \beta S + S^2\big)\mathbf{z}, \tag{6.18b}$$

$$\mathbf{q}_{opt} = \frac{1}{\sqrt{\gamma^2 + |\mathbf{x}|^2}}\begin{bmatrix}\mathbf{x}\\ \gamma\end{bmatrix}. \tag{6.18c}$$

Here

$$\alpha \equiv \lambda_{max}^2 - \big(\mathrm{tr}(B)\big)^2 + \mathrm{tr}\big(\mathrm{adj}(S)\big), \tag{6.19a}$$

$$\beta \equiv \lambda_{max} - \mathrm{tr}(B). \tag{6.19b}$$

When he was proposing the QUEST algorithm, Shuster's key observation was by choosing the $\lambda_0$ as the initial estimate the Newton–Raphson algorithm gives $\lambda_{max}$ within just few iterations. In fact, even $\lambda_{max} = \lambda_0$ approximation gives rather accurate results in many cases and a single iteration of Newton–Raphson algorithm is sufficient. Figure 6.1 gives the attitude estimation errors for QUEST algorithm comparing two cases, when Newton–Raphson algorithm is solved to get $\lambda_{max}$ and when $\lambda_{max} = \lambda_0$ approximation is used. Only Sun sensor and magnetometer measurements are used in this scenario. Reader can compare the errors considering the separation angle between the measurement vectors presented in Figure 5.2. Note that the approximation (and not implementing the Newton–Raphson algorithm) reduces the computational load approximately 24% on MATLAB environment.

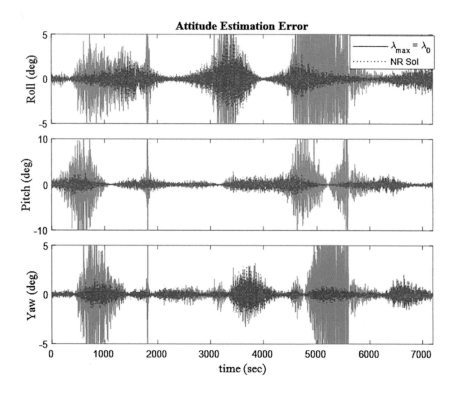

**FIGURE 6.1**    Estimation errors for QUEST algorithm for different values of $\lambda_{max}$.

Actually when there are only two vector measurements used in QUEST algorithm $\lambda_{max}$ can be calculated as below without needing any Newton–Raphson iteration [6]:

$$\lambda_{max} = \left\{ a_1^2 + a_2^2 + 2a_1a_2\left[\left(\boldsymbol{b}_1 \cdot \boldsymbol{b}_2\right)\left(\boldsymbol{r}_1 \cdot \boldsymbol{r}_2\right) + \|\boldsymbol{b}_1 \times \boldsymbol{b}_2\|\|\boldsymbol{r}_1 \times \boldsymbol{r}_2\|\right] \right\}^{1/2}. \tag{6.20}$$

It should be mentioned here that QUEST cannot give any attitude estimate when $(\lambda_{max} + \mathrm{tr}(B))I_{3 \times 3} - S$ is singular, which is actually showing that $\boldsymbol{q}_{opt}(4) = 0$ (since $\gamma = 0$ in Eq. 6.18). This deficiency can be easily overrun by redefining the attitude with respect to a rotated reference frame and rotating the reference vectors used for attitude estimation. Shuster uses the method of sequential rotations and performs rotation about one axis (of the reference frame) at a time until an acceptable reference frame is found. Of course search for such acceptable reference frame may increase the computational load of the algorithm and there are various other methods such as using an a priori quaternion to select the necessary single rotation as early as possible [2, 6].

Last, the covariance of the rotation angle error vector in the body frame for the QUEST is given as

$$P = \left[\sum_{i=1}^{n} a_i\left(I_{3 \times 3} - \boldsymbol{b}_i\boldsymbol{b}_i^T\right)\right]^{-1}. \tag{6.21}$$

This covariance matrix will be specifically needed when the estimated quaternion vector is used as measurement for an attitude filter in the integrated single-frame attitude filter concept, which will be discussed in detail in Chapter 10.

## 6.4 SVD METHOD

When we express the loss function as in Eq. (6.13), the problem reduces to the problem of maximizing the trace, $tr(AB^T)$. The matrix $B$ here has SVD of

$$B = U \Sigma^T V^T = U\text{diag}\left[\Sigma_{11} \ \Sigma_{22} \ \Sigma_{33} \ s\right]V^T, \tag{6.22}$$

where $U$ and $V$ are orthogonal and the singular values that obey $\Sigma_{11} \geq \Sigma_{22} \geq \Sigma_{33} \geq 0$. Then we can show that the trace is maximized for

$$U^T A_{opt} V = \text{diag}\begin{bmatrix} 1 & 1 & \det(U)\det(V) \end{bmatrix}, \tag{6.23}$$

and the optimal rotation matrix is [7],

$$A_{opt} = U\text{diag}\begin{bmatrix} 1 & 1 & \det(U)\det(V) \end{bmatrix}V^T. \tag{6.24}$$

The accuracy of the estimated $A_{opt}$ can be understood by examining the covariance matrix for rotation angle error. If we first define $s_1 = \Sigma_{11}$, $s_2 = \Sigma_{22}$, $s_3 = \det(U)\det(V)\Sigma_{33}$, then the covariance matrix $P$ is calculated as

$$P = U\text{diag}\left[(s_2 + s_3)^{-1} \ (s_3 + s_1)^{-1} \ (s_1 + s_2)^{-1}\right]U^T. \tag{6.25}$$

## 6.5 A BRIEF COMPARISON OF STATISTICAL METHODS FOR SMALL SATELLITE IMPLEMENTATIONS

In literature there are many studies comparing the performance of statistical methods for attitude estimation in terms of accuracy, robustness and computational load [2, 8, 9]. These include the recent researches that propose new algorithms such as Fast Linear Attitude Estimator [10] and Riemann–Newton method [11]. In common, they all show, Devenport's q-method, QUEST and SVD method performs same in term of accuracy as long as the measurement weights do not vary too widely and the measurements are represented with the right models. As to the robustness of the algorithms there are different claims specifically focusing on some extreme cases [2]. However, in our opinion, to be accounted as a robust algorithm for space applications, implementation of the algorithm for real missions is more important rather than testing on ground with the simulated data. In this sense, undoubtedly QUEST, which has been implemented for several mission as the sole attitude estimator or for preprocessing the star tracker data since 1979 [3] is a forerunner among all these algorithms.

Speed (or computational complexity) is of course another criterion that should be considered for implementation of the algorithms, especially for small satellites.

Advancement in the computer technology helps us implementing any of these algorithms with almost no issue when they are used as the sole attitude estimator. In this case, since a limited number of vector measurements is processed (as in case we are estimating the attitude using Sun sensor and magnetometer measurements) even the SVD method, which is computationally heavier than others, may be used in real-time. Nonetheless, if we think about using these algorithms for processing the vector star measurements, as a part of the star tracker algorithms for producing the quaternion measurements of the sensor, of course the speed becomes of more importance. A speedy algorithm is needed not just for capability to run the algorithm onboard but also for minimizing the delay in the measurements. QUEST is also advantageous in this sense. Variants of ESOQ are computationally light as well, but a long history of successful applications makes QUEST again preferable.

## 6.6 ATTITUDE DETERMINATION USING GNSS MEASUREMENTS

The Global Navigation Satellite System (GNSS) is commonly used to provide position, velocity and time information for the Earth orbiting LEO satellites. They are not one of the conventional attitude sensors. However, as discussed in several studies [12–14], having multiple receivers onboard the satellite gives us opportunity to use these sensors' measurements for attitude determination as well. Advantages in cost saving, size and mass make them an attractive alternative for attitude determination. Considering that the number of operational systems and the GNSS satellites are increasing recently, we may expect an increase in the number of GNSS-based attitude determination applications in practice as well.

Unlike the conventional sensors that provide vector measurements, the GNSS can only provide scalar measurements, such as pseudorange (or time of arrival), Doppler shift and carrier phase.

There are different methods for processing these scalar measurements to determine the attitude [12]. Here we will present the commonly used method based on measuring the carrier-phase single differences. In the end, we will show that the cost function to be minimized for attitude estimation has similarities to Wahba's function in Eq. (6.1).

Attitude determinations based on measuring the carrier-phase single differences requires at least 3 GNSS receivers onboard the satellite. The measurement geometry for two of these receivers is given Figure 6.2. Here, two GNSS receivers are mounted on the satellite. Their distance $b_i$ is known in the body frame. The line of sight vector to the $j^{th}$ GNSS satellite in the ECI, $s_j$, is also known once the position of the satellite is determined using the measurements from a minimum of four GNSS satellites. This vector is $As_j$ in the body frame once transformed using the attitude matrix of the satellite, which is unknown and to be determined. As a result, the range difference in the measurement of these two receivers can be written as

$$\Delta r_{ij} = b_i^T A s_j \tag{6.26}$$

However, what the receivers is measuring in practice is the carrier-phase single differences as mentioned. Phase difference due to the satellite's attitude is

$$\Delta \varphi_{ij} = 2\pi \left( N_{ij}^\varphi + \lambda^{-1} b_i^T A s_j \right) + \upsilon_{ij}. \tag{6.27}$$

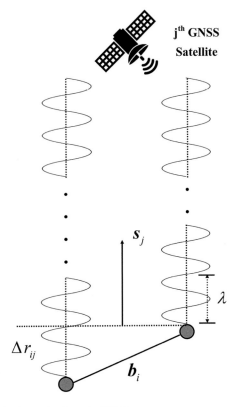

**FIGURE 6.2**    Attitude determination via GNSS measurements.

Here $\Delta\varphi_{ij}$ is the measured phase difference, $\lambda$ is the wavelength of the GNSS signal, $v_{ij}$ is the Gaussian white noise with standard deviation of $\sigma_{ij}$ and $N_{ij}^{\varphi}$ is the number of full wavelength differences of the paths from GNSS satellite $j$ to the receivers with baseline $i$. This last term is known as carrier phase integer ambiguity and there are several algorithms for its calculation (see Ch. 23 of [15]). Assuming it is calculated and corrected in the measurements, the normalized range differences to be used for attitude estimation can be given as

$$\Delta\bar{r}_{ij} = \lambda\left(\frac{\Delta\varphi_{ij}}{2\pi} - N_{ij}^{\varphi}\right). \tag{6.28}$$

Then the loss function to be minimized for attitude estimation is

$$J(A) = \frac{1}{2}\sum_{i=1}^{n}\sum_{j=1}^{m}a_{ij}\left(\Delta\bar{r}_{ij} - b_i^T A s_j\right)^2, \tag{6.29}$$

for a number of $m$ GNSS satellites in sight and $n$ base vectors. $a_{ij}$ is the normalized weight for each measurement. There is need for a minimum of 3 GNSS receivers

(2 baseline vectors) for three-axis attitude determination. Considering that minimum 4 GNSS satellites are in-sight[2] we will have at least 12 measurements to be used for attitude determination.

As can be seen Eq. (6.29) is similar to Wahba's loss function. Yet it is not as easy to solve. One method that is proposed to solve this problem by Crassidis and Markley [14] is the method of vectorized measurements. In this case we rather minimize the cost function for the baseline vectors in the inertial frame defined as $r_i = A^T b_i$. So the loss function to be minimized for each baseline vector $i = 1, 2, \ldots n$ is

$$J_i\left(r_i\right) = \frac{1}{2} \sum_{j=1}^{m} a_{ij} \left(\Delta \bar{r}_{ij} - r_i^T s_j\right)^2. \tag{6.30}$$

Once we have all these baseline vectors in the inertial frame ($r_i$) and their corresponding values in the body frame ($b_i$) the problem reduces to Wahba's problem and any of the statistical methods presented in this chapter can be used to determine the attitude. For other methods to determine the attitude using the GNSS measurements, the reader may refer to [12].

Although there is a solid theory for GNSS-based attitude determination methods with many different approaches, in practice the accuracy is limited due to several error sources. These include the reflections off the spacecraft (multipath effect), ephemeris error for the GNSS satellites and the satellite that we are estimating the attitude for, satellite clock errors, ionosphere errors and receiver errors. There are also inaccuracies in our knowledge for the satellite geometry and the baseline values in the body frame. Particularly for the small satellites as the baseline length shortens error tolerance decreases to few millimeters level if an attitude accuracy below 1° is aimed. Recent results for the Flying Laptop Satellite, which is a 120 kg of small satellite with a size of size 60 cm × 70 cm × 90 cm, show that the maximum error for GNSS based attitude estimations is about 2° [13]. Note that this is relative error compared to the star tracker measurements and before several calibration processes are applied.

## 6.7 CONCLUSION AND DISCUSSION

In the statistical method presented in this chapter, multiple – in fact more than two – vector measurements are integrated optimally to estimate the attitude. This is in contrast to the TRIAD, which does not produce optimal attitude estimates in conventional form and cannot be implemented for more than two measurements. To find the optimal attitude, the loss function that was proposed by Wahba can be minimized using different statistical methods such as Davenport's q-method, QUEST, SVD method, ESOQ and FLAM. These methods for single-frame attitude estimation need at least two vector measurements and knowledge of the relative precision of the attitude sensors.

The following are the main advantages of the statistical single-frame attitude determination methods:

- They provide means for computing an optimal three-axis attitude and can be sufficient for most small satellite missions as the standalone attitude estimator.

- They can be used for preprocessing the vector measurements and providing a single quaternion estimate to the attitude filter as the measurement. This is how today's star trackers provide quaternion measurements.

## NOTES

1  In [16] partially-factorized form of this QUEST characteristic polynomial is suggested to be used in the implementation of the QUEST algorithm.
2  This is the required number for position and velocity estimation using the GNSS measurements. Considering there are multiple operational GNSSs today, this number will be much higher.

## REFERENCES

1. G. Wahba, "A Least Squares Estimate of Satellite Attitude," *SIAM Rev.*, vol. 7, no. 3, pp. 409–409, July 1965.
2. F. L. Markley and D. Mortari, "Quaternion attitude estimation using vector observations," *J. Astronaut. Sci.*, vol. 48, no. 2–3, pp. 359–380, 2000.
3. M. D. Shuster, "The quest for better attitudes," *J. Astronaut. Sci.*, vol. 54, no. 3–4, pp. 657–683, 2006.
4. M. D. Shuster and S. D. Oh, "Three-axis attitude determination from vector observations," *J. Guid. Control. Dyn.*, vol. 4, no. 1, pp. 70–77, 1981.
5. P. Davenport, "A vector approach to the algebra of rotations with applications," 1968.
6. F. L. Markley and J. L. Crassidis, *Fundamentals of Spacecraft Attitude Determination and Control*. New York, NY: Springer New York, 2014.
7. F. L. Markley, "Attitude determination using vector observations and the singular value decomposition," *J. Astronaut. Sci.*, vol. 36, no. 3, pp. 245–258, 1988.
8. F. L. Markley and D. Mortari, "How to estimate attitude from vector observations," *Adv. Astronaut. Sci.*, vol. 103, no. PART III, pp. 1979–1996, 2000.
9. C. Hajiyev and D. Cilden Guler, "Review on gyroless attitude determination methods for small satellites," *Prog. Aerosp. Sci.*, vol. 90, no. November 2016, pp. 54–66, 2017.
10. J. Wu, Z. Zhou, B. Gao, R. Li, Y. Cheng, and H. Fourati, "Fast linear quaternion attitude estimator using vector observations," *IEEE Trans. Autom. Sci. Eng.*, vol. 15, no. 1, pp. 307–319, Jan. 2018.
11. Y. Yang, "Attitude determination using Newton's method on Riemannian manifold," *Proc. Inst. Mech. Eng. Part G J. Aerosp. Eng.*, vol. 229, no. 14, pp. 2737–2742, 2015.
12. S. T. Goh and K. S. Low, "Survey of global-positioning-system-based attitude determination algorithms," *J. Guid. Control. Dyn.*, vol. 40, no. 6, pp. 1321–1335, 2017.
13. A. Hauschild, U. Mohr, M. Markgraf, and O. Montenbruck, "Flight results of GPS-based attitude determination for the microsatellite Flying Laptop," *Navig. J. Inst. Navig.*, vol. 66, no. 2, pp. 277–287, 2019.
14. J. L. Crassidis and F. L. Markley, "New algorithm for attitude determination using global positioning system signals," *J. Guid. Control. Dyn.*, vol. 20, no. 5, pp. 891–896, 1997.
15. P. J. G. Teunissen and O. Montenbruck, Eds., *Springer Handbook of Global Navigation Satellite Systems*. Cham: Springer International Publishing, 2017.
16. Y. Cheng and M. D. Shuster, "Improvement to the implementation of the quest algorithm," *J. Guid. Control. Dyn.*, vol. 37, no. 1, pp. 301–305, 2014.

# 7 Kalman Filtering

In this chapter, we will consider the solution to a class of estimation problems originally developed by Kalman [1] and Kalman and Bucy [2] as extensions of Wiener's classical work [3]. Wiener described an optimal finite impulse response (FIR) filter in the mean squared error sense. His solution uses both the auto correlation and the cross correlation of the received signal with the original data, in order to derive an impulse response for the filter. Kalman also presented the optimal minimum squared error filter. However, Kalman's method has some superiorities when compared with Weiner's filter. In Kalman's method the impulse response of the filter is not determined and that's something advantageous when the numerical computations regarded. Since Kalman's filter is described by state space techniques, the filter can be used as a smoother, a filter or a predictor.

Our interest is in a class of linear minimum-error-variance sequential state estimation algorithms, referred to as "Wiener-Kalman filters," "Kalman Bucy filters" or, more commonly, just "Kalman filters," in recognition of their initial impetus to the theoretical development of the area which has seen extensive development in recent years.

Mathematically, the Kalman filter is a system of first-order ordinary differential equations with quadratic nonlinearities which are solved on digital computers. Kalman filter algorithms, or one of the many extensions and variations of them, have been applied in numerous practical situations, including navigation, space guidance and orbit determination. The Kalman filter is used for the following tasks:

1. Minimizing the measurement errors and finding a better value of a measured parameter;
2. Sensor fusion;
3. Estimating the unknown parameters or the state vector of the system;
4. Fault detection and diagnosis.

Depending on the application, one might want to obtain an estimate of the state at a certain time. If the state is estimated for some future time, the process is called prediction. If the estimate is made using all measurements up to and including the current moment, it is called filtering. If an estimate is made for some time in the past using measurements until the current moment, the process is called smoothing.

In this chapter, we limit ourselves to prediction and filtering. Kalman filter process will now be introduced. It basically consists of two parts:

1. **Time update (propagation)** – the prediction of the state vector and its error covariance using the system model and its statistics.
2. **Measurement update** – the improvements of the prediction (both the state and its error covariance) using the available measurements, which gives the filtered state in the end.

The Linear Kalman Filter (LKF) process has been designed to estimate the state vector in a linear model. If the model turns out to be nonlinear, a linearization procedure is usually performed in deriving the filtering equations.

In this chapter, concluding studies of estimation theory will be presented by giving explicit consideration to problems in nonlinear estimation. Many navigation, space guidance and orbit determination methods have inherently nonlinear observation models, and the dynamic process models for the majority of realistic vehicle-guidance and control problems are inherently nonlinear. In this chapter, we shall study the methods of linearization about a nominal trajectory, as well as methods to treat the nonlinear estimation problem directly. We will also consider a real-time linear Taylor approximation of the system function at the previous state estimate and that of the observation function at the corresponding predicted position. The Kalman filter so obtained will be called the extended Kalman filter (EKF). In this case, the filtering procedure is very simple and efficient. Furthermore, it has found many real-time applications. One such application is adaptive system identification. The details for the Unscented Kalman Filter (UKF), which is another type of nonlinear filtering algorithm, will be also presented in this chapter. In contrast to the EKF, the UKF does not linearize the system and approximates the nonlinear distribution using the sigma points. Applications of all these filters, specifically for small satellite attitude estimation, will be introduced in further chapters.

We shall begin our treatment of the Kalman filter with the discrete-time version of the problem, i.e. observation of a discrete dynamic system. For simple problems, the discrete algorithms can be manipulated by hand and considerable insight may be gained. The step-by-step processing of information lends itself to a simple development.

## 7.1 THE OPTIMAL DISCRETE LKF DERIVATION

In this section, we introduce and briefly discuss the mathematical model and the relations of the linear discrete-time Kalman filter.

Let us take a linear discrete dynamic system. The system's dynamic state equation defines the dynamics of the system, whereas the measurement equation defines the generation mechanism of the measurement. The equations for a linear system are written as follows:

State equation:

$$x(k+1) = \Phi(k+1,k)x(k) + G(k+1,k)w(k). \tag{7.1}$$

Measurement equation:

$$z(k) = H(k)x(k) + v(k), \tag{7.2}$$

where $x(k)$ is the $n$-dimensional state vector of the system, $\Phi(k+1,k)$ is the $n \times n$ transition matrix of the system, $w(k)$ is the $r$-dimensional random Gaussian noise vector (system noise) with zero mean and covariance matrix $E[w(k)w^T(j)] = Q(k)\delta(kj)$, $E$ is the statistic average operator, $\delta(kj)$ is the Kronecker symbol

$$\delta(kj) = \begin{cases} 1, & k = j \\ 0, & k \neq j \end{cases}.$$

$G(k + 1, k)$ is the $n \times r$ dimensional transfer matrix of the system noise, $z(k)$ is the $s$ dimensional measurement vector, $H(k)$ is the $s \times n$ dimensional measurement matrix of the system, $v(k)$ is the $s$-dimensional measurement noise vector, with zero mean and covariance matrix $E[v(k)v^T(j)] = R(k)\delta(kj)$.

The mean of the initial condition $x(0)$ is $\bar{x}(0)$, and the covariance matrix is $P(0)$. There is no correlation between the system noise $w(k)$ and the measurement noise $v(k)$: $E[w(k)v^T(j)] = 0$, $\forall\, k, j$.

It is desired to find the value of the state vector according to the sequence of the $z(k)$ measurement vectors. Linear filters theory based on Kalman filter must be used for that.

Let us examine the derivation of Kalman filter equations and make some comments briefly about the obtained basic results. The filtering problem for (7.1) system with set of measurements until $k$,

$$\{z(1), z(2), \ldots, z(k)\} = Z_1^k,$$

consists of finding the optimum estimation value of $x(k)$ state vector that corresponds to the minimum value of the error's mean quadratic value according to the measurement series. Let us show this value with $\hat{x}(k/k)$ and the error of the estimation value with:

$$\tilde{x}(k/k) = x(k) - \hat{x}(k/k). \tag{7.3}$$

As shown in the estimation theory [4], the estimation value of the state vector, which meets this criterion, is the conditional mean:

$$\hat{x}(k/k) = E\{X(k)/Z_1^k\}. \tag{7.4}$$

As it is obvious, a posterior probability distribution density is needed to be determined to calculate this value. This distribution density $f[x(k)/Z_1^k]$ can be found by using the Bayes formula:

$$f\left[x(k)/Z_1^k\right] = f\left[x(k)/Z_1^{k-1}, z(k)\right] = \frac{f\left[x(k)/Z_1^{k-1}\right]f\left[z(k)/x(k), Z_1^{k-1}\right]}{f\left[z(k)/Z_1^{k-1}\right]} \tag{7.5}$$

It is possible to show that distribution densities, which are included by expression (7.5), are all subject to Gaussian distribution by examining them separately. The parameters of these distributions are appointed in the basis of the state Eq. (7.1) and the observation Eq. (7.2). First,

$$f\left[z(k)/x(k), Z_1^{k-1}\right] = f\left[z(k)/x(k)\right] \tag{7.6}$$

equality can be written. Because, $v(k)$ is the white Gaussian noise and so in $x(k)$ value, $z(k)$ vector is not related to a priori $Z_1^{k-1}$ measurement sequence.

Simply, by using Eq. (7.2) the following result is obtained:

$$f\left[y(k)/x(k)\right] = N\left\{H(k)x(k), R(k)\right\},\tag{7.7}$$

Here, $N\{a,b\}$ representation is used for the Gaussian possibility distribution density with $a$ mean vector and $b$ covariance matrix.

In order to determine the parameters of the conditional distribution density $f\left[x(k)/Z_1^{k-1}\right]$, which are included by expression (7.5), let us adopt the form shown below:

$$E\left\{x(k)/Z_1^{k-1}\right\} = \hat{x}(k/k-1),\tag{7.8}$$

$$E\left\{\left[x(k)-\hat{x}(k/k-1)\right]\left[x(k)-\hat{x}(k/k-1)\right]^T / Z_1^{k-1}\right\}$$
$$= E\left\{\tilde{x}(k/k-1)\tilde{x}^T(k/k-1)/Z_1^{k-1}\right\} = P(k/k-1)\tag{7.9}$$

According to the physical meaning, $\hat{x}(k/k-1)$ variable is a step extrapolation (prediction) value and $P(k/k-1)$ is the covariance matrix of the extrapolation error. By substituting the definition of $x(k)$ from formula (7.1) to the expressions (7.8) and (7.9) [5]:

$$\hat{x}(k/k-1) = E\left\{\Phi(k,k-1)x(k-1)+G(k,k-1)w(k-1)/Z_1^{k-1}\right\}$$
$$= \Phi(k,k-1)E\left\{x(k-1)/Z_1^{k-1}\right\}\tag{7.10}$$
$$= \Phi(k,k-1)\hat{x}(k-1/k-1)$$

$$P(k/k-1) = E\left\{\tilde{x}(k/k-1)\tilde{x}^T(k/k-1)/Z_1^{k-1}\right\}$$
$$= \Phi(k,k-1)P(k-1/k-1)\Phi^T(k,k-1)+G(k,k-1)Q(k-1)G^T(k,k-1)\tag{7.11}$$

Here,

$$P(k-1/k-1) = E\left\{\tilde{x}(k-1/k-1)\tilde{x}^T(k-1/k-1)/Z_1^{k-1}\right\}\tag{7.12}$$

is the covariance matrix of the filtering error in a priori step.

When deriving expressions (7.10) and (7.11), it is regarded that, the mean value of $w(k)$ system noise vector is zero and there is not any correlation between this vector and $x(k)$ state vector. In conclusion,

$$f\left[x(k)/Z_1^{k-1}\right] = N\left\{\hat{x}(k/k-1), \Phi(k, k-1)P(k-1/k-1)\Phi^T(k, k-1)\right.$$
$$\left. + G(k, k-1)Q(k-1)G^T(k, k-1)\right\}. \tag{7.13}$$

Let us determine the parameters of the Gaussian distribution placed at the denominator of Eq. (7.5). By using observation Eq. (7.2) and taking into account that there is not any correlation between state vector $x(k)$ and observation error vector $v(k)$, it can be written that

$$E\left\{z(k)/Z_1^{k-1}\right\} = E\left\{H(k)x(k)+v(k)/Z_1^{k-1}{}_1\right\} = H(k)\hat{x}(k/k-1) \tag{7.14}$$

$$E\left\{\left[z(k)-H(k)\hat{x}(k/k-1)\right]\left[z(k)-H(k)\hat{x}(k/k-1)\right]^T\right\}$$
$$= E\left\{\tilde{z}(k/k-1)\tilde{z}^T(k/k-1)\right\} = P_{\tilde{z}} = H(k)P(k/k-1)H^T(k)+R(k). \tag{7.15}$$

Here,

$$\tilde{z}(k/k-1) = z(k)-H(k)\hat{x}(k/k-1) \tag{7.16}$$

is the "innovation" process which plays an important role in Kalman filter theory. Thus;

$$f\left[z(k)/Z_1^{k-1}\right] = N\left\{H(k)\hat{x}(k/k-1), H(k)P(k/k-1)H^T(k)+R(k)\right\}. \tag{7.17}$$

By substituting the expressions (7.7), (7.13) and (7.17) in formula (7.5) for $f\left[x(k)/Z_1^k\right]$ a posterior possibility density and using the addition procedure to determine the full square, it can be shown that, this density is the Gaussian density with $\hat{x}(k/k)$ mean value. The covariance matrix of that density is the covariance matrix of the filtering error.

Thus, possibility estimation algorithm of linear discrete dynamic system is expressed by recurrent equation system as given below:

The estimation value equation:

$$\hat{x}(k/k) = \Phi(k, k-1)\hat{x}(k-1/k-1)+K(k)$$
$$\times\left[z(k)-H(k)\Phi(k, k-1)\hat{x}(k-1/k-1)\right] \tag{7.18}$$
$$= \hat{x}(k/k-1)+K(k)\tilde{z}(k/k-1)$$

Here, $K(k/k)$ is the optimal gain matrix of the filter.

$$
\begin{aligned}
K(k) &= P(k/k)H^{T}(k)R^{-1}(k) \\
&= P(k/k-1)H^{T}(k)\left[H(k)P(k/k-1)H^{T}(k)+R(k)\right]^{-1}
\end{aligned}
\tag{7.19}
$$

The covariance matrix of the filtering error:

$$
\begin{aligned}
P(k/k) &= P(k/k-1)-P(k/k-1)H^{T}(k) \\
&\times\left[H(k)P(k/k-1)H^{T}(k)+R(k)\right]^{-1}H(k)P(k/k-1)
\end{aligned}
\tag{7.20}
$$

The covariance matrix of the extrapolation error:

$$
\begin{aligned}
P(k/k-1) &= \Phi(k,k-1)P(k-1/k-1)\Phi^{T}(k,k-1) \\
&+ G(k,k-1)Q(k-1)G^{T}(k,k-1)
\end{aligned}
\tag{7.21}
$$

The initial conditions:

$$
\hat{x}(0/0)=\bar{x}(0),\quad P(0/0)=P(0).
\tag{7.22}
$$

Optimal filter, which has calculation algorithm expressed by (7.18–7.22) equations, is called Kalman filter. $k/k-1$ series show the values that are predicted by using a priori estimates, $k/k$ series show the estimation values for which all measurements are used – including $z(k)$ - at $k^{th}$ time step.

$K(k)$ and $P(k/k)$ can be also shown with the expressions given below [6]:

$$
K(k)=P(k/k-1)H^{T}(k)\left[H(k)P(k/k-1)H^{T}(k)+R(k)\right]^{-1}
\tag{7.23}
$$

$$
P(k/k)=\left[I-K(k)H(k)\right]P(k/k-1).
\tag{7.24}
$$

Alternatively, we have the following expressions for the $P(k/k)$ and $K(k)$:

$$
P(k/k)=\left[P^{-1}(k/k-1)+H^{T}(k)R^{-1}(k)H(k)\right]^{-1}.
\tag{7.25a}
$$

$$
P(k/k)=P(k/k-1)\left[I+H^{T}(k)R^{-1}(k)H(k)P(k/k-1)\right]^{-1}
\tag{7.25b}
$$

$$
K(k)=P(k/k)H^{T}(k)R^{-1}(k),
\tag{7.26}
$$

where $I$ is the unit matrix.

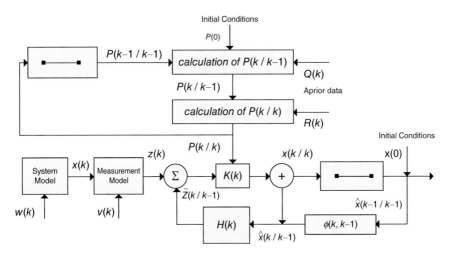

**FIGURE 7.1**    The structural diagram of the Kalman filter.

The initial values $\bar{x}(0)$ and $P(0)$, covariance matrix of the system noise $Q(k)$ and covariance matrix of the measurement noise $R(k)$ must be known previously for the filter to work.

The structural diagram of the Kalman filter is given in Figure 7.1.

According to formula (7.18), the estimation value is equal to the sum of $\hat{x}(k/k-1)$ extrapolation value and $K(k)\tilde{z}(k/k-1)$ correction term. The extrapolation (prediction) value is calculated by multiplying the estimation value found in the previous step by the transition matrix of the system. After that, correction is made to the extrapolation value. Hence, Kalman filter works on the principle of correction (update) of the prediction.

The Kalman filter algorithm includes the following steps [7]:

1. One step further prediction of the states (determination of the extrapolation value): $\hat{x}(k/k-1) = \Phi(k,k-1)\hat{x}(k-1/k-1)$ ;
2. One step prediction of the covariance (predicted state covariance)

$$P(k/k-1) = \Phi(k,k-1)P(k-1/k-1)\Phi^{T}(k,k-1) + G(k,k-1)Q(k-1)G^{T}(k,k-1);$$

3. Determination of the difference of the measurement and the extrapolation value (innovation sequence)

$$\tilde{z}(k/k-1) = y(k) - H(k)\hat{x}(k/k-1);$$

4. Calculation of the Kalman Gain

$$K(k) = P(k/k-1)H^{T}(k)\left[H(k)P(k/k-1)H^{T}(k) + R(k)\right]^{-1};$$

5. State and covariance correction (update)

$$\hat{x}(k/k) = \hat{x}(k/k-1) + K(k)\tilde{z}(k/k-1)$$

$$P(k/k) = [I - K(k)H(k)]P(k/k-1);$$

6. Continue to the next recursive step.

The important features of the Kalman filter are given below [8]:

1. The filter can be easily realized in a computer once the model of the dynamic system is known;
2. The value found by means of the filter is linear with respect to the measurement;
3. The covariance matrix of the filtering error $P(k/k)$ is not related to the measurements of $z(k)$ since the filter is linear, and it can be calculated separately out of the recursive loops for especially saving the computational load;
4. Filtering algorithms can be easily applied to multi-dimensional case;
5. Kalman filter is equivalent to Wiener filter in steady state for stationary dynamic filters.

Adaptive filters become more important when the mathematic model of the system is not known previously or it has changed during the operation process. In this case the evaluation operation is combined with the identification of the parameters and/or the model structure. To do so the nonlinear equations must be linearized first.

## 7.2  STABILITY OF OPTIMAL LKF

As it is seen above the Kalman filter is a time-domain filter, because the design is done in the time domain rather than the frequency domain. One of the advantages of the Kalman filter is its ability to estimate time-variable parameters. When Kalman filter is used, convergence of estimated values to the actual values of parameters depends on stability condition of the filter. The stability of conventional digital filters is easily analyzed with $z$-transform methods [9]. We shall do the same for the optimal Kalman filter.

It is assumed that the optimal discrete Kalman filter (7.18–7.21) has reached a constant-gain condition. The estimation Eq. (7.18) can also be expressed for convenience:

$$\hat{x}(k/k) = \hat{x}(k/k-1) + K(k)[z(k) - H(k)\hat{x}(k/k-1)]. \qquad (7.27)$$

Replace $\hat{x}(k/k-1)$ with $\Phi(k,k-1)\hat{x}(k-1/k-1)$ in Eq. (7.27). After simplifications the following equation is obtained [8]:

$$\hat{x}(k/k) = [\Phi(k,k-1) - K(k)H(k)\Phi(k,k-1)]\hat{x}(k-1/k-1) + K(k)z(k). \qquad (7.28)$$

Taking the $z$-transform of both sides of Eq. (7.28) and retarding $\hat{x}(k/k)$ by one step in the time domain is equivalent to multiplying $\hat{X}_z(k/k)$ by $z^{-1}$ in the $z$-domain. In the $z$-domain this means:

$$\hat{X}_z(k/k) = \left[\Phi(k,k-1) - K(k)H(k)\Phi(k,k-1)\right]z^{-1}\hat{X}_z(k/k) + K(k)Z_z(k). \quad (7.29)$$

After rearranging terms, we have

$$\left[zI - \left\{\Phi(k,k-1) - K(k)H(k)\Phi(k,k-1)\right\}\right]\hat{X}_z(k/k) = zK(k)Z_z(k), \quad (7.30)$$

where $z$ denotes the usual z-transform variable, $Z_z(k)$ refers to the z-transform of the measurement vector.

It is well known from linear system theory that the bracketed quantity on the left side of Eq. (7.30) describes the natural modes of the system. The determinant of the bracketed $n \times n$ matrix gives the characteristic polynomial for the system, i.e. [8]

$$\text{Characteristic polynomial} = \left[zI - \left\{\Phi(k,k-1) - K(k)H(k)\Phi(k,k-1)\right\}\right]. \quad (7.31)$$

The roots of this polynomial provide information about the filter stability. If all the roots lie inside the unit circle in the z-plane, the filter is stable; conversely, if any root lies on or outside the unit circle, the filter is unstable. As a matter of terminology, the roots of the characteristic polynomial are the same as the eigenvalues of $\{\Phi(k, k-1) - K(k)H(k)\Phi(k, k-1)\}$.

## 7.3 LKF IN CASE OF CORRELATED SYSTEM AND MEASUREMENT NOISE

Optimum Kalman filter equations can be modified to accommodate correlated system and measurement noise.

Let the process and measurement models be defined as

$$x(k+1) = \Phi(k+1,k)x(k) + w(k) \quad (7.32)$$

$$z(k) = H(k)x(k) + v(k), \quad (7.33)$$

where

$$E\left[w(k)w^T(i)\right] = Q(k)\delta(ki) \quad (7.34)$$

$$E\left[v(k)v^T(i)\right] = R(k)\delta(ki) \quad (7.35)$$

$$E\left[w(k-1)v^T(i)\right] = C(k). \quad (7.36)$$

An explanation why we are concerned with the cross correlation of $v(k)$ with $w(k-1)$, rather than $w(k)$ is given in [4]: "$w(k-1)$ - not $w(k)$- represents the cumulative effect of the white forcing function in the continuous model in the interval of $t(k-1)$ to $t(k)$. Similarly, $v(k)$ represents the cumulative effect of the white measurement noise in the continuous model when averaged over the same interval of $t(k-1)$ to $t(k)$".

Rewriting Eq. (7.32) with $k$-retarded one-step will help in this regard:

$$x(k) = \Phi(k,k-1)x(k-1) + w(k-1). \tag{7.37}$$

Therefore, if we wish to have a correspondence between the continuous and discrete models for small $\Delta t$, it is the cross correlation between $v(k)$ and $w(k-1)$ that we need to include in the discrete model.

Using the estimation Eq. (7.27) and the measurement Eq. (7.33) we express the estimation error:

$$
\begin{aligned}
e(k) &= x(k) - \hat{x}(k/k) = x(k) - \left[\hat{x}(k/k-1) + K(k)\{z(k) - H(k)\hat{x}(k/k-1)\}\right] \\
&= \left[I - K(k)H(k)\right]e(k/k-1) - K(k)v(k),
\end{aligned}
\tag{7.38}
$$

where

$$e(k/k-1) = x(k) - \hat{x}(k/k-1). \tag{7.39}$$

$e(k/k-1)$ and $v(k)$ will be correlated, so we will work this out as an additional problem:

$$
\begin{aligned}
E\left[e(k/k-1)v(k)^T\right] &= E\left[\left(x(k) - \hat{x}(k/k-1)\right)v(k)^T\right] \\
&= E\left[\left(\Phi(k,k-1)x(k-1) + w(k-1)\right.\right. \tag{7.40} \\
&\quad \left.\left. -\Phi(k,k-1)\hat{x}(k-1/k-1)\right)v(k)^T\right]
\end{aligned}
$$

Note that $v(k)$ will not be correlated with either $x(k-1)$ or $\hat{x}(k-1/k-1)$, because of its whiteness. Therefore, Eq. (7.40) reduces to

$$E\left[e(k/k-1)v(k)^T\right] = E\left[w(k-1)v(k)^T\right] = C(k). \tag{7.41}$$

We form the expression for the error covariance matrix $P(k)$ by using Eq. (7.38):

$$
\begin{aligned}
P(k) &= E\left[e(k)e(k)^T\right] \\
&= E\left\{\left[(I - K(k)H(k))e(k/k-1) - K(k)v(k)\right]\right. \tag{7.42} \\
&\quad \left.\left[(I - K(k)H(k))e(k/k-1) - K(k)v(k)\right]^T\right\}
\end{aligned}
$$

Taking into account (7.41) and expanding (7.42), we get [4]:

$$P(k) = \left[ I - K(k)H(k) \right] P(k/k-1) \left[ I - K(k)H(k) \right]^T + K(k)R(k)K(k)^T$$
$$- \left[ I - K(k)H(k) \right] C(k)K(k)^T - K(k)C(k)^T \left[ I - K(k)H(k) \right]^T.$$

$$(7.43)$$

As it is seen from (7.43), the error covariance matrix $P(k)$ involves the cross-correlation parameter $C(k)$. To find the optimal gain, we differentiate trace $P(k)$ with respect to $K(k)$ and set the result equal to zero.

After necessary mathematical operations, the resulting optimal gain is obtained as

$$K(k) = \left( P(k/k-1)H(k)^T + C(k) \right)$$
$$\times \left[ H(k)P(k/k-1)H(k)^T + R(k) + H(k)C(k) + C(k)^T H(k)^T \right]^{-1}.$$

$$(7.44)$$

Substituting the optimal gain expression (7.44) into the general $P(k)$ equation (7.43) yields a posteriori equation for $P(k)$. After some algebraic manipulations, this leads to either of the two forms [4]:

$$P(k) = P(k/k-1) - K(k)$$
$$\times \left[ H(k)P(k/k-1)H(k)^T + R(k) + H(k)C(k) + C(k)^T H(k)^T \right] K(k)^T$$

$$(7.45)$$

or

$$P(k) = \left[ I - K(k)H(k) \right] P(k/k-1) - K(k)C(k)^T. \qquad (7.46)$$

The extrapolation and the extrapolation error covariance equations are not affected by the cross correlation between $w(k-1)$ and $v(k)$ because of the whiteness property of each. Therefore, they are expressed as

$$\hat{x}(k+1/k) = \Phi(k+1,k)\hat{x}(k/k) \qquad (7.47)$$

$$P(k+1/k) = \Phi(k+1,k)P(k/k)\Phi(k+1,k)^T + Q(k). \qquad (7.48)$$

Equations (7.27), (7.44), (7.46), (7.47) and (7.48) comprise the complete set of LKF equations for the correlated system and measurement noise case [5].

## 7.4 DISCRETE KALMAN FILTERING WHEN SYSTEM AND MEASUREMENT NOISES ARE NOT ZERO-MEAN PROCESSES

In the previous sections, it was assumed that system noise $w(k)$ and measurement noise $v(k)$ were zero-mean processes. Now let us consider $w(k)$ is not a zero-mean process, but

$$E\{w(k)\} = \mu_w(k). \tag{7.49}$$

We try to make corrections in the Kalman filter equations in order to have the estimate that remains unbiased, since the known mean must be incorporated into the estimation equations.

The equation for the propagation of the mean of the state vector is [4]

$$\mu_x(k+1) = \Phi(k+1,k)\mu_x(k) + G(k)\mu_x(k) \tag{7.50}$$

with $\mu_x(0) = E\{x(0)\}$. Subtract this equation from the system-model equation and define $x^0(k) = x(k) - \mu_x(k)$ and $w^0(k) = w(k) - \mu_w(k)$, we obtain

$$x^0(k+1) = \Phi(k+1,k)x^0(k) + G(k)w^0(k) \tag{7.51}$$

with $E\{x^0(0)\} = E\{x(0) - \mu_x(0)\} = 0$. We note that both $x^0(k)$ and $w^0(k)$ are zero-mean processes. Similarly, we can write the observation model as

$$z^0(k) = H(k)x^0(k) + v(k), \tag{7.52}$$

where $z^0(k) = z(k) - H(k)\mu_x(k)$.

The Kalman filter may now be directly applied to the modified zero-mean system and observation models to obtain the estimate of $x^0(k)$ as

$$\hat{x}^0(k/k) = \Phi(k,k-1)\hat{x}^0(k-1/k-1) + K(k) \\ \times \left[ z^0(k) - H(k)\Phi(k,k-1)\hat{x}^0(k-1/k-1) \right]. \tag{7.53}$$

Since $\mu_x(k)$ is known, the best (in the sense of minimum variance) estimate of $x(k)$, given $z(k)$, must be $\hat{x}^0(k) + \mu_x(k)$. Let us add the equation for the mean,

$$\mu_x(k) = \Phi(k,k-1)\mu_x(k-1) + G(k-1)\mu_w(k-1) \tag{7.54}$$

to Eq. (7.53). Hence we obtain

$$\hat{x}(k/k) = \Phi(k,k-1)\hat{x}(k-1/k-1) + G(k-1)\mu_w(k-1) + K(k) \\ \times \left[ z^0(k) - H(k)\Phi(k,k-1)\hat{x}^0(k-1/k-1) \right]. \tag{7.55}$$

Substituting for $z^0(k)$ yields

$$\hat{x}(k/k) = \Phi(k,k-1)\hat{x}(k-1/k-1) + G(k-1)\mu_w(k-1)$$
$$+ K(k)\Big[z(k) - H(k)\mu_x(k) - H(k)\Phi(k,k-1)\hat{x}^0(k-1/k-1)\Big]. \tag{7.56}$$

Substituting the expression for $\mu_x(k)$, we have

$$\hat{x}(k/k) = \hat{x}(k/k-1) + K(k)\Big[z(k) - H(k)\hat{x}(k/k-1)\Big], \tag{7.57}$$

where the extrapolation value is

$$\hat{x}(k/k-1) = \Phi(k,k-1)\hat{x}(k-1/k-1) + G(k-1)\mu_w(k-1). \tag{7.58}$$

Although the foregoing development involved the case, where $\hat{\mu}_w \neq 0$, the same approach can also be used to handle the case where there is a known control input, so that the system model becomes

$$x(k+1) = \Phi(k+1,k)x(k) + G(k)w(k) + B(k)u(k). \tag{7.59}$$

Here $u(k)$ is a known deterministic control input. In this case the extrapolation value $\hat{x}(k/k-1)$ turns out

$$\hat{x}(k/k-1) = \Phi(k,k-1)\hat{x}(k-1/k-1) + B(k-1)u(k-1) + G(k-1)\mu_w(k-1). \tag{7.60}$$

This relation is particularly important in the stochastic control problems.

In a similar manner, we can handle the case in which $v(k)$ is not a zero-mean process or in which a known deterministic input y is added to the output so that the observation model is

$$z(k) = H(k)x(k) + v(k) + y(k), \tag{7.61}$$

where $E\{v(k)\} = \mu_v(k)$. For this problem, the filter Eq. (7.57) becomes [4]

$$\hat{x}(k/k) = \hat{x}(k/k-1) + K(k)\Big[z(k) - \mu_v(k) - y(k) - H(k)\hat{x}(k/k-1)\Big]. \tag{7.62}$$

The remaining filtration equations are the same as the optimal LKF (7.19–7.22).

## 7.5 DIVERGENCE IN THE KALMAN FILTER AND THE METHODS AGAINST DIVERGENCE

In some applications, one finds that the actual estimation errors greatly exceed the values which would be theoretically predicted by the error variance. In fact, the actual error may become unbounded, even though the error variance in the Kalman filter algorithm is vanishingly small. This phenomenon, referred as *divergence*, or *data saturation* [4], can seriously affect the usefulness of the Kalman filter. The possibility of such unstable behavior was first suggested by Kalman in [1] and later noted by Knoll and Edelstein in [10] and others in the application of Kalman filter algorithms to space navigation and orbit determination.

One of the major causes of divergence is inaccuracies in the modeling process used to determine the process or measurement model, due to failure of linearization, lack of complete knowledge of the physical problem or the simplifying assumptions necessary for mathematical tractability. Errors in the statistical modeling of noise variances and mean or unknown inputs may also lead to divergence. Another source of divergence is round-off errors, inherent in any digital implementation of the filter algorithm, which may cause the error-variance matrix to lose its positive definiteness or symmetry.

The correction term of the Kalman filter $K(k)\tilde{z}(k/k-1)$ depends on measuring fault level and the accuracy of this value at that time. The correction term will decrease with increasing accuracy of $\hat{x}(k/k)$ value. If the error of the system is small, the optimal gain coefficient $K(k)$ of the filter approaches to zero as $k \to \infty$. At last, the filter loses its sensitivity to new measurements and gives the extrapolation value as the final estimate. One of the failures of the Kalman filter occurs when the mathematical model of the system is not true. This causes the value of the state vector to collapse. In order to avoid this, the values of parameters of the filter $(K(k), P(k/k), Q(k))$ are limited from below or finite memory filter is used. In this case, the obstacle protection specialty of the filter would decrease.

When the input noise is small compared with the measurement, the error variance, as computed by the Kalman filter algorithm, and hence the gain, tends to become very small as time increases. In the case where there is no system noise, the gain asymptotically approaches zero. The divergence problem also occurs when the system noise is small compared with the measurement noise.

Based on the cause of the divergence, it is possible to modify the Kalman filter to prevent divergence. The basic concept of these approaches is to limit the decrease in the gain in order to avoid the rejection of the measurements. These procedures can be placed into three broad classes [11]:

1. Direct increase of the gain matrix;
2. Limiting of the error covariance;
3. Artificial increase of the variance of system noise.

Averaging the non-diagonal terms of the covariance and checking its positive definiteness are two other methods that are used when Kalman filter is implemented on computers. For example, Riccati equation is solved according to $P(k/k)$ matrix. $P(k/k)$ matrix may not be symmetric and positive definite. This occurs due to truncation error in computer (every calculation device has limited word length). The stabilization of the filter is done to remove significant truncation error and degeneration of the matrix. The matrix must be periodically made symmetric to avoid it from losing its symmetricity and positive definiteness. The terms $P_{ij}$ and $P_{ji}$ must be calculated and their half sum $0.5(P_{ij} + P_{ji})$ must be used. Furthermore, the negative diagonal terms are changed to positive. Also, binary logical computer arithmetic and symmetric calculation schemes can be used. All this reduces the truncation error; however, the required computer memory and the run time of the algorithm increase [8].

## 7.6 LINEARIZED KALMAN FILTER

The linearized Kalman filter, a heuristic extension of earlier work by Kalman and Bucy [2] is a minimum variance filter based on linearization about a nominal trajectory.

We will consider the class of nonlinear systems driven by white noise with white noise-corrupted measurements defined by

$$x(k+1) = \Phi\big[x(k),k\big] + G\big[x(k),k\big]w(k) \tag{7.63}$$

$$z(k) = h\big[x(k),k\big] + v(k), \tag{7.64}$$

where $x(k+1)$ is the state vector, $z(k)$ is the measurement at time $k$, $w(k)$ is the system noise, $v(k)$ is the measurement noise, $\Phi[x(k),k]$ is the nonlinear state transition function mapping the previous state to the current state, $h[x(k),k]$ is a nonlinear measurement model mapping current state to measurements, $G[x(k),k]$ is the transition matrix of the system noise.

It is assumed that both noise vectors $v(k)$ and $w(k)$ are linearly additive Gaussian, temporally uncorrelated with zero mean, which means

$$E\big[w(k)\big] = E\big[v(k)\big] = 0, \quad \forall k, \tag{7.65}$$

with the corresponding covariances:

$$E\big[w(i)w^T(j)\big] = Q(i)\delta(ij), \quad E\big[v(i)v^T(j)\big] = R(i)\delta(ij). \tag{7.66}$$

It is assumed that process and measurement noises are uncorrelated, i.e.

$$E\big[w(i)v^T(j)\big] = 0, \quad \forall i,j. \tag{7.67}$$

We can write the following expressions for nominal trajectory

$$\hat{x}_n(k+1) = \Phi\big[\hat{x}_n(k),k\big], \quad \hat{x}_n(k_0) = \mu_{x_0} \tag{7.68}$$

and the nominal measurement

$$z_n(k) = h\big[\hat{x}_n(k),k\big]. \tag{7.69}$$

It is assumed that deviation from the nominal measurement

$$\delta z(k) = z(k) - z_n(k) \tag{7.70}$$

and deviation from the nominal trajectory $\delta x(k) = x(k) - x_n(k)$ are small enough.

The linearized system model and the measurement model, respectively, are described as

$$\delta x(k+1) = \frac{\partial \Phi\left[\hat{x}_n(k), k\right]}{\partial \hat{x}_n(k)} \delta x(k) + G\left[\hat{x}_n(k), k\right] w(k) \tag{7.71}$$

$$\delta z(k) = \frac{\partial h\left[\hat{x}_n(k), k\right]}{\partial \hat{x}_n(k)} \delta x(k) + v(k). \tag{7.72}$$

Initial conditions are

$$\delta \hat{x}(k_0) = 0, \quad P_{\delta \bar{x}}(k_0) = P_{x0}.$$

In this case the estimate of $\delta x(k)$ is given by following linearized filter algorithm: Linearized one-stage prediction

$$\delta \hat{x}(k+1/k) = \frac{\partial \Phi\left[\hat{x}_n(k), k\right]}{\partial \hat{x}_n(k)} \delta \hat{x}(k). \tag{7.73}$$

Estimate for the linearized filter is

$$\delta \hat{x}(k+1) = \delta \hat{x}(k+1/k) + K(k+1)\left[\delta z(k+1) - \frac{\partial h\left[\hat{x}_n(k+1), k+1\right]}{\partial \hat{x}_n(k+1)} \delta \hat{x}(k+1/k)\right]. \tag{7.74}$$

Filter-gain for the linearized filter

$$K(k+1) = P_{\delta \bar{x}}(k+1/k) \frac{\partial h^T\left[\hat{x}_n(k+1), k+1\right]}{\partial \hat{x}_n(k+1)}$$

$$\times \left[\frac{\partial h\left[\hat{x}_n(k+1), k+1\right]}{\partial \hat{x}_n(k+1)} P_{\delta \bar{x}}(k+1) \frac{\partial h^T\left[\hat{x}_n(k+1), k+1\right]}{\partial \hat{x}_n(k+1)} + R(k+1)\right]^{-1}. \tag{7.75}$$

The linearized extrapolation error

$$P_{\delta \bar{x}}(k+1/k) = \frac{\partial \Phi\left[\hat{x}_n(k), k\right]}{\partial \hat{x}_n(k)} P_{\delta \bar{x}}(k) \frac{\partial \Phi^T\left[\hat{x}_n(k), k\right]}{\partial \hat{x}_n(k)}$$

$$+ G\left[\hat{x}_n(k), k\right] Q(k) G^T\left[\hat{x}_n(k), k\right]. \tag{7.76}$$

Linearized estimation error is

$$P_{\delta \bar{x}}(k+1) = P_{\delta \bar{x}}(k+1/k) - P_{\delta \bar{x}}(k+1/k) \frac{\partial h^T \left[ \hat{x}_n(k+1), k+1 \right]}{\partial \hat{x}_n(k+1)}$$

$$\times \left[ \frac{\partial h \left[ \hat{x}_n(k+1), k+1 \right]}{\partial \hat{x}_n(k+1)} P_{\delta \bar{x}}(k+1/k) \frac{\partial h^T \left[ \hat{x}_n(k+1), k+1 \right]}{\partial \hat{x}_n(k+1)} + R(k+1) \right]^{-1}$$

$$\times \frac{\partial h \left[ \hat{x}_n(k+1), k+1 \right]}{\partial \hat{x}_n(k+1)} P_{\delta \bar{x}}(k+1/k).$$

$$(7.77)$$

Complete linearized state estimate in this case can be written as

$$\hat{x}(k) = \hat{x}_n(k) + \delta \hat{x}(k). \qquad (7.78)$$

Consequently, the linearized discrete Kalman filter algorithms are presented by the formulas (7.73–7.78).

## 7.7 EXTENDED KALMAN FILTER

The Kalman filtering process has been designed to estimate the state vector in a linear model. If the model turns out to be nonlinear, a linearization procedure is usually performed in deriving the filtering equations. We will consider a real-time linear Taylor approximation of the system function at the previous state estimate and that of the observation function at the corresponding predicted position. The Kalman filter so obtained will be called the Extended Kalman Filter (EKF). This idea to handle a nonlinear model is quite natural, and the filtering procedure is fairly simple and efficient.

A state space description of a system which is not necessarily linear will be called a nonlinear model of the system. In this section, we will consider nonlinear process and measurements models of the form (7.63) and (7.64) with the prior statistics (7.65–7.67).

Initial conditions are

$$\hat{x}(k_0) = \mu_{x_0}, P_{\hat{x}}(k_0) = P_{x_0}.$$

The derivation of the EKF follows from that of the linear Kalman filter, by linearizing state and observation models using Taylor's series expansion [4].

Filtering algorithm in this case is as given below:

$$\hat{x}(k+1) = \hat{x}(k+1/k) + K(k+1)\left\{ z(k+1) - h\left[ \hat{x}(k+1/k), k+1 \right] \right\}. \quad (7.79)$$

One-stage prediction

$$\hat{x}(k+1/k) = \Phi \left[ \hat{x}(k), k \right]. \qquad (7.80)$$

Filter-gain

$$
K(k+1) = P_{\tilde{x}}(k+1/k) \frac{\partial h^T \left[ \hat{x}(k+1/k), k+1 \right]}{\partial \hat{x}(k+1/k)}
$$

$$
\times \left[ \frac{\partial h \left[ \hat{x}(k+1/k), k+1 \right]}{\partial \hat{x}(k+1/k)} P_{\tilde{x}}(k+1/k) \frac{\partial h^T \left[ \hat{x}(k+1/k), k+1 \right]}{\partial \hat{x}(k+1/k)} + R(k+1) \right]^{-1}
$$

(7.81)

Extrapolation error

$$
P_{\tilde{x}}(k+1/k) = \frac{\partial \Phi \left[ \hat{x}(k), k \right]}{\partial \hat{x}(k)} P_{\tilde{x}}(k) \frac{\partial \Phi^T \left[ \hat{x}(k), k \right]}{\partial \hat{x}(k)}
$$
$$
+ G \left[ \hat{x}(k), k \right] Q(k) G^T \left[ \hat{x}(k), k \right].
$$

(7.82)

Estimation error

$$
P_{\tilde{x}}(k+1) = P_{\tilde{x}}(k+1/k) - P_{\tilde{x}}(k+1/k) \frac{\partial h^T \left[ \hat{x}(k+1/k), k+1 \right]}{\partial \hat{x}(k+1/k)}
$$

$$
\times \left[ \frac{\partial h \left[ \hat{x}(k+1/k), k+1 \right]}{\partial \hat{x}(k+1/k)} P_{\tilde{x}}(k+1/k) \frac{\partial h^T \left[ \hat{x}(k+1/k), k+1 \right]}{\partial \hat{x}(k+1/k)} + R(k+1) \right]^{-1}
$$

$$
\times \frac{\partial h \left[ \hat{x}(k+1/k), k+1 \right]}{\partial \hat{x}(k+1/k)} P_{\tilde{x}}(k+1/k).
$$

(7.83)

The filter expressed by the formulas (7.79–7.83) is called the Extended Kalman Filter.

It is prudent to note a number of problematic issues specific to the EKF. Unlike the linear filter, the covariances and gain matrix must be computed online as estimates and predictions are made available, and will not, in general, tend to constant values. This significantly increases the amount of computation that must be performed online by the algorithm. Also, if the predicted trajectory is too far away from the true trajectory, then the true covariance will be much larger than the estimated covariance and the filter will become poorly matched. This might lead to severe filter instabilities. Last, the EKF employs a linearized model that must be computed from an approximate knowledge of the state. Unlike the linear algorithm, this means that the filter must be accurately initialized at the start of operation to ensure that the linearized models obtained are valid. All these issues must be taken into account to achieve acceptable performance for the EKF [12].

## 7.8 UNSCENTED KALMAN FILTER

Another variant of Kalman filter for nonlinear systems is the UKF. The essence of the UKF is the unscented transform, a deterministic sampling technique that we use for obtaining a minimal set of sample points (or sigma points) from the a priori mean and covariance of the states. These sigma points go through nonlinear transformation. The posterior mean and the covariance are determined using the transformed sigma points [12–14]:

The UKF is derived for discrete-time nonlinear equations, so the system model is given by:

$$x(k+1) = f\big(x(k),k\big) + w(k),\tag{7.84}$$

$$y_k = h\big(x_k,k\big) + v_k.\tag{7.85}$$

Here, $x(k)$ is the state vector and $y(k)$ is the measurement vector. Moreover $w(k)$ and $v(k)$ are the process and measurement error noises, which are assumed to be Gaussian white noise processes with the covariance of $Q(k)$ and $R(k)$, respectively.

The initial step of the UKF algorithm is determining the $2n + 1$ sigma points with a mean of $x\big(k|k\big)$ and a covariance of $P(k|k)$. For an $n$-dimensional state vector, these sigma points are obtained by

$$x_0\big(k|k\big) = \hat{x}\big(k|k\big)\tag{7.86a}$$

$$x_\gamma\big(k|k\big) = \hat{x}\big(k|k\big) + \left(\sqrt{(n+\kappa)P\big(k|k\big)}\right)_\gamma,\tag{7.86b}$$

$$x_{\gamma+n}\big(k|k\big) = \hat{x}\big(k|k\big) - \left(\sqrt{(n+\kappa)P\big(k|k\big)}\right)_\gamma,\tag{7.86c}$$

where $x_0\big(k|k\big)$, $x_\gamma\big(k|k\big)$, and $x_{\gamma+n}\big(k|k\big)$ are sigma points, $n$ is the state number, and $\kappa$ is the scaling parameter which is used for fine tuning. $\left(\sqrt{(n+\kappa)P\big(k|k\big)}\right)_\gamma$ corresponds to the $\gamma^{th}$ column of the indicated matrix and $\gamma$ is given as $\gamma = 1\ldots n$.

The next step of the UKF procedure is evaluating the transformed set of sigma points for each of the points by

$$x_l\big(k+1|k\big) = f\Big[x_l\big(k|k\big), k\Big]. \quad l = 0\ldots 2n.\tag{7.87}$$

Thereafter, these transformed values are used for calculating the predicted mean, $\hat{x}\big(k+1|k\big)$, and covariance, $P(k + 1|k)$

$$\hat{x}\big(k+1|k\big) = \sum_{l=0}^{2n} \lambda_l x_l\big(k+1|k\big),\tag{7.88a}$$

$$P\left(k+1|k\right)=\left\{\sum_{l=0}^{2n}\lambda_l\left[x_l\left(k+1|k\right)-\hat{x}\left(k+1|k\right)\right]\left[x_l\left(k+1|k\right)-\hat{x}\left(k+1|k\right)\right]^T\right\}+Q(k).$$

(7.88b)

The weights are defined as

$$\lambda_0=\frac{\kappa}{n+\kappa};\quad \lambda_l=\frac{1}{2\left(n+\kappa\right)}l=1...2n.$$

(7.89)

The following part is the update phase of the UKF algorithm. In that phase, the predicted observation vector is calculated as

$$\hat{y}\left(k+1|k\right)=\sum_{l=0}^{2n}\lambda_l y_l\left(k+1|k\right),$$

(7.90)

where

$$y_l\left(k+1|k\right)=h\left[x_l\left(k+1|k\right),k\right]\quad l=0...2n.$$

(7.91)

After that, the observation covariance matrix is determined as

$$P_{yy}\left(k+1|k\right)=\sum_{l=0}^{2n}\lambda_l\left[y_l\left(k+1|k\right)-\hat{y}\left(k+1|k\right)\right]\left[y_l\left(k+1|k\right)-\hat{y}\left(k+1|k\right)\right]^T,$$

(7.92)

where the innovation covariance is

$$P_{vv}\left(k+1|k\right)=P_{yy}\left(k+1|k\right)+R\left(k+1\right).$$

(7.93)

Here, $R(k+1)$ is the measurement noise covariance matrix. Moreover, the cross-correlation matrix can be obtained as

$$P_{xy}\left(k+1|k\right)=\sum_{l=0}^{2n}\lambda_l\left[x_l\left(k+1|k\right)-\hat{x}\left(k+1|k\right)\right]\left[y_l\left(k+1|k\right)-\hat{y}\left(k+1|k\right)\right]^T.$$

(7.94)

For update using the measurements, $y(k+1)$, an innovation sequence is found as

$$e\left(k+1\right)=y\left(k+1\right)-\hat{y}\left(k+1|k\right),$$

(7.95)

and then the Kalman gain is computed by

$$K\left(k+1\right)=P_{xy}\left(k+1|k\right)P_{vv}^{-1}\left(k+1|k\right).$$

(7.96)

Finally, the updated states and covariance matrix are determined,

$$\hat{x}\left(k+1|k+1\right) = \hat{x}\left(k+1|k\right) + K\left(k+1\right)e\left(k+1\right), \tag{7.97}$$

$$P\left(k+1|k+1\right) = P\left(k+1|k\right) - K\left(k+1\right)P_{vv}\left(k+1|k\right)K^{T}\left(k+1\right). \tag{7.98}$$

Here, $\hat{x}\left(k+1|k+1\right)$ is the estimated state vector and $P(k+1|k+1)$ is the estimated covariance matrix.

Even though still being a popular method as an estimator for nonlinear systems, the EKF has some disadvantages, especially in case of high nonlinearity, which appears to be a common problem in the many of the applications, including the attitude estimation. Generally, this is caused by the mandatory linearization phase of the EKF procedure and so the Jacobians derived with that purpose. For most of the applications, generation of the Jacobians is difficult, time consuming and prone to human errors [12]. Nonetheless, the linearization brings about an unstable filter performance when the time step intervals for the update are not sufficiently small and the system is not almost linear on the timescale of the update intervals. As a result, the filter may diverge [13]. Of course, when the filter is propagated for shorter time steps computational load increases. These facts show that the EKF may be efficient only if the system is almost linear on the timescale of update intervals.

In contrast to the EKF, UKF does not need any linearization and is valid to higher order expansions than the EKF. Compared with the EKF's first-order accuracy, the estimation accuracy of the UKF is improved to the third-order for Gaussian data and at least second-order for non-Gaussian data.

## 7.9 OTHER NONLINEAR FILTERING ALGORITHMS

Apart from EKF and UKF, there are various other nonlinear filtering algorithms, proposed especially in last decade. Some examples to these nonlinear filtering algorithms are, a second-order EKF (EKF2), which resembles the UKF [15], cubature Kalman filter (CKF), which is a special form of the UKF [16], Quadrature and Square-Root Quadrature Kalman Filters [17] and Gauss-Hermite Kalman Filter [18]. They all are results of search for having more accurate estimation results for the nonlinear systems and possess several advantages even over UKF. Yet their advantages/disadvantages should be considered on the application basis. There are also variants of these filtering algorithms that are derived as a result of the search for decreasing their computational load for practical applicability.

Presenting all these different nonlinear filtering algorithms is out of the scope of this book. We are mainly interested in applicability of the filters for attitude estimation. In this sense, EKF and UKF are the ones which have a more solid background; EKF has been used for several decades as an attitude estimation algorithm and UKF also has application examples as an attitude estimator, especially for recent small satellite missions. On the other hand, although there are several studies in theory, these other nonlinear filters do not have so many examples for being used as the attitude estimation algorithm for an actual mission. This is why we kept the nonlinear filters presented in this section limited to EKF and UKF.

## 7.10 CONCLUSION AND DISCUSSION

The Kalman filtering algorithms suitable for estimations for linear and nonlinear systems are considered in this chapter. The optimal discrete Kalman filter stability, correlated system and measurement noise processes, divergence in the Kalman filter, the methods for numerical stabilization of the filters and other filtering problems are discussed. The case in which the system and measurement noises are not a zero-mean process is considered.

Three popular approaches to solve nonlinear estimation problems are the linearized Kalman filter, extended Kalman filter (EKF) and unscented Kalman filter (UKF). The linearized Kalman filter and EKF are linear estimators for a nonlinear system derived by linearization of the nonlinear state and observations equations. Among these, the EKF is an efficient extension to the Kalman filter that was introduced regarding the inherent nonlinearity for many real-world systems and since then it has been widely preferred for various applications including the attitude estimation.

To implement the Kalman filter for nonlinear systems without any linearization step, the unscented transform and so UKF can be used. The UKF algorithm is a more recent filtering method which has many advantages over the well-known EKF. The essence of the UKF is the fact that a nonlinear distribution can be approximated more easily than a nonlinear function or transformation. Instead of the analytical linearization used in EKF, UKF statistically captures the system's nonlinearities. The UKF avoids the linearization step of the EKF by introducing sigma points to catch higher order statistic of the system. As a result it satisfies both better estimation accuracy and convergence characteristic. The EKF is more sensitive to measurement faults than the UKF, especially in case of additional bias and noise in the measurements.

## REFERENCES

1. R. E. Kalman, "A new approach to linear filtering and prediction problems," *J. Basic Eng.*, vol. 82, no. 1, pp. 35–45, 1960.
2. R. E. Kalman and R. S. Bucy, "New results in linear filtering and prediction theory," *J. Fluids Eng. Trans. ASME*, vol. 83, no. 1, pp. 95–108, 1961.
3. N. Wiener, *Extrapolation, Interpolation, and Smoothing of Stationary Time Series.* Cambridge, USA: The MIT Press, 1964.
4. A. P. Sage and J. L. Melsa, *Estimation Theory with Applications to Communications and Control.* New Yor, USA: McGraw-Hill, 1971.
5. C. Hajiyev, *Experimental Data Processing Methods and Engineering Applications (in Turkish).* Ankara: Nobel Publishing and Distribution Inc., 2010.
6. C. Hajiyev, *Radio Navigation (in Turkish).* Istanbul: Istanbul Technical University Press, 1999.
7. C. Hajiyev, H. Ersin Soken, and S. Yenal Vural, *State Estimation and Control for Low-cost Unmanned Aerial Vehicles.* Cham: Springer International Publishing, 2015.
8. C. Hajiyev and F. Caliskan, *Fault Diagnosis and Reconfiguration in Flight Control Systems*, vol. 2. Boston, MA: Springer US, 2003.
9. R. G. Brown and P. Y. C. Hwang, *Introduction to Random Signals and Applied Kalman Filtering with Matlab Exercises.* Wiley, 2012.

10. A. L. Knoll and M. M. Edelstein, "Estimation of local vertical and orbital parameters for an earth satellite using horizon sensor measurements," *AIAA J.*, vol. 3, no. 2, pp. 338–345, Feb. 1965.

11. F. H. Schlee, C. J. Standish, and N. F. Toda, "Divergence in the Kalman filter.," *AIAA J.*, vol. 5, no. 6, pp. 1114–1120, Jun. 1967.

12. S. J. Julier, J. K. Uhlmann, and H. F. Durrant-Whyte, "A new approach for filtering nonlinear systems," *Proc. 1995 Am. Control Conf. - ACC'95*, vol. 3, no. June, pp. 0–4, 1995.

13. S. Julier, J. Uhlmann, and H. F. Durrant-Whyte, "A new method for the nonlinear transformation of means and covariances in filters and estimators," *IEEE Trans. Automat. Contr.*, vol. 45, no. 3, pp. 477–482, Mar. 2000.

14. S. J. Julier and J. K. Uhlmann, "Unscented Filtering and Nonlinear Estimation," *Proc. IEEE*, vol. 92, no. 3, pp. 401–422, Mar. 2004.

15. F. Gustafsson and G. Hendeby, "Some relations between extended and unscented Kalman filters," *IEEE Trans. Signal Process.*, vol. 60, no. 2, pp. 545–555, 2012.

16. I. Arasaratnam and S. Haykin, "Cubature kalman filters," *IEEE Trans. Automat. Contr.*, vol. 54, no. 6, pp. 1254–1269, 2009.

17. I. Arasaratnam and S. Haykin, "Square-root quadrature Kalman filtering," *IEEE Trans. Signal Process.*, vol. 56, no. 6, pp. 2589–2593, 2008.

18. S. Särkkä, "Bayesian filtering and smoothing," *Bayesian Filter. Smoothing*, pp. 1–232, 2010.

# 8 Adaptive Kalman Filtering

The Kalman filter (KF) approach to the state estimation is quite sensitive to any uncertainties and malfunctions. If the condition of the real system does not correspond to the models, used in the synthesis of the filter, then these differences, which may be due to some possible failures in the sensors and actuators or any other uncertainties in the models, significantly decrease the performance of the estimation system. In such cases the KF can be adapted and Adaptive or Robust Kalman Filters can be used to compensate the uncertainties and recover the possible faults.

In literature there is not a single convention for naming the adaptive Kalman filters and using the terms "adaptive" and "robust." One preference is to use the term "adaptive" whenever the filter's time-varying covariance matrices are tuned and use the term "robust" when the filter is made insensitive to the outliers and faults, especially in measurements. We think, whenever the filter's covariance matrices are tuned, whether this is against the time-varying uncertainties or faults, it becomes an adaptive Kalman filter since the Kalman gain is no more optimal. So, in our terminology the adaptive Kalman filter somehow encapsulates the robust Kalman filters. Specifically, for filters, which are designed to compensate the sensor/actuator faults (e.g. outliers, abrupt faults etc.), we use both naming conventions.

The KF can be made adaptive and hence insensitive to the uncertainties and faults by using various different techniques (Figure 8.1) [1, 2]. We will be interested with mainly four of these: The Multiple-Model-based Adaptive Estimation (MMAE) [3, 4], joint state and covariance estimation [5–7] and autocorrelation and covariance matching techniques for Innovation-based Adaptive Estimation (IAE) [8, 9] and Residual-based Adaptive Estimation (RAE) [10, 11]. While in the first technique a bank of Kalman filters run in parallel under different models for the filter's statistical information, in the rest the adaptation is done by directly estimating the covariance matrices of the measurement and/or system noises. For joint state and covariance estimation methods, the covariance matrices are estimated together with the states using the Bayesian approach. For the innovation/residual-based adaptive Kalman filters, the filter is adapted by either directly estimating the covariance matrices (autocorrelation techniques) or scaled using the calculated variables (covariance matching techniques).

The MMAE methods described in [3, 4] assume the faults (and/or modes of uncertainties) are known. The parallel running KFs are designed for known conditions (or faults) and the algorithm works within these limits. For example, for an attitude estimation filter different filters may be running with different inertia values

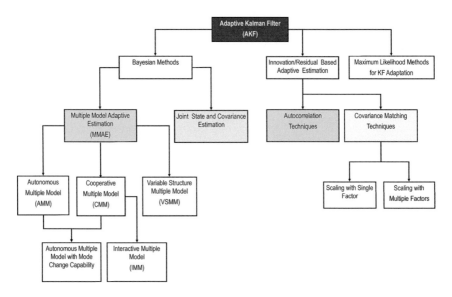

**FIGURE 8.1**   Different Methods for Adaptive Kalman Filtering.

for the spacecraft to account for uncertainties in these parameters. The MMAE approach requires high computational load due to several KFs running parallel. As a result of these drawbacks the MMAE method can be used for only a limited number of applications and are not widely applied for satellite attitude estimation, at least in practice.

Joint state and covariance estimation adapt the filter using Bayesian methods. In this method, the propagation stage is similar to that in a conventional KF. The update stage incorporates the variations in the covariance matrices. These matrices are calculated with an iterative process for each recursive step of the filter. There are adaptive Kalman filter algorithms derived based on Bayesian methods for both linear and nonlinear systems [12]. Due to the iterations and repeated matrix calculations, the computational load is higher compared to conventional KF.

Estimation of the covariance matrices requires using the innovation or residual vectors for $N$ epoch. This increases the storage burden and presents the setting the moving window width, $N$, as another problem. Furthermore, IAE and RAE estimators require that the number, type and distribution of the measurements for all epochs within a window should be consistent. If they do not, the covariance matrices cannot be estimated based on the innovation or residual vectors.

Another concept for KF adaptation is to scale the noise covariance matrix by multiplying it with a time-dependent variable. One of the methods for constructing such algorithm is to use a scale factor as a multiplier to the process or measurement noise covariance matrices [13]. These types of algorithms are also named as fading memory algorithms and the Adaptive Kalman Filter (AKF) is referred as Adaptive Fading Kalman Filter (AFKF) [14]. When the process noise covariance matrix is scaled (Q-adaptation), that means there is an actuator fault or any uncertainty in the process model. When the measurement noise covariance matrix is scaled (R-adaptation), a sensor fault or an uncertainty in the measurements is the problem. Scaling can be done using both a single scale factor or multiple scale factors [15]. The essence of using multiple scale factors is the performance of the KF that varies for each variable. Unlike the adaptation with single scale factor, the multiple scale factor-based method scales only the required terms of the process/measurement noise covariances, so any unnecessary information loss is prevented by disregarding only the data of the faulty sensor/actuator. This is especially advantageous for compensating the sensor faults. If there is fault in any of the sensors (e.g. one of the several magnetometers), the RKF algorithm can be utilized and by the use of multiple measurement noise scale factors as a multiplier on the measurement noise covariance matrix, insensitiveness of the filter to the measurements of only this specific sensor can be ensured.

Moreover there are also published works in literature which are interested in adapting both the R and Q matrices all the time [16–18]. In these works, the simultaneous R- and Q-adaptations by using different combinations of these adaptation methods in case of sensor/actuator faults and model uncertainties are presented.

Especially when dealing with sensor/actuator faults as well as detecting the fault, it is also important to isolate it. When a KF is used, the decision statistics that are used for fault detection change regardless the fault is in the sensors or in the actuators. To solve this problem, a Robust Kalman Filter (or one of its nonlinear variants such as Robust Unscented Kalman filter) based on the Doyle-Stein condition can be used [19, 20]. In this case distinguishing the sensor and actuator faults is easy by using this RKF.

In this chapter, we present some of the widely used adaptive Kalman filtering algorithms. The adaptation algorithms are applicable to linear Kalman filter so the resulting algorithms are all for linear systems. Their variants suitable for adapting the nonlinear Kalman filters will be given in future chapters together with application examples for the small satellite attitude estimation.

## 8.1 *A PRIORI* UNCERTAINTY AND ADAPTATION

### 8.1.1 *A Priori* Uncertainty

A priori uncertainty degree may vary as follows:

a. Complete a priori statistical uncertainty: In this case, neither shape nor parameters of the probability distribution rule for the components of the measured and estimated random processes are known. However, the limited variation areas for the components of the random processes are given. In such a case, the estimation algorithms can be formed only on the basis of a guaranteed approach.

b. Partial a priori statistical uncertainty: In this case, probability distribution rule is known for some of the components of the measured and estimated random processes. The number for the parameters with unknown probability characteristics should not be high. Increase in this number deteriorates the quality of the solution for the problem. In this sense, when there is a priori uncertainty in the parameters, distribution set is given rather than the probability distribution rule for the random processes. Estimation algorithm should be chosen such that it satisfies the optimality criteria for the given distribution set. It means that the estimation algorithm must be adaptive.

## 8.1.2 ADAPTATION

There are three approaches for the adaptation problem: the parametric approach, invariant principle-based approach and structural approach. The parametric approach is the most common one. In this approach the algorithm uses the measurement data for both estimating the required components of the random processes and restoring a priori statistical characteristics of the dynamical system and measurements.

For most of the parametric adaptive estimation methods, a self-tuning circuit is added to the regular KF algorithm. When the KF is designed, it is assumed that the statistical characteristics of the system model, measurements and noises are known accurately. Yet this assumption is not often ensured and the filter becomes sub-optimal. Sub-optimality might be also caused by the simplified process calculations in the filter algorithm. As a result, the adaptation of the noise covariances becomes a necessity. Such adaptation procedure should be also followed for determining the type of the fault caused by the sub-optimality of the filter. These types of KFs with adapted noise covariances are called the AKF.

Some possible methods for designing the AKF are as follows:

- The MMAE method is used. In the MMAE approach, a bank of KFs run in parallel under different models for the statistical filter information matrices, i.e. the process noise covariance matrix Q and/or the measurement noise covariance matrix R.
- Unknown noise covariances of the KF are determined by the statistical analysis of the innovation or residual series. This might be either as direct estimation of the covariance matrices or as an adaptation performed on the basis of covariance scaling (Figure 8.2).
- Probability methods such as the maximum likelihood are used for the noise estimation. Estimated values for the unknown noise covariances are periodically used to renew the noise information in the filtering algorithm.
- The filter is adapted using Bayesian methods. An iterative procedure is followed for determining the Q and R matrices. The unknown covariances are estimated by analyzing the residual error which was caused by using the same covariance values at the previous iteration step. This method is generally used when the required computational burden for performing a specific test is high.

In the next section, we review some of the existing common techniques for the KF adaptation.

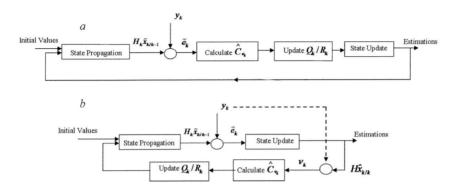

**FIGURE 8.2** The KF adaptation by the statistical analysis of the innovation and residual series: a. Innovation-based KF adaptation; b. Residual-based KF adaptation. For both adaptation techniques update for the process and measurement noise covariances might be either as estimation or scaling.

## 8.2 MULTIPLE MODEL-BASED ADAPTIVE ESTIMATION

The MMAE method is based on implementing a bank of KFs with different models and merging the estimates of all filters by using each model's probabilities for being the true one (Figure 8.3). The state estimate might be either the output of the model with the highest probability or a weighted sum of all the outputs.

The MMAE methods are usually categorized into three [3, 21, 22]. The Autonomous Multiple Model (AMM) estimator is the type where each KF run independently with the assumption that the true mode is time invariant. This is a highly

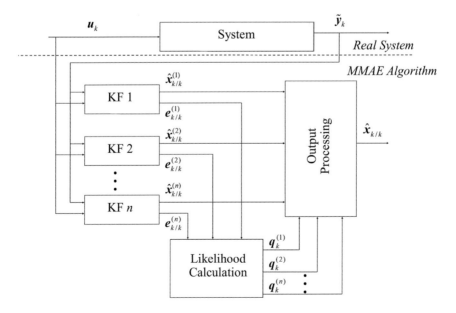

**FIGURE 8.3** Multiple Model Adaptive Estimation: AMM Approach.

appreciated method especially for fault detection and isolation. For fault tolerant control, each model in the bank represents a different sensor/actuator failure condition and the fault can be detected when the probability for any model increases above an alarm threshold. In contrast, the Cooperative Multiple Model (CMM) estimation method that can be used for the hybrid estimation has optimal performance potential when the true mode changes. In this case, the AMM method's assumption of one true mode is no more valid. To propose a feasible solution for the hybrid estimation, all the KFs in the CMM estimator interact with each other rather than running autonomously. The latest type of the MM method is the variable structure MM (VSMM), for which the number of models (so the KFs) is not fixed but changes by time.

Here we will give further details of first two types of MM estimators, the AMM and CMM.

The classical AMM algorithm works assuming that only one of the models in the filter bank is true and all the KFs run independently. In the classical approach, the Bayesian likelihood for each KF in a bank of $n$ KFs running parallel is

$$q_k^{(i)} = \frac{\text{pdf}_k^{(i)} q_{k-1}^{(i)}}{\sum\limits_{j=1}^{n} \text{pdf}_k^{(j)} q_{k-1}^{(j)}}, \tag{8.1}$$

where

$$\text{pdf}_k^{(i)} = \frac{1}{\left(2\pi\right)^{m/2} \left|S_k^{(i)}\right|^{1/2}} \exp\left(\frac{-\left[\tilde{\mathbf{y}}_k - \hat{\mathbf{y}}_k^{(i)}\right]^T \left(S_k^{(i)}\right)^{-1} \left[\tilde{\mathbf{y}}_k - \hat{\mathbf{y}}_k^{(i)}\right]}{2}\right), \tag{8.2}$$

and

$$S_k^{(i)} = H_k P_{k/k-1}^{(i)} H_k^T + R_k. \tag{8.3}$$

Here, $\hat{\mathbf{y}}_k^{(i)}$ is the predicted measurement, $S_k^{(i)}$ is the innovation covariance and $P_{k/k-1}^{(i)}$ is the predicted covariance for KFs. Besides, $H_k$ is the measurement matrix, $R_k$ is the measurement noise covariance matrix and $m$ is the number of measurements.

When we use the classical Bayesian likelihood in the AMM algorithm the likelihood for one of the models, which is assumed to be the true model with the lowest residual, goes to 1 as the other likelihoods diminish and become 0 eventually. In essence, this is as a result of including the memory of previous $q_{k-1}^{(i)}$ values in the calculation of the current likelihood, $q_k^{(i)}$.

Although there are documented examples for using the AMM for systems with multiple modes [23] (in contrast to single true mode), the CMM-based estimation algorithms have been used for similar problems more often. In this context, certainly the IMM estimation method has proved itself via a significant number of successful applications, specifically for the target tracking. The IMM method reduces the ever-growing number of hypothesis in the underlying CMM method by introducing the interaction (mixing) step. In this interaction step, all the KFs are reinitialized depending on the mode transition probabilities such that

$$\bar{x}_{k-1/k-1}^{(i)} = \sum_{j=1}^{n} \hat{x}_{k-1/k-1}^{(j)} q_{k-1}^{j|i},$$  (8.4)

where

$$q_{k-1}^{j|i} = \frac{\pi_{ji} q_{k-1}^{(j)}}{\sum_{j=1}^{n} \pi_{ji} q_{k-1}^{(j)}}.$$  (8.5)

Here, $\hat{x}_{k-1/k-1}^{(j)}$ are state estimates for each KF from the previous filtering step, $\bar{x}_{k-1/k-1}^{(i)}$ are the reinitialized state values for the KFs and $\pi_{ji}$ are the mode transition probabilities for $\forall i, j \in N$. N denotes the number of possible models, which are represented with $n$ filters in the algorithm. Likewise the estimation covariance for each filter $P_{k-1/k-1}^{(i)}$ must be reinitialized. The mode transition probabilities and each filter's models can be decided after Monte Carlo simulations for the problem.

Two types of output processing are considered for the MMAE algorithms. In the first, the estimates are fused considering each KF's likelihood. This is also called soft-switching. Overall estimate and estimation covariance become

$$\hat{x}_{k/k} = \sum_{j=1}^{n} \hat{x}_{k/k}^{(j)} q_k^{(j)},$$  (8.6a)

$$P_{k/k} = \sum_{j=1}^{n} \left\{ P_{k/k}^{(j)} + \left[ \hat{x}_{k/k}^{(j)} - \hat{x}_{k/k} \right] \times \left[ \hat{x}_{k/k}^{(j)} - \hat{x}_{k/k} \right]^T \right\} q_k^{(j)}.$$  (8.6b)

Another option is to accept the estimate with the maximum likelihood as the overall estimate,

$$\left. \begin{aligned} \hat{x}_{k/k} &= \hat{x}_{k/k}^{(i)} \\ P_{k/k} &= P_{k/k}^{(i)} \end{aligned} \right\}, \quad \text{for} \quad i = \arg\max q_k^{(i)}.$$  (8.7)

This method of output processing is also called hard-switching.

The MMAE methods in general have several drawbacks. First of all, when these methods are used, it is assumed the faults or the modes are known. The parallel running KFs can represent only a limited number of faults or uncertainty levels. For example, suppose that the AMM algorithm is run with different inertia values assigned to each KF in the bank for spacecraft attitude estimation problem. Eventually the algorithm will converge to the estimates of one of these KFs, but this is still the

KF with the highest likelihood and it is not known how much the final inertia estimate matches with the real values. Moreover, of course, the MMAE approach
requires high computational load due to several KFs running parallel.

The IMM method has some other drawbacks in addition to those are common to
all MMAE method. These are the introduced complexity over the AMM estimation
and the necessity to know the mode transition probabilities. Indeed, how to define
each mode is still important. As discussed in [24] when this algorithm is used for
fault tolerant estimation, it is impossible to associate each fault with unknown magnitudes as a different mode in the IMM algorithm as indefinitely many modes would
be required in this case.

## 8.3 ADAPTIVE KALMAN FILTERING WITH NOISE COVARIANCE ESTIMATION

In this section, the existing methods for the adaptive Kalman filtering, which are
based on noise covariance estimation, are reviewed and their drawbacks are
discussed.

### 8.3.1 INNOVATION-BASED ADAPTIVE ESTIMATION

Among the adaptive Kalman filtering methods, the IAE is the mostly used method
for the filter adaptation [8, 9]. The essence of this approach is to investigate the
behavior of the innovation sequence and determine whether the real characteristics of
the noise match with its a priori characteristics. When the real innovation sequence is
different than the white noise, adaptation of the process noise covariance (Q) or measurement noise covariance (R) is necessary. We check the innovation sequence to see
if it has zero mean expected value and given covariance.

The covariance of the innovation sequence ($\tilde{e}_j$) is calculated by the formula

$$\hat{C}_{e_k} = \frac{1}{N} \sum_{j=k-N+1}^{k} \tilde{e}_j \tilde{e}_j^T \tag{8.8}$$

Here $N$ is the size of the moving window.

If a priori and real characteristics of the noise do not match, the R and Q covariances should be estimated, respectively. These estimates are used as the updated
values for the covariance matrices in the Kalman filter algorithm. Hence for the AKF
with noise covariance estimation, the filter is used for both estimating the states and
characteristics of the noise covariances.

In the IAE, the R and Q matrices should be updated at each iteration in accordance
with the estimations. The equations for estimating the matrices in the IAE-based
AKF are

$$\hat{R}_k = \hat{C}_{e_k} - H_k P_{k/k-1} H_k^T, \tag{8.9a}$$

$$\hat{Q}_k = K_k \hat{C}_{e_k} K_k^T. \tag{8.9b}$$

### 8.3.2 INNOVATION-BASED ADAPTIVE FILTRATION ALGORITHM
### FOR STATIONARY SYSTEMS

The adaptive algorithm that predicts the state vector of a static dynamic system together with the covariance matrices of the system and measurements noises was proposed by Mehra and it is named after him [25]. The problem is as follows:

The mathematical model of system is given as

$$X_{k+1} = \Phi X_k + \Gamma W_k; \quad \Phi \quad \text{and} \quad \Gamma \to const, \tag{8.10}$$

and the measurement model is

$$y_k = Hx_k + V_k; \quad H \to const \tag{8.11}$$

Let $\nu_k = y_k - Hx(k/k-1)$ be the innovation process and $C_i = E\left(v_k v_{k-i}^T\right)$ the innovation covariance, where $E$ is the statistic average operator. Then the estimation of the innovation covariance $\hat{C}_i$ can be found using formula below:

$$\hat{C}_i = \frac{1}{N} \sum_{k=1}^{N} v_k v_{k-i}^T, \tag{8.12}$$

where $x(R/r-1) = E\left(x_R / Y_1^{R-1}\right)$ and $N$ - is the size of the moving window.

We have the a priori data as

$$w_k \approx N(0, Q); \quad v_k \approx N(0, R); \quad x_0 \approx N\left[x(0/0), P_0\right]. \tag{8.13}$$

The matrices $R$ and $Q$ are constant and unknown; $R_0$ and $Q_0$ are the a priori values of the $R$ and $Q$, respectively;

$$\text{cov}\left(w_R, w_j\right) = \text{cov}\left(v_R, v_j\right) = \text{cov}\left(x_0, w_k\right) = \text{cov}\left(x_0, v_k\right) = 0 \tag{8.14}$$

The system is fully observable and controllable, i.e.

$$\text{rank}\left[H^T, \left(H\Phi\right)^T, ..., \left(H\Phi^{n-1}\right)^T\right] = n$$

$$\text{rank}\left[\Gamma, \Phi\Gamma, ..., \Phi^{n-1}\Gamma\right] = n. \tag{8.15}$$

The stationary filter for this case can be built with the below steps:

$$x_{k+1/k} = \Phi x_{k/k}, \tag{8.16a}$$

$$x_{k/k} = x_{k/k-1} + Kv_k \tag{8.16b}$$

$$v_k = y_k - H x_{k/k-1} \tag{8.16c}$$

$$K = \hat{P}H^T \left( H\hat{P}H^T + \hat{R} \right)^{-1} \tag{8.16d}$$

$$P = \Phi\left[ E - KH \right]P\Phi^T + \Gamma\hat{Q}\Gamma^T, \tag{8.16e}$$

where $x_{k+1/k} = E(x_{k+1}/Y^k)$, $x_{k/k} = E\left( x_k / Y_1^k \right)$, $P = \text{cov}\left( x_{k+1}, x_{k+1} / Y_1^k \right)$ .

When $k \to \infty$, the estimation algorithms for the matrices $R$ and $Q$ are given via expressions below.

*R matrix estimation algorithm*:

$$\hat{R} = \begin{cases} R_0 \\ \hat{C}_0 - H\left( \hat{P}H^T \right) \end{cases}; \quad \hat{P}H^T = \begin{cases} P_0 H^T \\ K\hat{C}_0 + A^+ \begin{bmatrix} \hat{C}_1 \\ \hat{C}_n \end{bmatrix} \end{cases} \tag{8.17a}$$

$$A = \begin{bmatrix} H\Phi \\ H\Phi\left( E - KH \right)\Phi \\ \dots \\ H\left[ \Phi\left( E - KH \right) \right]^{n-1}\Phi \end{bmatrix} \tag{8.17b}$$

$A^+ = (A^T A)^{-1}A^T$ is the pseudo inverse matrix for A.

*Q matrix estimation algorithm*:

$$\sum_{j=0}^{i-1} H\Phi^j \Gamma\hat{Q}\Gamma^T \left( \Phi^{-i} \right)^T = H\hat{P}\left( \Phi^{-i} \right)^T H^T - H\Phi^i \hat{P}H^T - \sum_{j=0}^{i-1} H\Phi^j \hat{\Omega}\left( \Phi^{-i} \right)^T H^T \tag{8.18a}$$

for $i = 1, \dots, n$. Here

$$\hat{\Omega} = \Phi\left[ -KH\hat{P} - \hat{P}H^T K^T + K\hat{C}_0 K^T \right]\Phi^T. \tag{8.18b}$$

Number of unknown elements of the matrix $Q$ is.
The disadvantages of the Mehra algorithm can be listed as follows:

1. $\hat{Q}$ estimation requires a large volume of computation when we are getting the solution by (8.18);
2. The estimation values $\hat{Q}$ and $\hat{R}$ cannot be found at the rate of measurements. For estimation, the measurements must be accumulated;
3. The estimation values $\hat{Q}$ and $\hat{R}$ are highly dependent on their apriori values $Q_0$ and $R_0$.

### 8.3.3 Innovation-Based Adaptive Filtration Algorithm with Feedback

Let's examine the adaptive filtering algorithm for the dynamic system with constant and unknown $Q_k$ and $R_k$ matrices and nonstationary noises [26]. At this point, assume that, $Q_k$ and $R_k$ matrices, are time-varying. The problem is as follows:

The mathematical model of system is

$$X_{k+1} = \Phi X_k + \Gamma W_k; \quad \Phi, \quad \Gamma \to const \tag{8.19}$$

and the measurement model is

$$y_{k+1} = H x_{k+1} + v_{k+1}; \quad H \to const. \tag{8.20}$$

We have the a priori data as

$$w_R \approx N\left(0, Q_k\right); \quad v_{k+1} \approx N\left(0, R_{k+1}\right); \quad x_0 \approx N\left(x\left(0/0\right), P_0\right), \tag{8.21}$$

where $Q_k, R_{k+1}, P_0$ are unknown matrices and

$$\text{cov}\left(w_k, w_x\right) = 0; \quad \text{cov}\left(v_k, v_j\right) = 0; \quad \text{cov}\left(x_0, w_k\right) = \text{cov}\left(x_0, v_k\right) = 0. \tag{8.22}$$

In this case the adaptive filtration algorithm is given by the following equations:

Equation for the extrapolation value

$$x_{k+1/k} = E\left(X_{k+1} / Y_1^k\right) \tag{8.23}$$

can be written as

$$x_{k+1/k} = \Phi x_{k/k}. \tag{8.24}$$

The filtering algorithm for estimating the states of

$$x_{k+1/k+1} = E\left(X_{k+1} / Y_1^{k+1}\right) \tag{8.25}$$

is presented in the form of recurrent expressions as:

$$\hat{x}_{k+1/k+1} = \hat{x}_{k+1/k} + K_{k+1} V_{k+1/k} \tag{8.26a}$$

$$V_{k+1/k} = y_{k+1} - H x_{k+1/k} \tag{8.26b}$$

$$K_{k+1} = S^+ E_v\left(V_{k+1/k} V_{k+1/k}^T\right) \tag{8.26c}$$

$$S^+ = \left(S^T S\right)^{-1} S^T \tag{8.26d}$$

$$E_v = \begin{bmatrix} V_{k+1/k} & V^T_{k+1/k} \\ \cdots\cdots\cdots\cdots\cdots\cdots\cdots \\ V_{k+l/k} & V^T_{k+l/k} \end{bmatrix}; \quad S = \begin{bmatrix} H \\ H\Phi \\ \cdots\cdots \\ H\Phi^{l-1} \end{bmatrix} \qquad (8.26e)$$

### 8.3.4 Residual-Based Adaptive Estimation

In the RAE approach, the adaptation is applied on the measurement and process noise covariances based on the changes in the residual sequence [10, 11]. The residual sequence is formulated as

$$v_k = y_k - H\hat{x}_{k/k}, \qquad (8.27)$$

where $\hat{x}_{k/k}$ is the estimated state vector. The residual covariance is given as

$$\hat{C}_{v_k} = \frac{1}{N} \sum_{j=k-N+1}^{k} v_j v_j^T. \qquad (8.28)$$

In order to estimate the $R$ and $Q$ matrices the following equations are used:

$$\hat{R}_k = \hat{C}_{v_k} + H_k P_{k/k} H_k^T \qquad (8.29a)$$

$$\hat{Q}_k = \frac{1}{N} \sum_{j=k-N+1}^{k} \Delta x_j \Delta x_j^T + P_{k/k} - F P_{k-1/k-1} F^T \qquad (8.29b)$$

Here $P_{k/k}$ is the estimated covariance for the recent step ($k$), $P_{k-1/k-1}$ is the estimated covariance for the previous step ($k-1$) and $\Delta x_k$ is the state correction sequence, which is the difference between the estimated and predicted states as

$$\Delta x_k = \hat{x}_{k/k} - \hat{x}_{k/k-1}. \qquad (8.30)$$

In the steady state we may consider just the first term of (8.29b). Then regarding $\Delta x_k = K_k v_k$ equality, (8.29b) may be approximated with (8.9b) [10].

For some application areas such as the integrated INS/GPS system, it is experienced that the RAE is more appropriate than the IAE for adaptive estimation.

### 8.3.5 Drawbacks of Adaptive Noise Covariance Estimation Methods

Both the IAE and RAE have certain drawbacks when they are used for the KF adaptation. These are as follows:

1.   We need to use the innovation/residual sequences for $N$ epoch. In this case the computational load increases and determination of the moving window size $N$ appears as another problem.

2. The IAE and RAE estimators require that the number, type and distribution of the measurements for all epochs within a window should be consistent. If they do not, the covariance matrices of the measurement noises cannot be estimated based on the innovation or the residual vectors.

3. For the R estimation in the IAE, if $H_k P_{k/k-1} H_k^T > \hat{C}_{e_k}$ then $\hat{R}_k$ becomes negative definite and the KF may collapse.

## 8.4 ADAPTIVE KALMAN FILTERING WITH NOISE COVARIANCE SCALING

In this section, we present the novel KF adaptation techniques which adapt the noise covariance matrices by scaling. These algorithms are essentially a type of adaptive algorithm based on the covariance matching.

### 8.4.1 INNOVATION-BASED ADAPTIVE SCALING

#### 8.4.1.1 R-Adaptation

The essence of the innovation-based R-adaptation is the comparison of real and theoretical values of the innovation covariance [27]. When the operational condition of the measurement system mismatches with the model used in the synthesis of the filter, then the KF gain changes related to the differentiation in the innovation covariance matrix. Under these circumstances, the innovation covariance differs as

$$\hat{C}_{e_k} = H_k P_{k/k-1} H_k^T + S_k R_k \tag{8.31}$$

and so the Kalman gain becomes

$$K_k = P_{k/k-1} H_k^T \left( H_k P_{k/k-1} H_k^T + S_k R_k \right)^{-1} \tag{8.32}$$

Here, $S_k$ is the measurement noise scale factor (SF).

In this approach, the Kalman gain is changed when the predicted observation $H_k \hat{x}_{k/k-1}$ is considerably different from the actual observation, $y_k$, because of the significant changes in the operational condition of the measurement system. In other words, if the real value of the filtration error exceeds the theoretical error (or if a priori and real characteristics of the noise do not match) as

$$tr\left\{ \tilde{e}_k \tilde{e}_k^T \right\} \geq tr\left\{ H_k P_{k/k-1} H_k^T + R_k \right\}, \tag{8.33}$$

the filter must be run adaptively. Here, $tr\{\cdot\}$ denotes the trace of the related matrix.

There are two possible techniques for scaling the R matrix. The first one is to use just a single scale factor (SSF) and the second one is to scale with a matrix formed of multiple scale factors (MSF).

For obtaining the SSF, let us substitute (8.31) into (8.33) and regard that the adaptation begins at the point where the equality condition for (8.33) is satisfied,

$$tr\left\{\tilde{e}_k\tilde{e}_k^T\right\} = tr\left\{H_kP_{k/k-1}H_k^T\right\} + S_k tr\left\{R_k\right\}. \tag{8.34}$$

Then, in the light of $tr\left\{\tilde{e}_k\tilde{e}_k^T\right\} = \tilde{e}_k^T\tilde{e}_k$ equality, the SSF, $S_k$, can be written as

$$S_k = \frac{\tilde{e}_k^T\tilde{e}_k - tr\left\{H_kP_{k/k-1}H_k^T\right\}}{tr\left\{R_k\right\}}. \tag{8.35}$$

In case of fault in the measurements, the adaptation of the KF is performed via automatically correcting the Kalman gain. If the inequality condition for (8.33) is met, then the scale factor $S_k$ increases. Bigger $S_k$ causes a smaller Kalman gain (8.32) because of the covariance of the innovation sequence (8.31) which also increases. Consequently, small Kalman gain value reduces the effect of the faulty innovation sequence on the state estimation process. In all other cases, where measurement system operates normally, the SSF takes the value of $S_k = 1$ and so the filter runs optimally.

On the other hand, as clearly demonstrated with simulations in [28, 29], using the MSF is a healthier procedure for the adaptation since the filter disregards only the measurement of the faulty sensor in this case rather than disregarding all as for the SSF. When we scale the R matrix using the MSF, the relevant term of the Kalman gain matrix, which corresponds to the innovation channel of the faulty sensors, is fixed individually.

In order to determine the scale matrix formed of the MSF, the real and theoretical values of the innovation covariance matrix must be compared like we did for the SSF. When there is a measurement malfunction in the estimation system, the real error will exceed the theoretical one. Hence, if a scale matrix, $S_k$, is added in to the algorithm as [28],

$$\hat{C}_{e_k} = H_kP_{k/k-1}H_k^T + S_kR_k, \tag{8.36}$$

then, it can be determined by the formula of

$$S_k = \left(\hat{C}_{e_k} - H_kP_{k/k-1}H_k^T\right)R_k^{-1}. \tag{8.37}$$

In case of normal operation, the scale matrix will be a unit matrix as $S_k = I$. Here $I$ represents the unit matrix. Nonetheless, $S_k$ matrix, found by the use of (8.37) may be non-diagonal and have diagonal elements which are "negative" or less than "one." This is mainly because of the limited $N$ number and the approximation errors. $S_k$ matrix should be diagonal because only its diagonal terms have significance on the adaptation since each diagonal term corresponds to the noise covariance of each measurement (for the adaptation procedure $S_k$ matrix is multiplied with the diagonal R matrix). Besides the measurement noise covariance matrix must be positive definite (that is why the multiplier $S_k$ matrix cannot have negative terms) and also any term of this matrix cannot decrease in time for this specific problem since there is no possibility for increasing the performance of the on-board sensor (that is why the multiplier $S_k$ matrix cannot have terms less than one). Therefore, in order to avoid such situation, composing the scale matrix by the following rule is suggested:

$$S^* = diag\left(s_1^*, s_2^*, \ldots, s_n^*\right),$$ (8.38)

where

$$s_i^* = \max\left\{1, S_{ii}\right\} \quad i = 1, n.$$ (8.39)

Here, $S_{ii}$ represents the $i^{th}$ diagonal element of the matrix $S$. Apart from that point, if the measurements are faulty, $S_k^*$ will change and so affect the Kalman gain matrix;

$$K_k = P_{k/k-1}H_k^T\left(H_kP_{k/k-1}H_k^T + S_k^*R_k\right)^{-1}.$$ (8.40)

In case of any kind of malfunction, the element of the scale matrix, which corresponds to the faulty component of the innovation vector, increases and so the terms in the related column of the Kalman gain decreases. As a consequence, the effect of the faulty innovation term on the state update process reduces and accurate estimation results can be obtained even in case of measurement malfunctions.

We name the R-adapted KF as the RKF since it is robust against sensor faults.

### 8.4.1.2 Q-Adaptation

When there is an actuator fault in the system that ends up with changes in the control distribution matrix, the real characteristics of the noise do not match with its a priori characteristics. As a solution the Q-adaptation should be performed. The essence of the method is again comparing the real and theoretical values of the innovation covariance matrix. When there is an actuator fault, the real error will exceed the theoretical one.

Besides, it is known that using multiple factors for adaptation satisfies better estimation accuracy. Hence, a fading matrix is used for the adaptation rather than a fading factor. We name the matrix used for the Q-adaptation as fading matrix in order to distinguish from the scale matrix used for the R-adaptation, otherwise the function of scale and fading matrices are same. The fading matrix, $\Lambda_k$, built of multiple fading factors (MFF), is added in to the algorithm as

$$\hat{C}_{e_k} = H_k\left(F_kP_{k-1/k-1}F_k^T + \Lambda_kG_kQ_kG_k^T\right)H_k^T + R_k.$$ (8.41)

Then the fading matrix can be determined as,

$$\Lambda_k = \left(\hat{C}_{e_k} - H_kF_kP_{k-1/k-1}F_k^TH_k^T - R_k\right)\times\left(H_kG_kQ_kG_k^TH_k^T\right)^{-1}.$$ (8.42)

In a similar manner with the R-adaptation, gained fading matrix should be diagonalized since the $Q$ matrix must be a diagonal, positive definite matrix.

$$\Lambda^* = diag\left(\lambda_1^*, \lambda_2^*, \dots, \lambda_n^*\right), \tag{8.43a}$$

$$\lambda_i^* = \max\left\{1, \Lambda_{ii}\right\} \quad i = 1, n. \tag{8.43b}$$

Here, $\Lambda_{ii}$ represents the $i^{th}$ diagonal element of the matrix $\Lambda$. Apart from that point, if there is a fault in the system, $\Lambda_k^*$ must be put into process as,

$$P_{k/k-1} = F_k P_{k-1/k-1} F_k^T + \Lambda_k^* G_k Q_k G_k^T. \tag{8.44}$$

## 8.4.2 RESIDUAL-BASED ADAPTIVE SCALING

We may also perform a residual-based scaling for the adaptation of the R matrix. In this case the scale matrix formed of the MSF is

$$\tilde{S}_k = \left(\hat{C}_{v_k} + H_k P_{k/k} H_k^T\right) R_k^{-1}. \tag{8.45}$$

Same as the proposed method for the innovation-based R-scaling, $\tilde{S}_k$ should be diagonalized using the following rule:

$$\tilde{S}^* = diag\left(\tilde{s}_1^*, \tilde{s}_2^*, \dots, \tilde{s}_n^*\right), \tag{8.46a}$$

$$\tilde{s}_i^* = \max\left\{1, \tilde{S}_{ii}\right\} \quad i = 1, n. \tag{8.46b}$$

Here, $\tilde{S}_{ii}$ represents the $i^{th}$ diagonal element of the matrix $\tilde{S}$.

On the other hand, a residual-based scaling for the process noise covariance adaptation is rather complicated because of the inverse matrix calculations and the analytical solution for obtaining the residual-based MFF, $\tilde{\Lambda}_k$, is difficult. Therefore, it is not presented and discussed here.

## 8.5 SIMPLIFIED RKF AGAINST MEASUREMENT FAULTS

To build a simplified RKF against the measurement faults, we first assume that the process model is described by the state equations for linear dynamic system

$$x(k+1) = \Phi(k+1,k)x(k) + w(k) \tag{8.47}$$

and the measurement model by the observation equation of the type

$$y(k) = H(k)x(k) + \gamma(k)v(k), \tag{8.48}$$

where $x(k)$ is the $n$-dimensional vector of state of the system, $\Phi(k + 1, k)$ is its transition matrix of dimension $n \times n$, $w(k)$ is the random $r$-dimensional Gaussian system noise with zero mean and covariance matrix $E[w(k)w^T(j)] = Q(k)\delta(kj)$, $y(k)$ is the $s$-dimensional measurement vector, $H(k)$ is the $s \times n$ dimensional measurement matrix of the system, $v(k)$ is the random s-dimensional Gaussian measurement noise with zero mean and covariance matrix $E[v(k)v^T(j)] = R(k)\delta(kj)$, $\gamma(k)$ is a random variable which takes two values, i.e.

$\gamma(k) = 1$ (normal operating sensors)

$\gamma(k) = \sigma \gg 1$ (faulty sensor measurements).

We assume that the $\gamma(k)$ is an independent random variable at each step, and $\gamma(k) = 1$, with a probability of $q(k)$ (probability of normal operating sensors); $\gamma(k) = \sigma$, with a probability of $1 - q(k)$ (probability of a fault).

When the variance of the measurement noise in the case of fault is significantly more than that in the normal functioning mode, i.e. $\gamma(k) = \sigma \gg 1$, the following expressions are obtained [30] for the suboptimum filter:

$$\hat{x}(k/k) = \hat{x}(k/k-1) + p(1/k)K(k)\left[ y(k) - H(k)\hat{x}(k/k-1) \right]. \quad (8.49a)$$

$$K(k) = P(k/k-1)H^T(k)\left[ H(k)P(k/k-1)H^T(k) + R(k) \right]^{-1} \quad (8.49b)$$

where $\hat{x}(k/k)$ is the estimate of the state vector, $\hat{x}(k/k-1)$ is the extrapolation (prediction) estimate for one step, $p(1/k)$ is the probability that the quantity $\gamma$ in the given step will take the value 1; $K(k)$ is the gain matrix of the filter, $P(k/k-1)$ is the covariance matrix of extrapolation errors.

The only difference between the given filter above and the Kalman filter is that the gain matrix is pre-multiplied by $p(1/k)$. For $p(1/k) = 1$, the filter coincides with the Kalman filter, and for $p(1/k) = 0$, it transforms to an extrapolator.

The calculations of the $p(1/k)$ values are very cumbersome because they depend on multi-dimensional functions. Accurate realization of the above filter requires considerable computing efforts and, besides, is not always necessary. From this viewpoint, it is necessary to work out methods to facilitate a reduction in the required volume of calculations for putting into effect the filtering algorithms, which are resistant to faults in the sensors.

It is shown in [30] on the basis of mathematical modeling that, in the case of scalar measurements, the quantity $p(1/k)$ can practically assume only two values in each specific realization, i.e. unity, if the assumed realization of $y(k)$ falls in the allowed range of values of this realization, and zero, in the opposite case. By accounting for this, in case of vector measurements for solving the stated problem, it is proposed to check, as the measurements flow in, whether the innovation vector $\Delta(k) = y(k) - H(k)\hat{x}(k/k-1)$ falls in the area $\Omega$ of its permissible values. In this, we assume that [31]

$$p(1/k) = 1 \quad \text{if} \quad \Delta(k) \in \Omega \quad (8.50a)$$

$$p(1/k) = 0 \quad \text{if} \quad \Delta(k) \notin \Omega. \quad (8.50b)$$

We verify the fulfillment of conditions (8.50) in accordance with the conditions discussed below.

On the strength of the center limiting theorem of the probability theory, and considering the many components affecting the value of the extrapolation estimate, we assume the distribution of its true value at any instant as multidimensional and normal. Considering this, in the case of normal operating measurement channel, the vector $\Delta(k)$ will also have a multidimensional normal distribution with the following density:

$$f\left(\Delta(k)\right) = \frac{1}{\sqrt{(2\pi)^s \left|P_\Delta(k)\right|}} \exp\left\{-\frac{1}{2}\left[y(k) - H(k)\hat{x}(k/k-1)\right]^T\right.$$

$$\left. P_\Delta^{-1}(k)\left[y(k) - H(k)\hat{x}(k/k-1)\right]\right\}, \tag{8.51}$$

where $P_\Delta(k) = H(k)P(k/k-1)H^T(k) + R(k)$ is the innovation covariance and positive definite matrix. The equation

$$\left[y(k) - H(k)\hat{x}(k/k-1)\right]^T P_\Delta^{-1}(k)\left[y(k) - H(k)\hat{x}(k/k-1)\right] = \rho^2 \tag{8.52}$$

in case of a matrix $P_\Delta(k)$ with real elements is an ellipsoid equation (ellipse equation when $s = 2$) on the strength of the positive definite of $P_\Delta(k)$ [27]. Representing $\rho$ forms the permissible ellipsoid variations in the parametric space.

Locating the innovation vector $\Delta(k)$ inside the permissible ellipsoid corresponds to condition (8.50a). On the other hand, locating the vector $\Delta(k)$ outside the ellipsoid corresponds to condition (8.50b). It is known that the probability of locating inside the ellipsoid is a function of only $\rho$ [32]. $\rho$ is the dimension of the semiaxes of the ellipsoid in the standard deviations. By setting a high probability such that the innovation vector $\Delta(k)$ falls in the permissible ellipsoid, we determine the permissible value of the parameter $\rho$ -- $\rho_{per}$ from the following condition:

If the dimension of the innovation vector $s \geq 2$, then for even numbers

$$P\left[\Delta(k) \in \Omega\right] = 1 - R\left(\frac{s}{2} - 1; \frac{\rho^2}{2}\right), \tag{8.53}$$

where $R(m, a_p) = \sum_{i=0}^{m} a_p^i e^{-a_p} / i$ is the function which describes the probability distribution of random variables subjected to the Poisson distribution with the parameter $a_p$.

If $s \geq 3$, then for odd numbers

$$P\left[S(k) \in \Omega\right] = 2\Phi(\rho) - \frac{\sqrt{2}}{\sqrt{\pi}} e^{-\rho^2/2} \sum_{m=1}^{s-2} \frac{\rho^m}{m!!}, \tag{8.54}$$

where the sum is defined only for the odd number indices, $m = 1, 3, 5, \ldots, (s-2)$, $(s \geq 3)$;

$$\Phi(\rho) = \frac{1}{\sqrt{2\pi}} \int_0^\rho e^{-t^2/2} dt \text{ is the Laplace function.}$$

Expressions (8.53) and (8.54) characterize the probability that the vector $\Delta(k)$ lies inside the permissible ellipsoid [31].

Substituting $\rho_{per}$ in (8.52), we obtain the equation of the corresponding permissible ellipsoid as follows:

$$\left[ y(k) - H(k)\hat{x}(k/k-1) \right]^T P_\Delta^{-1}(k) \left[ y(k) - H(k)\hat{x}(k/k-1) \right] = \rho_{per}^2 \quad (8.55)$$

We ascertain that in the case when the innovation vector $\Delta(k)$ is located outside the permissible ellipsoid, the left side of Eq. (8.55) is more than $\rho_{per}^2$. Actually, we connect a certain arbitrary point $y(m)$ in the Figure 8.4 outside the permissible area to the point $H(m)\hat{x}(m/m-1)$ with the vector $y(m) - H(m)\hat{x}(m/m-1)$, which intersects the permissible ellipse at the point $y^*(m)$. Since $y(m)$ is located outside the permissible ellipse, then

$$\left[ y(m) - H(m)\hat{x}(m/m-1) \right]^T = \mu \left[ y^*(m) - H(m)\hat{x}(m/m-1) \right], \quad (8.56)$$

where $\mu > 1$.

Hence,

$$\left[ y(m) - H(m)\hat{x}(m/m-1) \right]^T P_\Delta^{-1}(m) \left[ y(m) - H(m)\hat{x}(m/m-1) \right]$$
$$= \mu^2 \left[ y^*(m) - H(m)\hat{x}(m/m-1) \right]^T P_\Delta^{-1}(m) \left[ y^*(m) - H(m)\hat{x}(m/m-1) \right]$$
$$> \left[ y^*(m) - H(m)\hat{x}(m/m-1) \right]^T P_\Delta^{-1}(m) \left[ y^*(m) - H(m)\hat{x}(m/m-1) \right] = \rho_{per}^2$$

$$(8.57)$$

which itself was required to be proved.

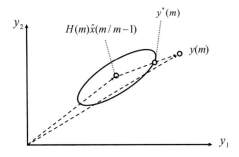

**FIGURE 8.4**  The determination of permissible ellipse.

We check that the innovation vector $\Delta(k)$ falls in the permissible area $\Omega$ by a simple substitution of the vector $y(k)$ in the equation of the ellipsoid (8.55). In case of abnormal measurements (in the presence of faults in the sensors), the left side of Eq. (8.55) is found to be more than $\rho_{per}^2$. As a result, the quantity $p(1/k)$ assumes the value zero, and the filter calculates the extrapolated value of the state vector without new data. If $\Delta(k) \in \Omega$, the left side of Eq. (8.55) becomes less than or equal to $\rho_{per}^2$. In this case, the quantity $p(1/k)$ takes the value of unity, and the examined filter transforms to a normal Kalman filter tuned to a normal operation mode ($\gamma(k) = 1$).

Hence, the covariance matrix of filtering errors $P(k/k)$ is calculated by the formulas:

$$P\left(k/k\right) = P\left(k/k-1\right) - K\left(k\right)H\left(k\right)P\left(k/k-1\right) \quad \text{for} \quad \Delta\left(k\right) \in \Omega \qquad (8.58a)$$

$$P\left(k/k\right) = P\left(k/k-1\right) \quad \text{for} \quad \Delta\left(k\right) \notin \Omega \qquad (8.58b)$$

Apart from the examined linear case, the given algorithm can be extended even to nonlinear systems. For this, it is necessary to linearize the equations of state or measurements.

## 8.6 CONCLUSION

The Kalman filter approach to the state estimation is quite sensitive to any malfunctions. If the condition of the real system does not correspond to the models, used in the synthesis of the filter, then these changes significantly decrease the performance of the estimation system.

In real conditions the statistical characteristics of measurement or system noises may change. If the nominal (predicted) trajectory is too far away from the true trajectory, then the true covariance will be much larger than the estimated covariance and the filter will become poorly matched. This might lead to severe filter instabilities. This is most often due to lack of careful modeling of sensors and environment. Failure to understand the limitations of the algorithm exacerbates the problem. It means that estimation algorithm should be adaptive. In such cases the Kalman filter can be adapted and adaptive or robust Kalman filters can be used to recover the possible faults.

In this chapter, first we reviewed the existing estimation-based adaptation techniques for the process and measurement noise covariance matrices in order to build an adaptive Kalman filter algorithms. In specific we presented the Multiple Model Adaptive Estimation (MMAE) algorithms and noise covariance estimation methods. As the MMAE approach requires several parallel Kalman filters, and the faults should be known, it can be used in limited applications. It may not be suitable for running for onboard attitude estimation, in practice. On the other hand, adaptive noise covariance estimation methods have their own drawbacks such as requiring buffered measurements within a window and consistency of the measurements/process within this window. In the solution of some practical problems (e.g. in the integrated INS/GPS system), the residual-based adaptive estimation is numerically more appropriate than innovation-based adaptive estimation.

Next in this chapter we introduced novel noise covariance scaling based methods for the filter adaptation. By the use of defined variables named as scale factor and fading factor procedures for adapting both the measurement and noise covariance matrices are given. It is shown that using multiple factors rather than a single factor for adaptation is more advantageous.

Finally, the simplified RKF algorithm against measurement faults is considered. In the case of measurement fault, this filter calculates the extrapolated value of the state vector without new data and when there is no fault, the examined RKF transforms to a normal Kalman filter tuned for a normal operation mode.

## REFERENCES

1. M. Ger, C. Van Ommeren, M. Westenkirchner, and G. Herbold, "Multiple model concepts in navigational applications," *2016 DGON Inert. Sensors Syst. ISS 2016 - Proc.*, Karlsruhe, Germany, pp. 1–20, 2016.
2. J. Duník, O. Straka, O. Kost, and J. Havlík, "Noise covariance matrices in state-space models: A survey and comparison of estimation methods—Part I," *Int. J. Adapt. Control Signal Process.*, vol. 31, no. 11, pp. 1505–1543, 2017.
3. X. R. Li and V. Jilkov, "A survey of Maneuvering Target tracking. Part V: Multiple Model Methods," *IEEE Trans. Aerosp. Electron. Syst.*, vol. 41, no. 4, pp. 1255–1321, 2005.
4. P. S. Maybeck, "Multiple model adaptive algorithms for detecting and compensating sensor and actuator/surface failures in aircraft flight control systems," *Int. J. Robust Nonlinear Control*, vol. 9, no. 14, pp. 1051–1070, 1999.
5. E. Özkan, V. Šmídl, S. Saha, C. Lundquist, and F. Gustafsson, "Marginalized adaptive particle filtering for nonlinear models with unknown time-varying noise parameters," *Automatica*, vol. 49, no. 6, pp. 1566–1575, Jun. 2013.
6. S. Sarkka and J. Hartikainen, "Non-linear noise adaptive Kalman filtering via variational Bayes," *2013 IEEE International Workshop on Machine Learning for Signal Processing (MLSP)*, Southampton, UK, pp. 1–6, 2013.
7. S. Sarkka and A. Nummenmaa, "Recursive noise adaptive Kalman filtering by variational Bayesian approximations," *IEEE Trans. Automat. Contr.*, vol. 54, no. 3, pp. 596–600, Mar. 2009.
8. R. Mehra, "On the identification of variances and adaptive Kalman filtering," *IEEE Trans. Automat. Contr.*, vol. 15, no. 2, pp. 175–184, Apr. 1970.
9. P. S. Maybeck, *Stochastic Models, Estimation, and Control*. Newyork, USA: Academic Press, 1982.
10. A. H. Mohamed and K. P. Schwarz, "Adaptive Kalman filtering for INS/GPS," *J. Geod.*, vol. 73, no. 4, pp. 193–203, 1999.
11. J. Wang, M. P. Stewart, and M. Tsakiri, "Adaptive Kalman filtering for integration of GPS with GLONASS and INS," pp. 325–330, 2000.
12. K. Li, L. Chang, and B. Hu, "A variational Bayesian-based unscented Kalman filter with both adaptivity and robustness," *IEEE Sens. J.*, vol. 16, no. 18, pp. 6966–6976, 2016.
13. C. Hide, T. Moore, and M. Smith, "Adaptive Kalman filtering algorithms for integrating GPS and low cost INS," PLANS 2004. Position Location and Navigation Symposium (IEEE Cat. No.04CH37556), Monterey, CA, USA, pp. 227–233, 2004.
14. C. Hu, W. Chen, Y. Chen, and D. Liu, "Adaptive Kalman filtering for vehicle navigation," *J. Glob. Position. Syst.*, vol. 2, no. 1, pp. 42–47, 2003.
15. H. E. Soken, C. Hajiyev, and S.-I. Sakai, "Robust Kalman filtering for small satellite attitude estimation in the presence of measurement faults," *Eur. J. Control*, vol. 20, no. 2, 2014.

16. H. E. Soken, UKF Adaptation and Filter Integration for Attitude Determination and Control of Nanosatellites with Magnetic Sensors and Actuators. Hayama, Japan: The Graduate University for Advanced Studies, 2013.

17. C. Hajiyev and H. E. Soken, "Fault tolerant estimation of UAV dynamics via robust adaptive Kalman filter," in Complex Systems. Relationship between Control, Communications and Computing, G. Dimirovski, Ed. Switzerland: Springer International Publishing, 2016, pp. 369–394.

18. D. Jwo and S. Chang, "Particle swarm optimization for GPS navigation Kalman filter adaptation," *Aircr. Eng. Aerosp. Technol.*, vol. 81, no. 4, pp. 343–352, Jul. 2009.

19. C. Hajiyev and H. E. Soken, "Robust adaptive unscented Kalman filter for attitude estimation of pico satellites," *Int. J. Adapt. Control Signal Process.*, vol. 28, no. 2, 2014.

20. C. Hajiyev and H. E. Soken, "Robust Adaptive Kalman Filter for estimation of UAV dynamics in the presence of sensor/actuator faults," *Aerosp. Sci. Technol.*, vol. 28, no. 1, pp. 376–383, 2013.

21. T. Kirubarajan and Y. Bar-Shalom, "Kalman filter versus IMM estimator: When do we need the latter?," *IEEE Trans. Aerosp. Electron. Syst.*, vol. 39, no. 4, pp. 1452–1457, 2003.

22. Y. Bar-Shalom, X. R. Li, and T. Kirubarajan, *Estimation with Applications to Tracking and Navigation: Theory Algorithms and Software*. New Jersey, USA: Wiley, 2004.

23. H. E. Soken and S. Sakai, "A new likelihood approach to autonomous multiple model estimation," *ISA Trans.*, vol. 99, pp. 50–58, Apr. 2020.

24. Q. Zhang, "Stochastic hybrid system actuator fault diagnosis by adaptive estimation," *IFAC-PapersOnLine*, vol. 48, no. 21, pp. 150–155, 2015.

25. R. Mehra, "Identification and adaptive Kalman filtering," *Mechanics*, vol. 3, pp. 34–52, 1971.

26. M. A. Ogarkov, *Statistic Estimation Methods for Parameters of Random Processes (in Russian)*. Moscow: Energoatomizdat, 1990.

27. C. Hajiyev, "Adaptive filtration algorithm with the filter gain correction applied to integrated INS/radar altimeter," *Proc. Inst. Mech. Eng. Part G J. Aerosp. Eng.*, vol. 221, no. 5, pp. 847–855, May 2007.

28. C. Hajiyev and H.E. Soken, "Adaptive Kalman Filter with Multiple Fading Factors for UAV State Estimation," 7th IFAC International Symposium oh Fault Detection, Supervision and Safety of Technical Processes (SAFEPROCESS-2009)-Proc., Barcelona, Spain, pp. 77-82, 2009.

29. C. Hajiyev and H. E. Soken, "Robust Estimation of UAV Dynamics in the Presence of Measurement Faults," *J. Aerosp. Eng.*, vol. 25, no. 1, pp. 80–89, Jan. 2012.

30. Y. P. Grishin and Y. M. Kazarinov, *Fault-Resistant Dynamic Systems (in Russian)*. Moscow: Radio i Svyaz, 1985.

31. C. M. Gadzhiev, "Simplified filtering algorithm in the presence of fault in the measurement channel," *Meas. Tech.*, vol. 32, no. 6, pp. 505–507, 1989.

32. H. L. Van Trees, Detection, Estimation, and Modulation Theory, Part I. New York, USA: John Wiley & Sons, Inc., 2001.

# 9 Kalman Filtering for Small Satellite Attitude Estimation

Attitude filtering for spacecraft of low rotation rates is considered to be a mature research field with many publications for over 40 years [1–4] and applications to hundreds of missions [5–8]. Nonetheless, our search for more accurate attitude pointing is continuing and everyday new studies have been published. Apart from attitude filtering for specific missions or with specific (new) sensors, studies mostly aim at improving the pointing accuracy using new filtering algorithms. Few examples for different nonlinear filters for attitude estimation may be listed as Unscented Kalman Filter (UKF)[3, 9], Particle Filter [10, 11], Cubature Kalman Filter (CKF) [12, 13], Gauss-Hermite Quadrature Filter (GHQF) and Sparse Grid Quadrature Filter (SGQF) [14, 15].

There is also various research that have proposed a reduction of the computational load of the filtering algorithm without sacrificing the accuracy. This research is specifically for small satellite applications, for which the computational load is very limited. Steady-state Extended Kalman Filter [16], Gain-scheduled Kalman Filter [17], variants of sigma point filters with reduced number of sigma points [18] and sequential approaches with any of single-frame attitude estimators are examples for researches in this direction [19].

Another research field for attitude filtering is related to the attitude error representation within the filtering algorithm. These studies aim to increase the estimation accuracy by using different approaches to enforce the unit-norm constraint of the quaternion in the Kalman filter [20–22].

In this chapter, we briefly review the details of Kalman filtering for small satellite attitude estimation. Specifically, we discuss different approaches such as dynamics and gyro-based attitude filtering and present the methods to ensure the unit-norm constraint of the quaternions when these parameters are used for attitude representation in the filter. Moreover, we discuss about some small-satellite-originated problems and refer to the possible solutions.

Considering their popularity for real mission applications, we present here the algorithms based on Extended Kalman Filter (EKF) and Unscented Kalman Filter (UKF). However, they can be easily modified in accordance if one would like to use another filtering framework. Besides, this chapter is deduced to the traditional approach for the filtering, in which we use the sensor measurements without any preprocessing. In the next chapter we will discuss the nontraditional approach for attitude filtering, in which the filter is integrated with one of the single-frame methods and the vector measurements (e.g. from magnetometer) are first processed in the single-frame algorithm.

## 9.1 GYRO-BASED AND DYNAMICS-BASED ATTITUDE FILTERING

Before giving the details of attitude filtering algorithms, probably it is better to first briefly introduce two main approaches: gyro-based and dynamics-based attitude filtering. This will ease the understanding of further discussions within this chapter.

When there are gyros onboard the spacecraft providing the attitude rate measurements, then usually a gyro-based attitude filtering algorithm (or a reduced-order filter) is preferred. Such filter estimates the attitude and gyro biases and only needs the spacecraft kinematics as a part of the state propagation. The dynamics given with Eq. (2.6) is not necessary. Suppose that we are using the quaternions to represent the attitude. In this case for the gyro-based filtering algorithm, the state vector is

$$\hat{x} = \begin{bmatrix} \hat{q} \\ \hat{b}_g \end{bmatrix}. \tag{9.1}$$

Here $q$ is the quaternion vector including four attitude parameters, $q = [q_1 \ q_2 \ q_3 \ q_4]^T$. Leading three variables represent the vector part and the last one is the scalar term. So we can rewrite the quaternion as $q = \begin{bmatrix} g^T & q_4 \end{bmatrix}^T$, where $g = \begin{bmatrix} q_1 & q_2 & q_3 \end{bmatrix}^T$. Besides, $b_g$ is the gyro bias vector as $b_g = \begin{bmatrix} b_{gx} & b_{gy} & b_{gz} \end{bmatrix}^T$.

Considering the gyro-measurement model (Eq. 4.27), the angular rate of the satellite is estimated using the gyro measurements ($\tilde{\omega}_{bi}$), and the estimated gyro biases as

$$\hat{\omega}_{bi} = \tilde{\omega}_{bi} - \hat{b}_g. \tag{9.2}$$

The gyro-based attitude filtering algorithm has certain advantages such as [23]:

- Since the quaternion propagation can be performed using a discrete matrix and a closed-form solution to the state transition matrix (for EKF) has been developed for use in state-error propagation, the computational load for this method is low. This is specifically advantageous for small satellites.
- There is no need to know the moments of inertia of the satellite accurately.
- It is not necessary to build a realistic model for the disturbance torques such as gravity-gradient, sun pressure etc.

Especially having no uncertainty imposed by the dynamics, which is impossible to model perfectly in practice, makes the gyro-based attitude filter advantageous. For small satellites, due to the small moments of inertias, disturbance torques (e.g. the magnetic torque) can cause huge inaccuracies in the dynamics when not accounted for in the filter models. So a gyro-based attitude filter is more useful most of the time.

Of course such advantage of the gyro-based attitude filter is valid as long as we have rather accurate gyro measurements. If the inaccuracy in gyro measurements (due to the angular random noise, variation of the bias terms etc.) is much larger compared to that in spacecraft dynamics model, using a dynamics-based attitude filter (or full-order filter) would be more reasonable. In this case the state vector will be composed as

$$\hat{x} = \begin{bmatrix} \hat{q} \\ \hat{\omega}_{bi} \end{bmatrix}. \tag{9.3a}$$

There are also researchers who suggest using the dynamics even if we have gyros [24]. In this case the body angular rate vector, $\omega_{bi}$ and the gyro biases would be estimated alongside in a state vector of

$$\hat{x} = \begin{bmatrix} \hat{q} \\ \hat{\omega}_{bi} \\ \hat{b}_g \end{bmatrix}, \tag{9.3b}$$

as we see examples in [25, 26]. Larger state size and slightly higher computational demand is the explicit disadvantage of this approach.

## 9.2  ATTITUDE FILTERING USING EULER ANGLES

Euler angles are the most natural attitude representations for filtering, because of being minimal set with three angles for three rotations. However as discussed in Chapter 2, when Euler angles are used there is chance of singularity at certain angles depending on the used sequence. Thus, avoiding the singularity should be guaranteed and this may be possible if the attitude is defined with one of asymmetric sequences (such as 3-2-1) in terms of roll, pitch and yaw angles with respect to the orbit reference frame. This is specifically true for nadir-pointing small spacecraft, since the Euler angles should not deviate from small values throughout the mission.

One another disadvantage of using the Euler angles for attitude representation in the filter might be the trigonometric functions that appear both in kinematics and attitude matrix. Yet the nonlinearity is not a major issue especially for recent filtering algorithms such as UKF and even small satellite onboard computers are capable of easily evaluating these functions.

Examples for studies that use Euler angles to represent the attitude in the filtering algorithm can be seen in [12, 27–29].

## 9.3  ATTITUDE FILTERING USING QUATERNIONS

Quaternions are unarguably the most favored parameters to represent the attitude in the filtering algorithms for satellite attitude estimation. Because a quaternion vector is the lowest-dimensional attitude parameters free of any singularity [30]. As well as strong theoretical background, there are hundreds of publications demonstrating in-flight experience for attitude filters designed using the quaternions. These include validations with recent small satellite missions too [5, 6, 31].

Despite being used commonly, when the attitude is represented with quaternions the filtering algorithms given in Chapter 7 cannot be implemented straightforwardly as we do with Euler angles. The reason of such drawback is the constraint of quaternion unity given by $q^T q = 1$. An unbiased estimator cannot satisfy this constraint at the same time and if we somehow enforce the filter to produce norm-constrained quaternions, for instance by normalizing the filter estimates ($\hat{q}^+$) as

$$\hat{q} = \hat{q}^+ / \left\| \hat{q}^+ \right\|, \tag{9.4}$$

the estimates may be biased and the $7 \times 7$ covariance matrix may be ill-conditioned [30].

Next subsections will present alternate methods that have been used for satisfying the quaternion norm constraint in the attitude filter. Using a three-component local attitude error representation in the filter and updating the global quaternion estimate with a (quaternion) multiplicative approach is the most efficient method among them. Last two subsections are dedicated to the Multiplicative Extended Kalman Filter (MEFK) and the Unscented Attitude Filtering algorithms, which are produce of this approach.

### 9.3.1 Method of Quasi-measurement

One of the methods to have norm-constrained quaternion estimates is to include the constraint as a dummy measurement when updating the states [32]. Basic of the method is to augment the measurement output with the known algebraic constraint

$$\varsigma(x,t) = 0 \tag{9.5}$$

for a nonlinear system given as

$$\dot{x} = f(x,t), \tag{9.6}$$

$$y = h(x,t). \tag{9.7}$$

Hence, the augmented measurement vector becomes

$$y_A = \begin{bmatrix} h(x,t) \\ \varsigma(x,t) \end{bmatrix}. \tag{9.8}$$

Such method will still produce biased estimates.

The filter designer should be careful also with the covariance setting of this dummy measurement. The measurement noise covariance for the constraint measurement is theoretically 0. But including the measurement with 0 covariance is practically not possible so the filter should be relaxed with a small covariance value.

### 9.3.2 NORM-CONSTRAINED KALMAN FILTERING

Another method for satisfying the norm constraint is to introduce a Langrage multiplier in the filtering equations. In [33] the method is introduced for a linear Kalman filter. Here, we give the equations for the Lagrange multiplier method when it is implemented as a part of the UKF. For EKF the formulation will be similar to that given in the original publication [33]. The optimal Lagrange multiplier is defined as

$$\lambda_k = \frac{-1}{\bar{\varepsilon}_k} + \frac{\varepsilon_k^T P_{yy,k}^{-1} P_{xy,k}^T + \left(\hat{X}_k^-\right)^T}{\bar{\varepsilon}_k \sqrt{l}}, \tag{9.9}$$

and the UKF gain is modified as

$$K_k^* = \left( P_{xy,k}^T - \lambda_k \hat{x}_{k/k-1} \varepsilon_k^T \right)\left( P_{yy,k} + \lambda_k \varepsilon_k \varepsilon_k^T \right)^{-1} \tag{9.10}$$

to satisfy the norm constraint, whose defined value is $l = 1$. Here $\lambda_k$ is the Lagrange multiplier, $\varepsilon_k$ is the innovation vector, $\bar{\varepsilon}_k$ is the normalized innovation as $\bar{\varepsilon}_k = \varepsilon_k^T P_{yy,k}^{-1} \varepsilon_k$, $P_{yy,k}$ is the innovation covariance, $P_{xy,k}$ is the cross correlation matrix and $\hat{X}_k^-$ is the predicted state vector at discrete time step of $k$. The asterisk shows that the gain is modified and different from the unconstrained UKF gain $K_k$.

The state estimation covariance is given by the Joseph formula:

$$P_k^{+*} = P_k^- - K_k^* P_{xy,k}^T - P_{xy,k}\left(K_k^*\right)^T + K_k^*\left[ P_{xy,k}^T \left(P_k^-\right)^{-1} P_{xy,k} + R_k \right]\left(K_k^*\right)^T \tag{9.11}$$

Eqs. (9.9–9.11) are given for the norm constrained UKF in general case. In our problem only a part of the state, which is for quaternions, is subject to the constraint. Thus the state vector, $\hat{X}_k^-$, cross correlation matrix, $P_{xy,k}$, UKF gain, $K_k^*$ and state estimation covariance $P_k^{+*}$ must be partitioned. In this case, the Lagrange multiplier is calculated using only $\hat{q}$ instead of full state vector $\hat{X}$ and the corresponding first four rows of $P_{xy}$. The Kalman gain and state estimation covariance are calculated independently for the spin-axis direction terms and the rest of the states.

### 9.3.3 MULTIPLICATIVE EXTENDED KALMAN FILTER

The essence of the multiplicative EKF (MEKF) is to represent the attitude using a three-component local error representation in the filter and update the global attitude representation, which is in quaternions, using a multiplicative approach. The relation between the error quaternion ($\delta q$), the true quaternion ($q^{true}$) and the estimated quaternion ($\hat{q}$) is defined with the quaternion multiplication

$$q^{true} = \delta q \otimes \hat{q}, \tag{9.12}$$

which is defined as

$$q' \otimes q = \left[\Xi(q) \quad \vdots \quad q\right] q' \tag{9.13}$$

for two given quaternion vectors, $q$ and $q'$. Here $\Xi(q)$ matrix is given as

$$\Xi(q) = \begin{bmatrix} q_4 I_{3\times3} + [g\times] \\ -g^T \end{bmatrix} \tag{9.14}$$

for a quaternion vector rearranged as $q = \begin{bmatrix} g^T & q_4 \end{bmatrix}^T$. Thus $g = \begin{bmatrix} q_1 & q_2 & q_3 \end{bmatrix}^T$ is the vector part for quaternion. In (9.14) $[g\times]$ is the cross-product matrix as

$$[g\times] = \begin{bmatrix} 0 & -g_3 & g_2 \\ g_3 & 0 & -g_1 \\ -g_2 & g_1 & 0 \end{bmatrix}. \tag{9.15}$$

Regarding the three-component local error representation, one has multiple choices. The most straightforward approach might be using the three-component error quaternion defined as

$$\delta q_{1:3} = \delta g / \delta q_4 \tag{9.16}$$

for local-error quaternion vector defined as $\delta q = \begin{bmatrix} \delta g^T & \delta q_4 \end{bmatrix}^T$.

On the other hand, due to the simplifications it brings about in the filtering algorithms (such as eliminating multiplications with factors of 2) and having components of roll, pitch and yaw error angles for any rotation sequence [30], using the rotation error vector defined as $\delta\alpha \approx 2\delta q_{1:3}$ for representing the local error attitude is useful. As a result, the local three-component attitude error vector to be used in the filtering algorithm can be directly defined from the error quaternion vector as

$$\delta\alpha = 2\delta g / \delta q_4. \tag{9.17}$$

Then the obvious transformation back to the error quaternion is

$$\delta g = \frac{\delta\alpha}{2}, \tag{9.18a}$$

$$\delta q_4 = 1. \tag{9.18b}$$

The kinematic equation for the rotation error vector, which is useful in the derivation of the filter equations is

$$\delta\dot\alpha = -\hat\omega \times \delta\alpha + \delta\omega. \tag{9.19}$$

Depending on the gyro-measurement availability, the designed MEKF can be a gyro-based filter of a dynamics-based filter. In any case we need to include additional states (e.g. gyro bias terms for a gyro-based filter) in the state vector. Then the final error state, which will be used in the filter, is formed as

$$\Delta x = \begin{bmatrix} \delta\alpha \\ \Delta\beta \end{bmatrix}, \tag{9.20}$$

where $\Delta\beta$ is the error vector of other variables to be estimated. For these variables, an additional error definition as $\beta^{true} = \Delta\beta + \hat{\beta}$ can be used.

Now we can briefly explain the steps for the MEKF algorithm for attitude estimation:

1. The filtering algorithm starts with the propagation of states. For quaternions we use the kinematics equation (2.22). If this is a dynamics-based filter the satellite dynamics equation (2.6) is also needed. For a gyro-based filter, the assumed model for gyro biases can be $\dot{b}_g = 0$.
2. The $6 \times 6$ estimation covariance matrix defined for the error states is propagated with (7.82). The system transition matrix can be calculated in the continuous form as

$$F(t) = \begin{bmatrix} -[\hat{\omega}\times] & -I_3 \\ 0_3 & 0_3 \end{bmatrix} \tag{9.21}$$

for a gyro-based filter (i.e. $\Delta\beta = \Delta b_g$). Here it should be noted that $\delta\omega = -(\Delta b_g + \eta_{gn})$ in (9.19). On the other hand, for a dynamics-based filter (i.e. $\Delta\beta = \delta\omega$) the same system transition matrix is given as

$$F(t) = \begin{bmatrix} -[\hat{\omega}\times] & I_3 \\ 0_3 & J^{-1}\{[(J\omega)\times] - [\omega\times]J\} \end{bmatrix}, \tag{9.22}$$

considering that the rate error satisfies the linear differential equation of [34]

$$\delta\dot{\omega} = J^{-1}\{[(J\omega)\times] - [\omega\times]J\}\delta\omega. \tag{9.23}$$

The discrete form of the system transition matrix, which is needed in the filter's implementation, is approximated as

$$\phi_k \approx I_{6\times6} + \Delta t F(t). \tag{9.24}$$

For gyro-based filter both the state propagation equation and this system transition matrix can be calculated more efficiently with a power series approach [30]. In this case, the process noise covariance matrix which is formed of gyro-error specifications,

$$Q = \begin{bmatrix} \sigma_v^2 I_{3\times3} & 0_{3\times3} \\ 0_{3\times3} & \sigma_u^2 I_{3\times3} \end{bmatrix}, \tag{9.25}$$

can be approximated as

$$Q_k = \begin{bmatrix} \left(\sigma_v^2 \Delta t + \frac{1}{3}\sigma_u^2 \Delta t^3\right)I_{3\times3} & -\left(\frac{1}{2}\sigma_u^2 \Delta t^2\right)I_{3\times3} \\ -\left(\frac{1}{2}\sigma_u^2 \Delta t^2\right)I_{3\times3} & \left(\sigma_u^2 \Delta t\right)I_{3\times3} \end{bmatrix} \tag{9.26}$$

in discrete form.

3. The measurement update is performed using the vector measurements as

$$\Delta \hat{x}_k^+ = \begin{bmatrix} \delta \hat{\alpha}_k^+ \\ \Delta \hat{\beta}_k^+ \end{bmatrix} = K\left( z_k - h_k\left( \hat{x}_k^- \right) \right), \tag{9.27}$$

where

$$z_k = \begin{bmatrix} b_1 \\ b_2 \\ \vdots \\ b_N \end{bmatrix} \tag{9.28}$$

is the set of measurements for vector measurements $b_i$ in body frame ($i = 1 \ldots N$) and

$$h\left( \hat{x}_k^- \right) = \begin{bmatrix} \left[ A\left( \hat{q}_k^- \right) r_1 \right] \\ \left[ A\left( \hat{q}_k^- \right) r_2 \right] \\ \vdots \\ \left[ A\left( \hat{q}_k^- \right) r_N \right] \end{bmatrix} \tag{9.29}$$

is the set of predicted measurement for predicted quaternion vector $\hat{q}_k^-$ and reference vectors $r_i^1$.

Once the updated error states are obtained, the full quaternion vector is updated by first calculating the error quaternion using (9.18) and then with quaternion multiplication as

$$\hat{q}_k^* = \delta q\left( \delta \hat{\alpha}_k^+ \right) \otimes \hat{q}_k^-, \tag{9.30}$$

which is followed by normalization

$$\hat{q}_k^+ = \frac{\hat{q}_k^*}{\left\| \hat{q}_k^* \right\|}. \tag{9.31}$$

Note that, considering (9.13), two operations with (9.18) and (9.30) to get $\hat{q}_k^*$ from the updated three-component error state is mathematically equivalent to the single operation of

$$\hat{q}_k^* = \hat{q}_k^- + \frac{1}{2}\Xi\left( \hat{q}_k^- \right)\delta \hat{\alpha}_k^-. \tag{9.32}$$

The remaining states are updated simply as

$$\hat{\beta}_k^+ = \hat{\beta}_k^- + \Delta \hat{\beta}_k^+. \tag{9.33}$$

4. After the update, the error states $\delta \hat{\alpha}_k^+$ and $\Delta \hat{\beta}_k^+$ are reset.

The general flow of the MEKF algorithm is given in Figure 9.1. The final form with specific system transition matrix etc. depends on the availability of the gyro measurements, which is shown with a dashed border.

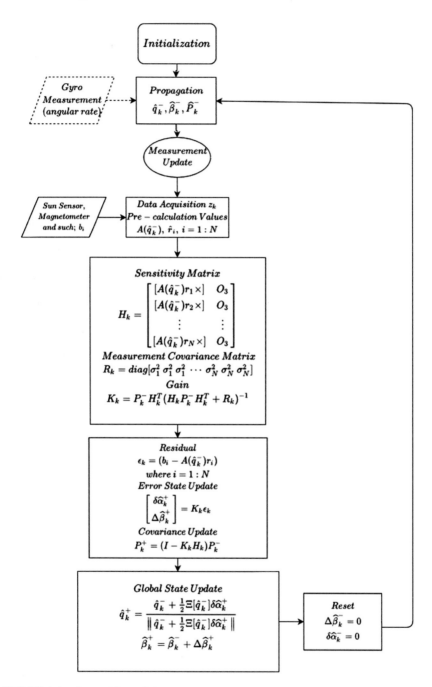

**FIGURE 9.1** General flow of the MEKF for attitude filtering.

Figures 9.2–9.3 give attitude estimation results for two different MEKF algorithms, in exactly same scenario, for a nanosatellite. One of these filters is gyro-based and the other one is dynamics-based. For measurements the magnetometer noise is zero mean Gaussian white noise with a standard deviation of $\sigma_m = 300$nT. Moreover, the standard deviation for the Sun sensor noise is taken as $\sigma_s = 0.1°$. The angular random walk and rate random walk values for the used gyros are $\sigma_v = 2.47$ arcsec $/ \sqrt{s}$ and $\sigma_u = 6.36 \times 10^{-4}$ arcsec $/ \sqrt{s^3}$, respectively. The spacecraft dynamics used for deriving the true simulated data include the magnetic and gravity gradient disturbance torques, which are not accounted for in the filter model for dynamics-based filter. The residual magnetic moment is assumed as $M = \begin{bmatrix} -0.51 & 0.042 & 0.11 \end{bmatrix}^T$ Am$^2$.

Advantages of using the gyro measurements whenever they are available is clearly seen when two results (Figures 9.2 and 9.3) are compared, especially in eclipse (gray areas) when the filter is left with only magnetometer measurements for update.

### 9.3.4 Unscented Attitude Filtering

The formulation for UKF for attitude estimation, when quaternions used for attitude representation, is similar to that of MEKF. Again we use the three-component local attitude error in the filter's state vector and update the global quaternion with a multiplicative approach. The main difference appears due to the sigma points, which we need in the UKF formulation for approximating the nonlinear distribution.

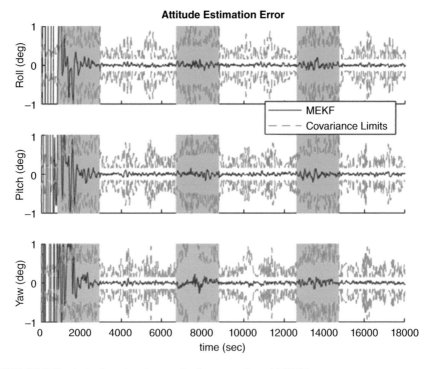

**FIGURE 9.2** Attitude estimation results for a gyro-based MEKF.

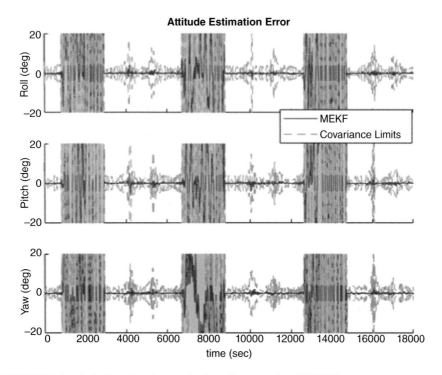

**FIGURE 9.3** Attitude estimation results for a dynamics-based MEKF.

The sigma-points, which are calculated for the three-component error states (see Eq. 7.86) and regard that we have $6 \times 6$ estimation covariance matrix as in MEKF), must be first transformed to the full quaternion sigma points for state propagation and then back to the three-component "propagated" sigma points for the rest of the estimation process. Thus the transformation in between the three-component error state and full quaternion vector is needed also in the propagation stage, not just for state update.

In line with the essential reference for Unscented Attitude Filtering [3], here we use generalized Rodrigues parameters to represent the three-component error attitude:

$$\delta \boldsymbol{p} = f\left[\delta \boldsymbol{g} / \left(a + \delta q_4\right)\right]. \tag{9.34}$$

Here, $a$ is a parameter from 0 to1 and $f$ is the scale factor. When $a = 0$ and $f = 1$ then Eq. (9.34) gives the Gibbs vector and when $a = 1$ and $f = 1$ then Eq. (9.34) gives the standard vector of modified Rodrigues parameters. In [3] $f$ is chosen as $f = 2(a + 1)$. See also if $a = 0$ and $f = 2$ then $\delta \boldsymbol{p}$ becomes the rotation error vector that we used in MEKF formulation as $\delta \boldsymbol{p} \approx \delta \boldsymbol{\alpha}$. The inverse transformation from $\delta \boldsymbol{p}$ to $\delta \boldsymbol{q}$ is given by

$$\delta q_4 = \frac{-a\|\delta \boldsymbol{p}\|^2 + f\sqrt{f^2 + \left(1 - a^2\right)\|\delta \boldsymbol{p}\|^2}}{f^2 + \|\delta \boldsymbol{p}\|^2}, \tag{9.35a}$$

$$\delta \boldsymbol{g} = f^{-1}\left(a + \delta q_4\right)\delta \boldsymbol{p}. \tag{9.35b}$$

Note that this inverse transformation includes also the normalization.

Now we can introduce the procedure for the UKF with attitude error representation. First let us define the following state vector:

$$x_0\left(k|k\right) = \hat{x}\left(k|k\right) = \begin{bmatrix} \delta\hat{p}\left(k|k\right) \\ \Delta\hat{\beta}\left(k|k\right) \end{bmatrix} \tag{9.36}$$

The vector for the sigma points can be partitioned into two parts:

$$x_l\left(k|k\right) = \begin{bmatrix} x_l^{\delta p}\left(k|k\right) \\ x_l^{\Delta\beta}\left(k|k\right) \end{bmatrix} \quad l = 0\ldots12, \tag{9.37}$$

where $x_l^{\delta p}\left(k|k\right)$ is the attitude error part and $x_l^{\Delta\beta}\left(k|k\right)$ is the part for rest of the states. The sigma points calculated using Eq. (9.37) can be called as error sigma points which correspond to the sigma points calculated for error quaternions. We need to calculate the full sigma points that correspond to the full state with four quaternion terms and six bias terms in order to propagate the model. Therefore,

$$\hat{q}_0\left(k|k\right) = \hat{q}\left(k|k\right) \tag{9.38a}$$

$$\hat{q}_l\left(k|k\right) = \delta\hat{q}_l\left(k|k\right) \otimes \hat{q}\left(k|k\right) \quad l = 0\ldots12, \tag{9.38b}$$

where $\delta\hat{q}_l\left(k|k\right) = \begin{bmatrix} \delta\hat{g}_l\left(k|k\right)^T & \delta\hat{q}_{4_l}\left(k|k\right) \end{bmatrix}^T$ is represented by Eq. (9.35):

$$\delta\hat{q}_{4_l}\left(k|k\right) = \frac{-a\left\|x_l^{\delta p}\left(k|k\right)\right\|^2 + f\sqrt{f^2 + \left(1 - a^2\right)\left\|x_l^{\delta p}\left(k|k\right)\right\|^2}}{f^2 + \left\|x_l^{\delta p}\left(k|k\right)\right\|^2} \quad l = 0\ldots12 \tag{9.39a}$$

$$\delta\hat{g}_l\left(k|k\right) = f^{-1}\left[a + \delta\hat{q}_{4_l}\left(k|k\right)\right]x_l^{\delta p}\left(k|k\right) \quad l = 0\ldots12 \tag{9.39b}$$

Eq. (9.38a) requires that $x_0^{\delta p}\left(k|k\right)$ be zero and this is due to the reset of the attitude error to zero after each estimation step.

Next, the quaternions are propagated using the kinematic model so we have $\hat{q}_l\left(k+1|k\right)$ as the propagated quaternions.

As a result, now we have the propagated full sigma points. The propagated error sigma points must be calculated to continue the filtering process. So as to do that, we use the representation in Eq. (9.34);

$$x_0^{\delta p}\left(k+1|k\right) = 0 \tag{9.40a}$$

$$x_l^{\delta p}\left(k+1|k\right)=f\,\frac{\delta\hat{g}_l\left(k+1|k\right)}{a+\delta\hat{q}_{4_l}\left(k+1|k\right)}\qquad l=0\ldots12,\qquad(9.40b)$$

where

$$\left[\delta\hat{g}_l\left(k+1|k\right)^T\quad\delta\hat{q}_{4_l}\left(k+1|k\right)\right]^T=\delta\hat{q}_l\left(k+1|k\right)$$

$$=\hat{q}_l\left(k+1|k\right)\otimes\left[\hat{q}_0\left(k+1|k\right)\right]^{-1}\quad l=0\ldots12$$

$$(9.41)$$

Note that $\delta\hat{q}_0\left(k+1|k\right)$ is the identity quaternion. Now using the propagated error sigma point $x_l\left(k+1|k\right)$, it is possible to calculate the predicted mean and covariance (Eq. 7.88a and 7.88b).

Next step is to calculate the predicted observation vector by using the propagated full sigma points (Eqs. 7.90 and 7.91). Then the observation covariance, innovation covariance and cross correlation matrices can be computed using Eqs (7.92–7.94). After calculating the innovation series and the Kalman gain (Eqs. 7.95–7.96), the state vector and the covariance can be updated by the use of Eqs. (7.97 and 7.98) with $\hat{x}\left(k+1|k+1\right)=\left[\delta\hat{p}\left(k+1|k+1\right)^T\quad\Delta\hat{\beta}\left(k+1|k+1\right)^T\right]^T$ as a result for the state update.

Now we have the updated state vector for error quaternions and we need to update the full quaternion vector by

$$\hat{q}_l\left(k+1|k+1\right)=\delta\hat{q}\left(k+1|k+1\right)\otimes\hat{q}_0\left(k+1|k\right),\qquad(9.42)$$

where $\delta\hat{q}\left(k+1|k+1\right)=\left[\delta\hat{g}\left(k+1|k+1\right)^T\quad\delta\hat{q}_4\left(k+1|k+1\right)\right]^T$ is represented by Eq. (9.35):

$$\delta\hat{q}_4\left(k+1|k+1\right)=\frac{-a\left\|\delta\hat{p}\left(k+1|k+1\right)\right\|^2+f\sqrt{f^2+\left(1-a^2\right)\left\|\delta\hat{p}\left(k+1|k+1\right)\right\|^2}}{f^2+\left\|\delta\hat{p}\left(k+1|k+1\right)\right\|^2}\qquad(9.43)$$

$$\delta\hat{g}\left(k+1|k+1\right)=f^{-1}\left[a+\delta\hat{q}_4\left(k+1|k+1\right)\right]\delta\hat{p}\left(k+1|k+1\right)\qquad(9.44)$$

Lastly $\delta\hat{p}\left(k+1|k+1\right)$ is reset to zero.

The flow to be followed for the rest of the states is similar but with an additive approach. For example, before the state propagation, the sigma points for the full states can be obtained by simply adding the error states with mean as

$$\hat{\beta}_l\left(k|k\right)=\hat{\beta}\left(k|k\right)+x_l^{\Delta\beta}\left(k|k\right)\qquad(9.45)$$

Figures 9.4–9.5 give attitude estimation results for two different Unscented Attitude Filtering algorithms, in exactly same scenario that we described for the MEKF in the previous section. As comprehensively discussed in [3], the UKF is more capable of handling the initial estimation error so has a quicker convergence to the true values. Besides, UKF gives more accurate estimation results especially for dynamics-based case, since it is more robust to uncertainties and nonlinearities.

## 9.4  ESTIMATION OF ADDITIONAL DYNAMICS PARAMETERS

In this section, we provide demonstration results for two algorithms that estimate the additional dynamics parameters, which is specifically useful for improving the attitude estimation accuracy of a dynamics-based filter and the attitude control accuracy in general. First of these algorithms, estimate the total disturbance torque acting on the satellite. The other algorithm focuses at estimating the residual magnetic moment terms to compensate the effects of magnetic disturbance torque only.

A general flow for both algorithms can be given as in Figure 9.6. Algorithms are used in a two-stage configuration with a gyro-based attitude filter. The attitude filter estimates the attitude and gyro biases. The estimated gyro biases are used to correct the angular rate measurements by gyros and this information is provided to the attitude dynamics estimation algorithm. Details of two different algorithms will be provided in the following subsections.

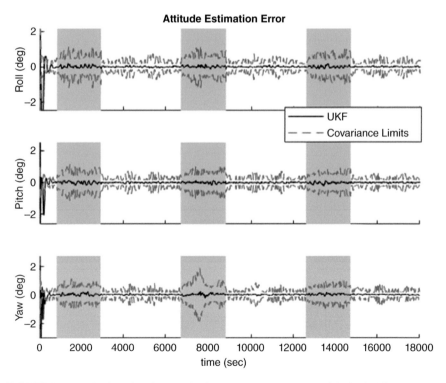

**FIGURE 9.4**  Attitude estimation results for a gyro-based Unscented Attitude Filter.

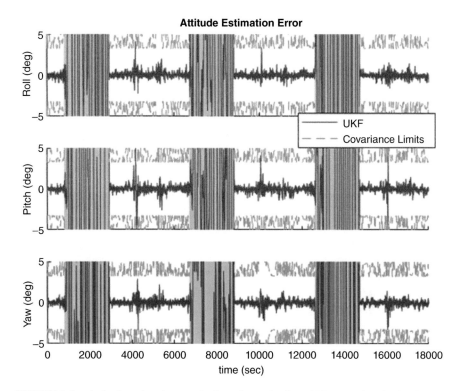

**FIGURE 9.5**    Attitude estimation results for a dynamics-based Unscented Attitude Filter.

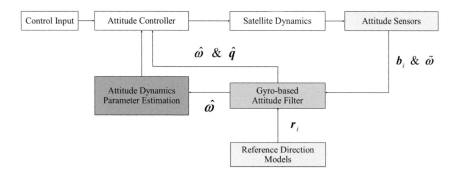

**FIGURE 9.6**    Flow for Estimation of Attitude Dynamics Parameters.

### 9.4.1 Disturbance Torque Estimation

In this estimation scenario, unknown constant components of the external torques affecting the satellite are estimated in addition to the attitude and angular rates [29, 35].

We have the satellite dynamics, accounting for only the disturbance torques as

$$\frac{d\omega_{bi}}{dt} = J^{-1} \left[ N_d - \omega_{bi} \times \left( J\omega_{bi} \right) \right]. \tag{9.46}$$

The disturbance torques are assumed to be constant, so the model used in the filter is simply,

$$\dot{N}_d = 0. \tag{9.47}$$

The UKF is used as the estimator. The inertia matrix of the satellite is $J = diag(2.1 \times 10^{-3}, 2.0 \times 10^{-3}, 1.9 \times 10^{-3})\text{kg.m}^2$ that corresponds to a 10cm cubic satellite with an approximate mass of 1.2 kg. The constant disturbance torque terms are assumed as $N_d = \begin{bmatrix} 5 & -3 & 4 \end{bmatrix}^T \text{Nm}$.

The magnetometer sensor noise is characterized by zero mean Gaussian white noise with a standard deviation of $\sigma_m = 300nT$.

The algorithm was tested in two cases. We are running a dynamics-based model for the estimations so there might be uncertainties mainly caused by the mismatch between the real dynamics and the model used in the filter. The only source for such uncertainty in our simulations might be the inertia terms that are not exactly known. Hence the two scenarios for the simulations are the cases with and without the uncertainty in the inertia terms.

In the first scenario there is no uncertainty in the inertia terms. Figure 9.7 presents an example for the estimation of the external disturbance torques. As clearly seen, the UKF converges to the real values after 10000th s and gives sufficiently accurate estimations for the torque terms.

The second simulation is performed for a scenario where we do not know the inertia of the satellite accurately and there are 5% and 10% uncertainty in the inertia

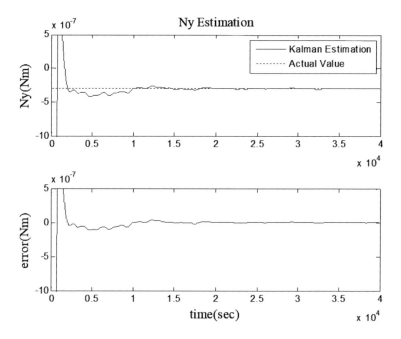

**FIGURE 9.7** Estimation of the constant external torque about "y" axis.

**TABLE 9.1**

**Root Mean Square Error for Disturbance Torque Estimation with and without Uncertainty in the Inertia Terms of the Satellite**

|                | Without Uncertainty | With 5% Uncertainty | With 10% Uncertainty |
|----------------|---------------------|---------------------|----------------------|
| $N_x$(Nm)      | 6.8174e-9           | 1.6826e-7           | 2.9664e-7            |
| $N_y$(Nm)      | 4.8459e-9           | 1.5259e-8           | 8.7539e-7            |
| $N_z$(Nm)      | 6.7456e-9           | 5.4487e-8           | 7.1730e-7            |

terms, respectively. For ease of understanding the results are given in table with comparison with the results obtained for the first scenario, beforehand. In Table 9.1, root mean square errors are tabulated for 10000 s between 20000th and 30000th s such that

$$RMSE_\gamma = \sqrt{\frac{1}{10000} \sum_{k=20001}^{30000} \left[ x_\gamma(k) - \hat{x}_\gamma(k) \right]^2} \quad \gamma = 1\ldots n. \qquad (9.48)$$

As can be seen, even for a small uncertainty in the inertia terms of the satellite the estimation error increases for all the estimated states. Yet for the simulation case with 5% uncertainty, estimation error is still within the acceptable bounds for cubesat applications. However, if the level of uncertainty is 10% then it is not possible to satisfy these requirements and the error becomes as large as the magnitude of the torque itself. Further examinations show that if there is even more uncertainty in the inertia terms of the satellite, the filter may diverge in the long term. Yet it should be considered that 10% inertia uncertainty is unlikely for especially nanosatellite missions with small mass and rigid body.

### 9.4.2 Residual Magnetic Moment Estimation

As discussed in Chapter 2, the magnetic disturbance torque is caused by the onboard electric current loop, small permanent magnet in some devices or some special material on the satellite, and does not strongly depend on the satellite size. Thus it is generally the main attitude disturbance source for the LEO small satellites [36, 37]. The residual magnetic moment (RMM) must be compensated in-orbit with an active control strategy and its accurate knowledge is also needed when a dynamics-based attitude filter is used.

In [35], an observer is proposed to estimate the RMM whereas in [38, 39] it is estimated using the Extended Kalman Filter (EKF). In [38], as well as the EKF, an UKF is designed for the RMM estimation and the estimation accuracies of these two different KFs are compared. In these studies, the RMM components are considered to be constant in time.

In [40, 41], considering that these parameters may change in practice with sudden shifts because of the instantaneous variations in the onboard electrical current, different adaptive Kalman filtering algorithms that are sensitive to the variations are proposed. Such instantaneous variations in the current may be caused by switching on/off of the onboard electronic devices or going into/out of eclipse. In such cases, a conventional KF cannot catch the new value of the parameter quickly if it is designed with a small process noise covariance in order to increase the steady-state estimation accuracy. The main issue is, especially if we use the feedforward cancellation technique for the RMM compensation, then the estimation accuracy is essential and so the KF must be designed with small process noise covariance. In other words, the inherent tracking capacity of the KF that can be assured by choosing a high process noise covariance must be sacrificed in order to increase the overall system performance. Therefore, if we want to design a KF with good tracking capability, as well as the high steady-state accuracy, then the filter should be adaptively designed such that it gives both good estimation results when there is no change in the parameter and good tracking performance when the parameter is changed.

The RMM estimation is based on the satellite dynamics equation, which is repeated here including only the magnetic disturbance torque:

$$\frac{d\omega_{bi}}{dt} = J^{-1}\left[ M \times B_b - \omega_{bi} \times \left( J\omega_{bi} \right) \right]. \tag{9.49}$$

Here $M$ is the RMM vector to be estimated and $B_b$ is the body frame magnetic field vector. There are two methods to obtain this body frame magnetic field vector values. The first, and the straightforward, method is to use the magnetometer measurements. The alternative second method may be transforming the calculated reference magnetic field values (in reference frame) to the body frame using the estimated attitude as $B_b = A\left(\hat{q}\right)b_i$. Due to the magnetometer errors, the second method will provide more accurate magnetic field data.

The RMM terms can be modeled constant in the filter as

$$\dot{M} = 0. \tag{9.50}$$

This is a fair assumption even if the terms are changing with sudden shifts. Even with this model an adaptive KF will be capable of quickly capturing the changes [40].

For this specific RMM estimation problem, the estimated state vector is composed of the body angular rates with respect to the inertial frame and RMM terms as given with

$$x = \begin{bmatrix} \omega_{BI} \\ M \end{bmatrix}. \tag{9.51}$$

The nonlinear process model is obtained by discrete-time integration of (9.49 and 9.50). Nevertheless, since the attitude filter supplies the angular rate, $\hat{\omega}_{bi}$, information, the measurement model may be represented with linear equation as

$$\tilde{y}_k = \begin{bmatrix} I_{3\times3} & 0_{3\times3} \end{bmatrix} x_k + v_k, \tag{9.52}$$

where $I_{3\times3}$ and $0_{3\times3}$ are $3 \times 3$ identity and null matrices, respectively.

Here we use the UKF. We preferred the UKF over the EKF since it is better in terms of estimation accuracy.

The inertia matrix of the examined satellite model is

$$J = \begin{bmatrix} 310 \text{ kg.m}^2 & 0 & 0 \\ 0 & 180 \text{ kg.m}^2 & 0 \\ 0 & 0 & 180 \text{ kg.m}^2 \end{bmatrix}. \tag{9.53}$$

We assume $B_b$ information is provided by the magnetometers. For the magnetometer measurements, the sensor noise is characterized by zero mean Gaussian white noise with a standard deviation of $\sigma_m = 300nT$. In the first scenario, where the RMM terms are constant and there is no abrupt change in time, the real RMM terms are $M = \begin{bmatrix} 0.1 & 0.02 & -0.05 \end{bmatrix}^T \text{ Am}^2$.

In Figure 9.8 the estimation result for the RMM in the $x$ axis is presented. In the top plot, the UKF estimation is given together with the actual value of the estimated parameter and in the lower plot the estimation error is shown. As seen the UKF accurately estimates the RMM terms. In order to make a better understanding of the performance the Root Mean Squared Error (RMSE) for the RMM terms of the state vector $(x_j(k)$ such that $j = 4...6)$ is calculated in between the 300th and 500th s. (for 2000 samples since $\Delta t = 0.1$ s):

$$RMSE_j = \sqrt{\frac{1}{2000} \sum_{k=3001}^{5000} \left[ x_j(k) - \hat{x}_j(k) \right]^2} \quad j = 4...6, \tag{9.54}$$

where $x_j(k)$ and $\hat{x}_j(k)$ are the real and estimated values of the $j^{th}$ state. As a result the RMSE is $2.626 \times 10^{-4}\text{Am}^2$, $2.584 \times 10^{-4}\text{Am}^2$ and $2.673 \times 10^{-4}\text{Am}^2$, respectively, for

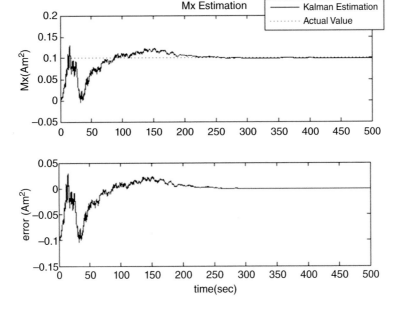

**FIGURE 9.8** Estimation of the RMM in x axis.

$M_x$, $M_y$ and $M_z$ estimations. That means the RMM estimation is accurate enough regarding the overall attitude control requirements and the magnetic disturbance can be compensated efficiently with a feed forward cancellation technique.

In the second scenario, this time the instantaneous change is realized as the change in the RMM terms at the 2000th s such that

$$M = \begin{cases} \begin{bmatrix} 0.1 & 0.02 & -0.05 \end{bmatrix}^T \text{Am}^2 & t < 2000 \text{ s} \\ \begin{bmatrix} 0.25 & 0.1 & -0.15 \end{bmatrix}^T \text{Am}^2 & t \geq 2000 \text{ s} \end{cases}, \tag{9.55}$$

and the same scenario is repeated for two different process noise levels, high and low, in order to clearly demonstrate the effect of the process noise covariance on the trade-off between the steady-state accuracy and tracking capability of the filter.

The Figure 9.9 presents the estimation results of the RMM in the $x$ axis for these two different noise levels. Obviously, the UKF tuned with low process noise has a poor tracking capability when the estimated parameter is changed and it takes almost one-third of the orbital period (the orbital period of the satellite for the performed simulation is 6400s.) for the filter to converge again to the required estimation accuracy (accepted as ±0.001 Am² for the RMM estimation). Conversely, the UKF with high noise is more agile to catch the new values of the RMM terms but performs noisy estimations with low accuracy during the steady-state regime. The Table 9.2 more clearly represents the fact that although the UKF with high process noise can agilely catch the new value of the RMM terms after the sudden change, the steady-state accuracy of the estimation is not high enough for satisfying good attitude control performance.

The investigations signify that if there is a necessity for both highly accurate RMM estimations during the steady state and good tracking speed when the states change suddenly, the KF must be built adaptively such that the filter parameters are tuned with respect to the requirements at that moment. The simplest method for achieving this is to use a change detector first for detecting the abrupt changes in the states and then increase the process noise covariance or estimation covariance of the filter in order to speed up the tracking process. The study in [40] introduces the novel method for change detection and KF adaptation, which can be referred for solving this problem.

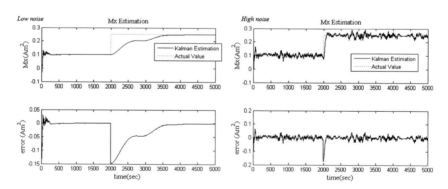

**FIGURE 9.9** Estimation of the RMM in x axis in case of sudden change for two different process noise levels.

**TABLE 9.2**

**Absolute Values of Error for the RMM Estimation in Case of Sudden Change**

| | Absolute Values of Estimation Error for the RMM Estimation | | | | | |
|---|---|---|---|---|---|---|
| | UKF with Low Noise | | | UKF with High Noise | | |
| | 1000th s | 2500th s | 4000th s | 1000th s | 2500th s | 4000th s |
| $M_x$ (Am²) | 0.000020 | 0.061680 | 0.002544 | 0.007702 | 0.002422 | 0.002908 |
| $M_y$ (Am²) | 0.000482 | 0.025425 | 0.000085 | 0.006528 | 0.003433 | 0.011384 |
| $M_z$ (Am²) | 0.000567 | 0.016087 | 0.004357 | 0.000547 | 0.036143 | 0.013680 |

## 9.5 ISSUES RELATED TO THE ATTITUDE FILTER'S COMPUTATIONAL LOAD

The computational load for the conventional MEKF algorithm is not high, and with several real-mission experiences its applicability even for nanosatellites has been demonstrated [6]. There are also nanosatellites implementing the Unscented Attitude Filter, successfully [7]. Yet, as introduced in the introduction there are several researches for decreasing the computational load of the attitude filtering algorithms. Steady-state Extended Kalman Filter [16], Gain-scheduled Kalman Filter [17], variants of sigma point filters with reduced number of sigma points [9, 18] and sequential approaches with any of single-frame attitude estimators are examples for researches in this direction [19]. We will discuss, specifically the computational advantages of integrating one of the single-frame methods with the filtering algorithms in Chapter 10.

One another important issue that should be mentioned here regarding the computational load of any filtering algorithm is the method preferred update using the measurements. In the general flowchart given for MEKF algorithm the measurement update is applied for all $N$ vector measurement at the same time. This may be computationally demanding, since it requires $3N \times 3N$ matrix inversion for gain calculation. In this case, using the Murrell's version for any of the filtering algorithm [30], which applies the vector update one-by-one, sequentially, may be used instead. In this case, the filter simply keeps on updating the error states and the filter's estimation covariance until all the vector measurements are processed as

$$\begin{bmatrix} \delta\hat{\alpha}_{k,i+1}^+ \\ \Delta\hat{\beta}_{k,i+1}^+ \end{bmatrix} = \begin{bmatrix} \delta\hat{\alpha}_{k,i}^+ \\ \Delta\hat{\beta}_{k,i}^+ \end{bmatrix} + K_{k,i}\left( z_{k,i} - h_k\left(\hat{x}_k^-\right) - H_k \begin{bmatrix} \delta\hat{\alpha}_{k,i}^+ \\ \Delta\hat{\beta}_{k,i}^+ \end{bmatrix} \right) \tag{9.56}$$

and

$$P_{k,i+1}^+ = \left( I - K_{k,i}H_k \right)P_{k,i}^+ \tag{9.57}$$

until $i$ reaches to $N$ with the last set of vector measurements. Note that for the very first vector measurement, i=1, + superscripts at the right side of the above equations should be −, so we use the propagated values as in original formulation.

## 9.6 CONCLUSION AND DISCUSSION

In this chapter, we briefly reviewed the details of Kalman filtering for small satellite attitude estimation. Specifically, we discussed different approaches such as dynamics and gyro-based attitude filtering and presented different methods to ensure the unit-norm constraint of the quaternions when these parameters are used for attitude representation in the filter. The Multiplicative Extended Kalman Filter and Unscented Kalman Filter algorithm, specifically designed to solve this problem are presented step-by-step to show the rationale of their design. These algorithms are given in their general form and can be easily modified by the designer, e.g. if there is need for augmenting the states.

In the last part of the chapter, we also introduced straightforward methods for estimating the attitude dynamics parameters for small satellites, especially to improve the attitude-pointing accuracy.

## NOTE

1  Note that when we have star tracker measurement, the measurement model becomes linear and we have simpler update sequence for the states. This will be discussed in Chapter 10 as a part of the filtering with nontraditional approach.

## REFERENCES

1. E. J. Lefferts, F. L. Markley, and M. D. Shuster, "Kalman filtering for spacecraft attitude estimation," *J. Guid. Control. Dyn.*, vol. 5, no. 5, pp. 417–429, Sep. 1982.
2. F. L. Markley and J. L. Crassidis, *Fundamentals of Spacecraft Attitude Determination and Control*. New York, NY: Springer New York, 2014.
3. J. Crassidis and F. L. Markley, "Unscented filtering for spacecraft attitude estimation," *J. Guid. Control. Dyn.*, vol. 26, no. 4, pp. 536–542, 2003.
4. J. L. Crassidis, F. L. Markley, and Y. Cheng, "Survey of nonlinear attitude estimation methods," *J. Guid. Control. Dyn.*, vol. 30, no. 1, pp. 12–28, 2006.
5. Y. Nakajima, N. Murakami, T. Ohtani, Y. Nakamura, K. Hirako, and K. Inoue, "SDS-4 attitude control system: In-Flight results of three axis attitude control for small satellites," *19th IFAC Symposium on Automatic Control in Aerospace*, 2013, vol. 46, no. 19, pp. 283–288.
6. J. C. Springmann and J. W. Cutler, "Flight results of a low-cost attitude determination system," *Acta Astronaut.*, vol. 99, no. 1, pp. 201–214, 2014.
7. A. Slavinskis *et al.*, "Flight results of ESTCube-1 attitude determination system," *J. Aerosp. Eng.*, vol. 29, no. 1, pp. 4015014–4015017, 2015.
8. S. Nakasuka, K. Miyata, Y. Tsuruda, Y. Aoyanagi, and T. Matsumoto, "Discussions on attitude determination and control system for micro/nano/pico-satellites considering survivability based on Hodoyoshi-3 and 4 experiences," *Acta Astronaut.*, vol. 145, no. December 2017, pp. 515–527, 2018.
9. S. K. Biswas, B. Southwell, and A. G. Dempster, "Performance analysis of fast unscented Kalman filters for attitude determination," *IFAC-PapersOnLine*, vol. 51, no. 1, pp. 697–701, 2018.
10. Y. Cheng and J. L. Crassidis, "Particle filtering for attitude estimation using a minimal local-error representation," *J. Guid. Control. Dyn.*, vol. 33, no. 4, pp. 1305–1310, 2010.

11. C. Zhang, A. Taghvaei, and P. G. Mehta, "Attitude estimation with feedback particle filter," *2016 IEEE 55th Conf. Decis. Control. CDC 2016*, no. Cdc, Las Vegas, USA, pp. 5440–5445, 2016.

12. R. V. Garcia, P. C. P. M. Pardal, H. K. Kuga, and M. C. Zanardi, "Nonlinear filtering for sequential spacecraft attitude estimation with real data: Cubature Kalman Filter, Unscented Kalman Filter and Extended Kalman Filter," *Adv. Sp. Res.*, vol. 63, no. 2, pp. 1038–1050, 2019.

13. X. Tang, Z. Liu, and J. Zhang, "Square-root quaternion cubature Kalman filtering for spacecraft attitude estimation," *Acta Astronaut.*, vol. 76, pp. 84–94, 2012.

14. B. Jia, M. Xin, and Y. Cheng, "Sparse Gauss-Hermite quadrature filter with application to spacecraft attitude estimation," *J. Guid. Control. Dyn.*, vol. 34, no. 2, pp. 367–379, 2011.

15. Y. Cheng, "Spacecraft attitude estimation using sparse grid quadrature filtering," in *Multisensor Attitude Estimation*, H. Fourati and D.E.C. Belkhiat (Eds.). Boca Raton, USA: CRC Press, 2016, pp. 315–329.

16. V. V. Unhelkar and H. B. Hablani, "Spacecraft attitude determination with sun sensors, horizon sensors and Gyros: Comparison of steady-state Kalman filter and extended Kalman filter," *Adv. Estim. Navig. Spacecr. Control*, pp. 413–437, 2015.

17. M. D. Pham, K. S. Low, S. T. Goh, and S. Chen, "Gain-scheduled extended kalman filter for nanosatellite attitude determination system," *IEEE Trans. Aerosp. Electron. Syst.*, vol. 51, no. 2, pp. 1017–1028, 2015.

18. M. Kiani and S. H. Pourtakdoust, "Concurrent orbit and attitude estimation using minimum sigma point unscented Kalman filter," *Proc. Inst. Mech. Eng. Part G J. Aerosp. Eng.*, vol. 228, no. 6, pp. 801–819, Mar. 2013.

19. H. E. Söken, "An attitude filtering and magnetometer calibration approach for nanosatellites," *Int. J. Aeronaut. Sp. Sci.*, vol. 19, no. 1, pp. 164–171, 2018.

20. R. Zanetti and K. J. DeMars, "Fully multiplicative unscented Kalman filter for attitude estimation," *J. Guid. Control. Dyn.*, vol. 41, no. 5, pp. 1–7, 2018.

21. S. A. O'Keefe and H. Schaub. "Shadow set considerations for modified rodrigues parameter attitude filtering," *Adv. Astronaut. Sci.*, vol. 150, no. 6, pp. 2777–2786, 2014.

22. M. S. Andrle and J. L. Crassidis, "Geometric integration of quaternions," *J. Guid. Control. Dyn.*, vol. 36, no. 6, pp. 1762–1767, 2013.

23. A. Fosbury, "Steady-state accuracy solutions of more spacecraft attitude estimators," in *AIAA Guidance, Navigation, and Control Conference*, Portland, USA, 2011, no. August.

24. Y. Yang and Z. Zhou, "Spacecraft dynamics should be considered in Kalman filter attitude estimation," *Adv. Astronaut. Sci.*, vol. 158, pp. 83–94, 2016.

25. O. J. Kim et al., "In-orbit results and attitude analysis of the snuglite cube-satellite," *Appl. Sci.*, vol. 10, no. 7, 2020.

26. G. F. De Oliveira et al., "A low-cost attitude determination and control system for the UYS-1 nanosatellite," *IEEE Aerosp. Conf. Proc.*, Big Sky, Montana, USA, 2013.

27. A. H. J. De Ruiter and C. J. Damaren, "Extended Kalman filtering and nonlinear predictive filtering for spacecraft attitude determination," *Can. Aeronaut. Sp. J.*, vol. 48, no. 1, pp. 13–23, 2002.

28. H. E. Soken and C. Hajiyev, "UKF-based reconfigurable attitude parameters estimation and magnetometer calibration," *IEEE Trans. Aerosp. Electron. Syst.*, vol. 48, no. 3, pp. 2614–2627, Jul. 2012.

29. H. E. Söken and C. Hajiyev, "UKF for the identification of the pico satellite attitude dynamics parameters and the external torques on IMU and magnetometer measurements," in *RAST 2009 - Proceedings of 4th International Conference on Recent Advances Space Technologies*, Istanbul, Turkey, 2009, pp. 547–552.

30. F. L. Markley and J. L. Crassidis, *Fundamentals of Spacecraft Attitude Determination and Control*. New York, NY: Springer New York, 2014.

31. D. Ran, T. Sheng, L. Cao, X. Chen, and Y. Zhao, "Attitude control system design and on-orbit performance analysis of nano-satellite - 'tian Tuo 1," *Chinese J. Aeronaut.*, vol. 27, no. 3, pp. 593–601, 2014.

32. P. Sekhavat, Q. G. Q. Gong, and I. M. Ross, "NPSAT1 parameter estimation using unscented Kalman filtering," *2007 Am. Control Conf.*, New York, USA, pp. 4445–4451, 2007.

33. R. Zanetti, M. Majji, R. H. Bishop, and D. Mortari, "Norm-constrained Kalman filtering," *J. Guid. Control. Dyn.*, vol. 32, no. 5, pp. 1458–1465, Sep. 2009.

34. M. N. Filipski and R. Varatharajoo, "Evaluation of a spacecraft attitude and rate estimation algorithm," *Aircr. Eng. Aerosp. Technol.*, vol. 82, no. 3, pp. 184–193, 2010.

35. H. E. Söken and C. Hajiyev, "Estimation of pico-satellite attitude dynamics and external torques via Unscented Kalman Filter," *J. Aerosp. Technol. Manag.*, vol. 6, no. 2, 2014.

36. S. I. Sakai, Y. Fukushima, and H. Saito, "Design and on-orbit evaluation of magnetic attitude control system for the REIMEI microsatellite," *Proc. 10th IEEE International Workshop on Advanced Motion Control*, Trento, Italy, 2008, pp. 584–589.

37. S. Busch, P. Bangert, S. Dombrovski, and K. Schilling, "UWE-3, in-orbit performance and lessons learned of a modular and flexible satellite bus for future pico-satellite formations," *Acta Astronaut.*, vol. 117, pp. 73–89, 2015.

38. T. Inamori, N. Sako, and S. Nakasuka, "Magnetic dipole moment estimation and compensation for an accurate attitude control in nano-satellite missions," *Acta Astronaut.*, vol. 68, no. 11–12, pp. 2038–2046, 2011.

39. W. H. Steyn and Y. Hashida. "In-Orbit attitude performance of the 3-axis stabilised SNAP-1 nanosatellite," *Proc. 15th AIAA/USU Conference on Small Satellites*, Utah, USA, 2001, pp. 1–10.

40. H. E. Soken, S. I. Sakai, and R. Wisniewski, "In-orbit estimation of time-varying residual magnetic moment," *IEEE Trans. Aerosp. Electron. Syst.*, vol. 50, no. 4, pp. 3126–3136, 2014.

41. H. E. Soken and S. Sakai, "Multiple-model adaptive estimation of time-varying residual magnetic moment for small satellites," in *Advances in Aerospace Guidance, Navigation and Control*, J. Bordeneuve-Guibe, A. Drouin and C. Roos (eds.). Cham: Springer International Publishing, 2015, pp. 303–321.

# 10 Integration of Single-Frame Methods with Filtering Algorithms for Attitude Estimation

Single-frame algorithms and filtering algorithms are two standalone methods that we can use for small satellite attitude estimation. Single-frame methods use the measurements obtained at a single point in time and fuse the information from sensors to optimally determine the attitude of the spacecraft. On the other hand, the filtering methods use the spacecraft dynamics information together with the measurements obtained over a period of time to sequentially estimate the spacecraft's attitude. Filtering methods are usually capable of providing more accurate attitude estimates than the single-frame methods and can provide estimates even when there are insufficient measurements for single-frame methods to work. Moreover, by using also gyro measurements or the spacecraft dynamics, they can provide the attitude rate estimate as well.

Integration of single-frame methods and the attitude filters provide several advantages, specifically for small satellites. These are as follows:

1. As a standalone technique the single frame works well as long as minimum two vector measurements are available and not parallel. However, if there is only one vector measurement (e.g. when the satellite is in eclipse and Sun sensor measurements are not available) or the attitude estimation problem is an underdetermined one because of parallel vectors, the single-frame method fails to provide any attitude estimate. On the other hand, a filter integrated to the overall algorithm can still provide attitude estimates by propagation using the gyro measurements or the attitude dynamics.
2. The single-frame method gives attitude estimates as frequent as the sampling rate of the sensor with lower measurement frequency (if there is no interpolation or propagation). The integrated method can provide attitude estimate with a higher frequency since it makes use of the gyro measurements or attitude dynamics.
3. The single-frame method does not estimate the attitude rates. For most of the cases, satellite attitude rates must be estimated especially for control purposes.

There are deterministic methods to estimate satellite's attitude rate from vector measurements, but usually a filtering-based method gives more accurate estimates.

4. When an integrated method is used, the measurement model for the attitude filter becomes linear. Especially if the attitude filter is an Unscented Kalman Filter (UKF), a linear measurement model means we have no need for sigma points at the measurement update phase and the filter reduces to a linear Kalman Filter. This reduces the computational load of the overall algorithm [1, 2], which is advantageous for small satellites.

5. Integrated algorithm has switching flexibility between different sensors, without requiring dramatic modification in the filtering algorithm. Considering the limited hardware capability of small satellites and the vulnerability of small satellite sensors, such flexibility ensures sustainability of the mission for longer periods [3].

6. In general, we have a more robust attitude estimation architecture when two methods are integrated. Specifically, the filter (especially EKF) converges faster since we have linear measurement. Moreover, the filter is inherently robust against the noise covariance increments in the measurements as will be discussed in Chapter 14.

7. With an integrated algorithm concurrent attitude estimation and attitude sensor calibration is easier (see TRIAD+UKF Approach for Attitude Estimation and Magnetometer Calibration in Chapter 16).

An earlier study on single-frame estimator aided attitude filtering was carried out in [4]. In this study, the authors integrate the algebraic method (TRIAD) and the EKF algorithms to estimate the attitude angles and angular velocities. The magnetometers, Sun sensors and horizon scanners/sensors are used as measurement devices and three different two-vector algorithms based on the Earth's magnetic field, Sun and nadir vectors are proposed. An EKF is designed to obtain the satellite's angular motion parameters with the desired accuracy. The measurement inputs for the EKF are the attitude estimates of the two-vector algorithms. Interest in "single-frame estimator aided attitude filtering" is higher in more recent literature [5–8]. The attitude determination concept of the Kyushu University mini-satellite QSAT is based on a combination of the Weighted-Least-Square (WLS) and KF 8-9. The WLS method produces the optimal attitude-angle observations at a single frame by using the Sun sensor and magnetometer measurements. The KF combines the WLS angular observations with the attitude rate measured by the gyros to produce the optimal attitude solution. In [8] an interlaced filtering method is presented for nanosatellite attitude determination. In this integrated system, the optimal-REQUEST and UKF algorithms are combined to estimate the attitude quaternion and gyro drifts. The optimal-REQUEST, which cannot estimate gyroscope drifts, is run for the attitude estimation. Then the UKF is used for the gyro-drift estimation on the basis of linear measurements obtained as optimal-REQUEST estimates. In the attitude determination system of the SDS-4 small spacecraft, the QUEST block provides quaternion measurements to the attitude filter using the magnetometer and Sun sensor measurements when the star tracker measurements are not available [6]. Last in [7], q-method is used for pre-processing the vector measurements before the predictive variable structure filter in the integrated algorithm.

Here, we may also refer to the studies where a single-frame attitude estimator used together with an attitude filter but not for providing the linear measurements [9, 10]. For linear measurements, it is equivalent to first update the attitude using the single-frame estimator and subsequently use this updated portion of the state to update the remainder of the state as it is to update the entire state at once. However, in [9, 10], the measurement model is nonlinear. A nonlinear update for the attitude is obtained solving the Wahba's problem and subsequently used to update the non-attitude states using the optimal gain for the linear measurement case. Therefore, in these studies, the attitude is updated using the single-frame estimator and all remaining non-attitude states using the standard nonlinear attitude filters.

In this chapter, we investigate different small satellite attitude estimation algorithms for which the single-frame and filtering methods are integrated and used together. First the single-frame method estimates the attitude using the vector measurements. Then the filtering algorithm uses the attitude estimates of the single-frame method as measurements and provides more accurate attitude information as well as estimates for the additional states.

## 10.1 INTEGRATION WHEN ATTITUDE IS REPRESENTED USING EULER ANGLES

The filter designer may prefer using the Euler angles for representing the attitude in the integrated architecture, especially if the attitude is defined with respect to the orbit reference frame. In this case the state vector in general form is

$$x = \begin{bmatrix} \varphi & \theta & \psi & \beta_x & \beta_y & \beta_z \end{bmatrix}^T. \tag{10.1}$$

Here, $\beta_x$, $\beta_y$ and $\beta_z$ are the additional parameters to be estimated such as gyro biases or angular rates depending on the design preferences.

For the UKF, the formulation is straightforward. The measurement model becomes linear and using a linear KF with an observation matrix of $H = \begin{bmatrix} I_{3\times3} & 0_{3\times3} \end{bmatrix}$, the states can be updated.

On the other hand, if an EKF is used the mathematical model of the satellite's rotational motion about its center of mass can be linearized using quasi-linearization method,

$$x_i = f\left(\hat{x}_{i-1}, \bar{\omega}_{o_{i-1}}\right) + F_{x_{i-1}}\left(x_{i-1} - \hat{x}_{i-1}\right) + F_o\left(\bar{\omega}_{o_{i-1}} - \omega_{o_{i-1}}^{comp}\right), \tag{10.2}$$

where $f\left(\hat{x}_{i-1}, \bar{\omega}_{o_{i-1}}\right)$ is the right-hand side of the satellite's rotational motion mathematical model based on estimated values; $F_o$ is the coefficient matrix accounting for the orbital angular velocity;

$$F_x = \left[\frac{\partial f}{\partial x}\right]_{\hat{x}_{i-1}, \bar{\omega}_{o_{i-1}}}, \quad F_o = \left[\frac{\partial f}{\partial \omega_o}\right]_{\hat{x}_{i-1}, \bar{\omega}_{o_{i-1}}} \tag{10.3}$$

Minimum of the error's standard deviation was selected as an optimum criterion. It is suggested to derive the satellite attitude estimation algorithm using Bayes' method.

The problem of finding the values of the system's parameters and output coordinates takes us to evaluation of $p\left(x_i \, / \, \overline{z}_i, \overline{\omega}_{o_i}\right)$ conditional probability density. To the Bayes' formula, this probability density can be written as [4]

$$p\left(x_i \, / \, Z^i, \overline{\omega}_{o_i}\right) = p\left(x_i \, / \, Z^{i-1}, z_i, \overline{\omega}_{o_i}\right) = \frac{p\left(x_i \, / \, Z^{i-1}, \overline{\omega}_{o_i}\right) p\left(z_i \, / \, x_i, Z^{i-1}, \overline{\omega}_{o_i}\right)}{p\left(z_i \, / \, Z^{i-1}\right)}, \qquad (10.4)$$

where $z_i^T = \left[z_{\varphi_i}, z_{\theta_i}, z_{\psi_i}\right]$ is the measurement vector.

Finding and substituting terms respectively into (10.4) and via taking into consideration that the minimum of the standard deviation, which was chosen as an optimum criterion (in this case the conditional mathematical expectation of the value's a posterior distribution will be the best value and as for the value's accuracy, the covariance matrix of this distribution will be used) and $p\left(x_i \, / \, Z^i, \overline{\omega}_{o_i}\right)$, is a Gauss distribution, the recursive algorithm for the satellite's attitude estimation is obtained as given below:

$$\hat{x}_i = f\left(\hat{x}_{i-1}, \overline{\omega}_{o_{i-1}}\right) + K_i\left[z_i - Hf\left(\hat{x}_{i-1}, \overline{\omega}_{o_{i-1}}\right)\right], \qquad (10.5)$$

$$P_i = P_i^- - P_i^- H^T\left[R + HP_i^- H^T\right]^{-1} HP_i^-, \qquad (10.6)$$

$$P_i^- = F_U P_{i-1} F_U^T + F_o D_{o_{i-1}} F_o^T + Q, \qquad (10.7)$$

where $P_i^-$ is the covariance matrix of the extrapolation error, $P_i$ is the covariance matrix of the estimation error, $K_i = P_i H^T R^{-1}$ is the gain matrix of Kalman filter, $D_{o_{i-1}}$ is the variance which characterizes uncertainty of the calculated values of satellite's orbital velocity, $R$ is the covariance matrix of measurement noise, which is diagonal matrix with diagonal elements built of the variances of angle and angle rate measurement noises and $Q$ is the covariance matrix of the system noises.

Equations given as (10.5–10.7) represent the extended Kalman filter, which fulfills recursive estimation of the satellite's rotational motion parameters about its mass center.

Estimation results when this method is used are given in Figures 10.1–10.3 for simulated data. Here three different TRIAD algorithms, based on the selected reference vectors (Earth's magnetic field vector and unit vectors in the direction of the Sun and the center of the Earth), are designed and redundant data processing algorithm based on Maximum Likelihood (ML) method is used to process individual outputs by these three TRIAD algorithms. The attitude angles, determined by the ML method, are used in EKF as the measurements. The attitude angle error covariances calculated for the estimations of the ML method are regarded as the measurement noise covariance for the EKF. In graphics, estimated values of Euler angles, the error between the actual values of the angles and their estimated values and estimation variances are shown.

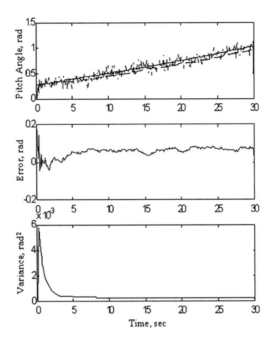

**FIGURE 10.1**    Results by using ML+EKF for pitch angle (solid line – actual value; dash and dotted line – measurement value; dashed line – EKF output).

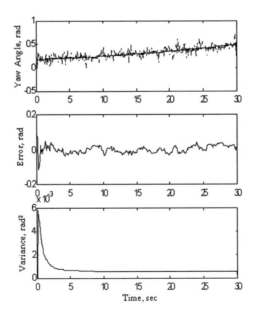

**FIGURE 10.2**    Results by using ML+EKF for yaw angle (solid line – actual value; dash and dotted line – measurement value; dashed line – EKF output).

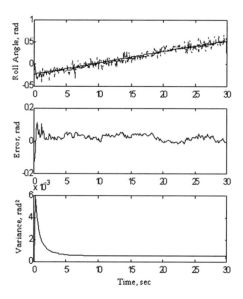

**FIGURE 10.3** Results by using ML+EKF for roll angle (solid line – actual value dash and dotted line – measurement value; dashed line – EKF output).

## 10.2 INTEGRATION WHEN ATTITUDE IS REPRESENTED USING QUATERNIONS

When the quaternions are used in the integrated architecture for attitude estimation, the overall process becomes computationally simpler specifically when QUEST or q-method is used as the single-frame algorithm. QUEST and q-method directly provides quaternion estimates and without any transformation these quaternion measurements can be used in the filter.

The quaternion measurements provided by the single-frame algorithm are treated in the same manner with star tracker measurements in the filtering algorithm. For MEKF or Unscented Attitude Filter, first we need to obtain the measurement of the component local attitude error. In terms of generalized Rodriguez parameters that we gave the formulation in Chapter 9 for the UKF,

$$\delta q_{obs} = q_{mes} \otimes \left[ \hat{q}_0 \left( k+1 | k \right) \right]^{-1}, \tag{10.8}$$

where, $q_{mes}$, the quaternion measurements from the SVD method are quaternion-multiplied with the predicted mean quaternion. Then regarding $\delta q_{obs} = \left[ \delta g_{obs}^T \quad \delta q_{4,obs} \right]^T$, measurement of the attitude error is calculated as

$$\delta p_{obs} = f \left[ \delta g_{obs} / \left( a + \delta q_{4,obs} \right) \right]. \tag{10.9}$$

Notice that if $f = 2$ and $a = 0$ we would have the measurements in terms of the rotation error vector $\delta \alpha$.

Figure 10.4 gives the general flow of the integrated algorithm for MEKF. We preferred to use QUEST as the single-frame method due to several advantages it

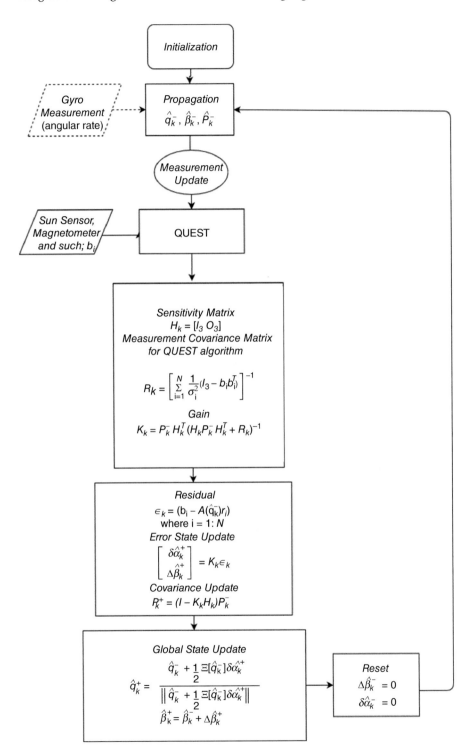

**FIGURE 10.4**  General flow of the QUEST+MEKF integrated algorithm for attitude filtering.

provides as we discussed in Chapter 6. However any other single-frame method as TRIAD, q-method of SVD can be used as well [1, 11–13]. Notice that measurement covariance matrix for the filter, $R$, is obtained via processing the $P_{QUEST}$, the covariance matrix for rotation angle errors for the QUEST algorithm.

Figures 10.5–10.6 give attitude estimation results for gyro-based MEKF and UKF algorithms when the vector measurements by Sun sensor and magnetometer are first processed in QUEST algorithm. The simulation scenario is exactly same with that we presented in Chapter 9 for a nanosatellite. The magnetometer noise is zero mean Gaussian white noise with a standard deviation of $\sigma_m = 300\text{nT}$. Moreover, the standard deviation for the Sun sensor noise is taken as $\sigma_s = 0.1°$. The angular random walk and rate random walk values for the used gyros are $\sigma_v = 2.47\,\text{arcsec}\,/\sqrt{s}$ and $\sigma_u = 6.36 \times 10^{-4}\,\text{arcsec}\,/\sqrt{s^3}$, respectively. The spacecraft dynamics used for deriving the true simulated data include the magnetic and gravity gradient disturbance torques, which are not accounted for in the filter model for dynamics-based filter. The residual magnetic moment is assumed as $M = \begin{bmatrix} -0.51 & 0.042 & 0.11 \end{bmatrix}^T \text{Am}^2$.

The results in Figures 10.5 and 10.6 show that there is almost no difference in the estimation when UKF or EKF is used. This is an expected situation since for a gyro-based filter with linear measurements, using UKF provides minimum advantage; we already have linear measurement model and the nonlinearity in the kinematics is not high.

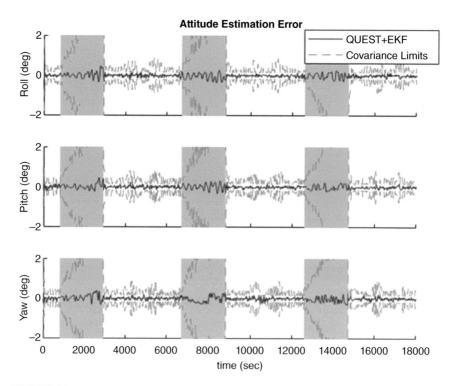

**FIGURE 10.5**   Attitude estimation results for a gyro-based filter when QUEST and MEKF are integrated.

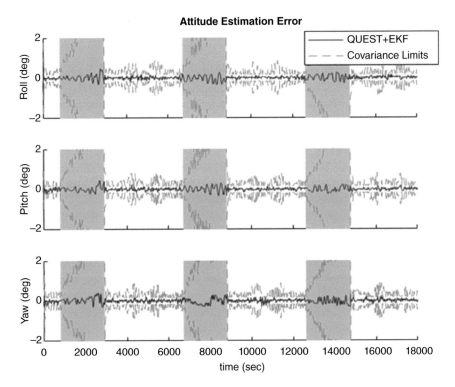

**FIGURE 10.6** Attitude estimation results for a gyro-based filter when QUEST and UKF are integrated.

On the other hand, please compare the results with Figures 9.2 and 9.4 in Chapter 9. First thing to notice will be much faster convergence of the QUEST+MEKF compared to the MEKF using the vector measurements. QUEST+UKF converges also faster than UKF with vector measurements but this is not as significant as we observe in MEKF case. Moreover, since we do not have any measurement coming from the QUEST block in eclipse, covariance bounds are naturally much larger for QUEST+filter algorithms for both MEKF and UKF. When vector measurements are used, covariance bounds in eclipse increases (since we have only magnetometer measurements) but they are still considerably lower than those of integrated algorithms.

## 10.3 CONCLUSION AND DISCUSSION

In this chapter, the single-frame and filtering algorithms are integrated to estimate the attitude of a small satellite. Single-frame method is used to pre-process the vector measurements and provide directly Euler angle or quaternion measurements to the filter block. Using this integrated architecture provides several advantages over the filtering-only methods especially for small satellite application. These are including but not limited to:

1. A linear measurement model and reduced computational load especially when the Unscented Kalman Filter is used. Depending on the exact architecture and the used single-frame method, it is possible to decrease the computational load more than 25%.
2. Switching flexibility between different sensors. When there are multiple vector measurements to be processed, we may need to modify the update phase of the filter depending on the available measurement at that time if vector-based update sequence is use. In contrast when measurements are pre-processed, we can run the algorithm without modifying the filter's structure.
3. A robust attitude estimation architecture. Integrated algorithm has much quicker convergence compared to the vector-based filtering algorithm. Moreover, since the measurement noise covariance of the filter is tuned depending on the estimation covariance of the single-frame algorithm, the overall algorithm is inherently robust against any kind of measurement uncertainties.

On the other hand, main disadvantage of integrated algorithms is, of course, being not capable of updating the states when the single-frame algorithm cannot provide measurements (e.g. when the satellite is in eclipse and there are only magnetometer measurements that can be used).

## REFERENCES

1. D. Cilden Guler, E. S. Conguroglu, and C. Hajiyev, "Single-frame attitude determination methods for nanosatellites," *Metrol. Meas. Syst.*, vol. 24, no. 2, pp. 313–324, 2017.
2. H. E. Soken and S. Sakai, "TRIAD+Filtering Approach for Complete Magnetometer Calibration," in *2019 9th International Conference on Recent Advances in Space Technologies (RAST)*, Istanbul, Turkey, 2019, pp. 703–708.
3. D. Y. Lee, H. Park, M. Romano, and J. Cutler, "Development and experimental validation of a multi-algorithmic hybrid attitude determination and control system for a small satellite," *Aerosp. Sci. Technol.*, vol. 78, pp. 494–509, 2018.
4. C. Hajiyev and M. Bahar, "Attitude determination and control system design of the ITU-UUBF LEO1 satellite," *Acta Astronaut.*, vol. 52, no. 2, pp. 493–499, 2003.
5. B. Y. Mimasu and J. C. Van Der Ha, "Attitude Determination Concept for QSAT," pp. 1–6, 2008.
6. Y. Nakajima, N. Murakami, T. Ohtani, Y. Nakamura, K. Hirako, and K. Inoue, "SDS-4 attitude control system: In-flight results of three axis attitude control for small satellites," *19th IFAC Symposium on Automatic Control in Aerospace*, Wurzburg, Germany, 2013, vol. 46, no. 19, pp. 283–288.
7. L. Cao and H. Li, "Unscented predictive variable structure filter for satellite attitude estimation with model errors when using low precision sensors," *Acta Astronaut.*, vol. 127, pp. 505–513, 2016.
8. W. Quan, L. Xu, H. Zhang, and J. Fang, "Interlaced optimal-REQUEST and unscented Kalman filtering for attitude determination," *Chinese J. Aeronaut.*, vol. 26, no. 2, pp. 449–455, Apr. 2013.
9. J. A. Christian and E. G. Lightsey, "Sequential optimal attitude recursion filter," *J. Guid. Control. Dyn.*, vol. 33, no. 6, pp. 1787–1800, Nov. 2010.

10. T. Ainscough, R. Zanetti, J. Christian, and P. D. Spanos, "Q-Method extended Kalman filter," *J. Guid. Control. Dyn.*, vol. 38, no. 4, pp. 752–760, Apr. 2015.
11. D. C. Guler and C. Hajiyev, *Gyroless Attitude Determination of Nanosatellites: Single-Frame Methods and Kalman Filtering Approach.* Saarbrucken, Germany: LAP LAMBERT Academic Publishing, 2017.
12. C. Hajiyev and D. Cilden Guler, "Review on gyroless attitude determination methods for small satellites," *Prog. Aerosp. Sci.*, vol. 90, no. November 2016, pp. 54–66, 2017.
13. H. E. Soken, D. Cilden, and C. Hajiyev, "Integration of single-frame and filtering methods for nanosatellite attitude estimation," in *Multisensor Attitude Estimation*, H. Fourati and D.E.C. Belkhiat (eds.). Boca Raton, USA: CRC Press, 2016, pp. 463–484.

# 11 Active Fault Tolerant Attitude Estimation

The attitude determination and control system of a small satellite is prone to several faults and uncertainties. Some of the reasons for such vulnerability can be listed as

- Using COTS devices, without any certain information regarding their in-orbit performance.
- Model uncertainties mainly due to disturbance torques.
- Temporal radiation and temperature effects on the sensors/actuators.
- Launching the satellite without comprehensive testing on-ground to meet the low-cost requirements and mass product demands.[1]

Thus, it is important to achieve fault tolerance in the design of attitude determination systems for both pointing accuracy and mission sustainability.

Similar with fault tolerant control systems in general [1], there are two possible methods for fault tolerance in an attitude estimator: Passive and active. In the passive category, the impaired attitude estimation system continues to operate with the same robust or adaptive estimator. The active category involves an on-line reconfiguration of the attitude estimator after the fault has occurred and has been detected. The active fault tolerant attitude estimation has two phases:

1. Fault detection and isolation (FDI)
2. Attitude estimator reconfiguration

Many fault detection methods have been developed to detect and identify faults in dynamic systems by using analytical redundancy [2–4]. In [2, 3], the algorithms for detection and diagnosis of multiple failures in the dynamic systems are described. They are based on the Interacting Multiple-Model (IMM) estimation algorithm, which is one of the most cost-effective adaptive estimation techniques for systems involving structural as well as parametric changes. The proposed algorithms provide an integrated framework for fault detection, diagnosis, and state estimation. In methods, described in these works, the faults are assumed to be known, and the Kalman filters are designed for the known types of faults. As the approach requires several parallel Kalman filters, and the faults should be known, it can be used in limited applications. Examples for IMM estimation-based FDI applications for satellite attitude determination and control can be seen in [5, 6].

In [4] an analytical redundancy-based approach for detecting and isolating sensor, actuator and component (i.e., plant) faults in complex dynamical systems, such as aircraft and spacecraft is developed. The method is based on the use of constrained

Kalman filters, which are able to detect and isolate such faults by exploiting functional relationships that exist among various subsets of available actuator input and sensor output data. A statistical change detection technique based on a modification of the standard generalized likelihood ratio (GLR) statistic is used to detect faults in real time. The GLR test, for which we see application examples for fault detection in attitude determination and control system in [7–9], requires the statistical characteristics of the system to be known before and after the fault occurs. As this information is usually not available after the fault, the method has limited applications in practice.

Fault-tolerant attitude control system architecture presented in [10] is based on the sensor reconfiguration. Part of the fault handling is dedicated to the duplicate components. Faults in non-duplicated sensors are detected using analytic redundancy methods based on different sensors. This approach deals with the hardware redundancy and it is very expensive.

In [11, 12], the neural network (NN) based methods to detect sensor and actuator failures are developed and discussed. In [11], several different neural networks were trained using either the resilient backpropagation, Levenberg-Marquardt, or Levenberg-Marquardt with Bayesian regularization training algorithms. Their results were analyzed to determine the accuracy of the networks with respect to isolating the faulty component or faulty subsystem within the attitude determination and control system. The performance of NN-based FDI methods is compared with that of other available FDI techniques. In [12] both actuator and sensor faults are considered. In the case study, it is assumed that the satellite has magnetorquers as actuator and magnetometers as the sensors. Two recurrent NNs are employed to identify the faults. The weights are updated based on a modified backpropagation scheme.

The methods based on artificial neural networks and genetic algorithms do not have physical bases. Therefore, according to the different data corresponding to the same event, the model gives different solutions. Thus, the model should continuously be trained by using the new data. This drawback of NN-based FDI method is clearly indicated in the conclusion section of [11] as "Though all of the networks were trained from the same example set, significant differences exist in the ability of the networks to positively detect and identify the faults with any consistency."

As another method, faults in multidimensional dynamic systems can be detected with the aid of innovation of the Kalman filter [13–16]. This approach does not require a priori statistical characteristics of the faults, and the computational burden is not very heavy. Generally, fault detection algorithms developed to check the statistical characteristics of innovation in real-time are based on the following fact. If a system of estimation operates normally, the normalized innovation in the Kalman filter coordinated with a dynamics model represents the white Gaussian noise with zero average value and unit covariance matrix. Change of indicated statistical characteristics of the normalized innovation is caused by a variety of problems: sensor faults, anomalous measurements, sudden shifts arising in the measurement channel, changes in the statistical characteristics of the process and/or measurements noises, computer malfunctions, deteriorated precision of instruments, increasing background noise of instruments, as well as divergence of real process trajectories and divergence of estimations generated by the Kalman filter. The task of efficiently detecting such changes has to be undertaken in real operating conditions in order to correct the estimations. It is also essential to take decisions in a timely manner to change test and operating conditions. Methods for checking the correspondence of the innovation to the white noise and revealing a change in its mathematical expectation are considered in [13, 14, 17–21]. The approaches that verify the innovation covariance are addressed in [13, 15–17, 22–25].

In this chapter, FDI in the attitude determination system of a small satellite is investigated. The attitude sensors used as a part of the system might be any of the sensors presented in Chapter 3. The satellite attitude is estimated via a nonlinear attitude filtering algorithm that was discussed in detail in Chapter 9.

## 11.1 THE INNOVATION AND ITS STATISTICAL PROPERTIES

The innovation is defined as the difference between the actual system output and the predicted output based on the predicted state. It is called as the innovation since it represents the new information brought in by the latest observation vector. Under normal conditions, the error signal is "small" and corresponds to random fluctuations in the output where all the systematic trends are eliminated by the model. However, under faulty conditions, the error signal is "large" and contains systematic trends since the model no longer represents the physical system adequately.

Let us take below linear dynamic system.
State equation:

$$x(k+1) = \phi(k+1,k)x(k) + G(k+1,k)w(k) \tag{11.1}$$

Measurement equation:

$$z(k) = H(k)x(k) + v(k), \tag{11.2}$$

where $x(k)$ is $n$ dimensional state vector of the system, $\phi(k+1,k)$ is $n \times n$ dimensional transition matrix of the system, $w(k)$ is $n$ dimensional random system noise, $G(k+1,k)$ is $n \times n$ dimensional transition matrix of system noise, $z(k)$ is $s$ dimensional measurement vector, $H(k)$ is $s \times n$ dimensional measurement matrix of the system, $v(k)$ is $s$ dimensional random measurement noise vector.

For the linear systems, when the output and the input are Gaussian, then the innovation is considered to be Gaussian. Hence only the mean and covariance of the innovation have to be specified to describe the statistical properties of the innovation completely. Kailath [26] has shown that if the filter is optimal, the error signal or the innovation is a white Gaussian noise with zero mean and known covariance (a sequence $\{x(n)\}$ is called white (or purely random) if it consists of a sequence of uncorrelated random variables).

Assume that random $w(k)$ and $v(k)$ vectors are white Gaussian noise, then their mathematical equations (mean) and covariance are as given below:

$$E\left[w(k)\right]=0; E\left[w(k)w^{T}(j)\right]=Q(k)\delta(kj);$$
$$E\left[v(k)\right]=0; E\left[v(k)v^{T}(j)\right]=R(k)\delta(kj);$$
$$E\left[w(k)v^{T}(j)\right]=0,$$

(11.3)

where $E$ is the statistical average operator, $T$ is transpose symbol and $\delta(kj)$ is Kronecker symbol.

$$\delta(kj)=\begin{cases}1; k=j\\0; k\neq j\end{cases}$$

The innovation of Kalman filter can be used to detect the faults that occur in such kind of systems. This sequence has the property presented below. If the system operates normally then innovation

$$\Delta(k)=z(k)-H(k)\hat{x}(k/k-1)$$

(11.4)

in Kalman filter which is adjusted according to the model of the system dynamics will be white Gaussian noise with zero-mean and covariance matrix of [13]:

$$P_{\Delta}(k)=H(k)P(k/k-1)H^{T}(k)+R(k).$$

(11.5)

Here,

$$P(k/k-1)=\phi(k,k-1)P(k-1/k-1)\phi^{T}(k,k-1)$$
$$+G(k,k-1)Q(k-1)G^{T}(k,k-1)$$

(11.6)

is the covariance matrix of extrapolation errors where $P(k-1/k-1)$ is the covariance matrix of estimation errors in the previous step.

The estimated value of state vector $\hat{x}(k/k)$ and the covariance matrix of the estimation error $P(k/k)$ can be found by the LKF as given:

$$\hat{x}(k/k)=\hat{x}(k/k-1)+K(k)\Delta(k)$$

(11.7)

$$K(k)=P(k/k-1)H^{T}(k)\left[H(k)P(k/k-1)H^{T}(k)+R(k)\right]^{-1}$$

(11.8)

$$P(k/k)=\left[I-K(k)H(k)\right]P(k/k-1)],$$

(11.9)

where $K(k)$ is the Kalman filter gain matrix and $I$ is the unit matrix.

It is more appropriate to use normalized innovation to detect the faults [1]:

$$\tilde{\Delta}(k) = \left[ H(k) P(k/k-1) H^T(k) + R(k) \right]^{-1/2} \Delta(k). \qquad (11.10)$$

Because in this case,

$$E\left[ \tilde{\Delta}(k) \tilde{\Delta}^T(j) \right] = P_{\tilde{\Delta}} = I\delta(kj).$$

The faults of measurement sensors (abnormal measurements, sudden shifts in the measurement channel and other difficulties such as a decrease in the instrument accuracy or an increase in the background noise) affect the characteristics of the normalized innovation by changing its white noise nature, displacing its zero mean and varying the unit covariance matrix. Thus, the problem is to detect any change in these parameters from their nominal value as quickly as possible.

The innovation covariance has been encountered in the derivation of the Kalman filter measurement update as well and thus it can be seen that the innovation is an integral part of the Kalman filter process.

Now the statistical properties of the innovation have been described and it is indicated how the innovations can be used for the performance analysis of Kalman filter. Since the properties of the innovation sequence are strictly defined when the filter is optimal, the innovation sequence resulting from an actually implemented filter can be monitored and compared to the faultless model description. Deviations from the theoretical characteristics may be caused by mismodelling of the dynamic and/or measurement model, failure of sensors, and outliers in data. Any kind of mismodelling will make the innovation depart from its theoretically defined nominal values.

The general performance of the filter can be monitored by analyzing the zero mean, Gaussianness, given covariance and whiteness of the innovation sequence. Furthermore, the innovation is the sole source of outlier detection. Outlier detection is also related to the wider field of the fault detection in dynamic systems. Finally, the innovation sequence offers a possible approach to adaptive filtering.

## 11.2 INNOVATION APPROACH-BASED SENSOR FDI

### 11.2.1 FAULT DETECTION VIA MATHEMATICAL EXPECTATION STATISTICS OF SPECTRAL NORM OF NORMALIZED INNOVATION MATRIX

In Eq. (10.5) of EKF for attitude estimation of linear measurements (10.5–10.7)

$$\Delta_i = \left[ z_i - Hf\left( \hat{x}_{i-1}, \bar{\omega}_{orb_{i-1}} \right) \right] \qquad (11.11)$$

is the innovation of EKF. If there is no fault in the estimation system, the normalized innovation

$$\tilde{\Delta}_i = \left[ HP_i^- H^T + R \right]^{-1/2} \Delta_i, \qquad (11.12)$$

of the EKF fits with the dynamic model representing the white Gaussian noise with zero average value and unitary covariance matrix [13];

$$E\left[\tilde{\Delta}_i\right] = 0; E\left[\tilde{\Delta}_k\tilde{\Delta}_j^T\right] = P_{\tilde{\Delta}_i} = I\delta_{kj}.$$

Faults causing abrupt changes in the characteristics of the measurement channel, malfunctions in the computer, as well as divergence of real process trajectories and estimations generated by the Kalman filter (which we shall hereafter name as faults of the estimation system) can cause changes in the demonstrated characteristics of the sequence $\tilde{\Delta}_i$ and can make it different from the white noise, displace the zero average value and change the unitary covariance matrix.

It is important to develop an efficient method for simultaneous checking of the mathematical expectation and the variance of the normalized innovation sequence (11.12), which does not require a priori information on variation values in the fault case and makes it possible to find the faults of the estimation system in real-time.

Let us introduce two hypotheses:

$\gamma_0$: the estimation system operates properly,

$\gamma_1$: there is a fault in the estimation system.

To find a fault, we build a matrix with the columns of innovation vectors of EKF and introduce the following definitions [27].

**Definition 11.1**    The innovation matrix of EKF is rectangular $n \times m$ matrix ($n$ dimension of innovation vector; $n \geq 2$; $m \geq 2$), with columns formed of innovation vectors $\Delta_i$ corresponding to $m$ different moments of time.

**Definition 11.2**    The innovation matrix, made up from normalized innovation vectors $\tilde{\Delta}_i$, is named as the normalized innovation matrix of EKF.

Hereafter for simplicity, we shall use the normalized innovation matrices $A$ consisting of a finite number of innovation vectors. If the check is realized in real time, it is reasonable to form the matrix $A_i$ at the $i$ th instant ($i \geq m$) of time from finite number $m$ ($m \geq 2$) of sequential innovation vectors $\underbrace{\cdots\tilde{\Delta}_{i-2},\tilde{\Delta}_{i-1},\tilde{\Delta}_i}_{m}$. In order to check the hypotheses $\gamma_0$ and $\gamma_1$, we use the spectral norm of the matrix $A_i$ built in this way. As it is known [28], the spectral norm $\|.\|_2$ of the real matrix $A_i$ is defined by the formula $\|A_i\|_2 \equiv \max\{(\lambda_i[A^T A_i])^{1/2}\}$, where $\lambda_i[A^T A_i]$ are eigenvalues of the matrix $A^T A_i$. Square roots from eigenvalues of the matrix $A^T A_i$ i.e. values $(\lambda_i[A^T A_i])^{1/2}$ are named as singular values of the matrix $A_i$. Hence the spectral norm of the matrix $A_i$ is equal to its maximum singular value. The singular values are real and non-negative [28]. Because of that determination of singular values and consequently, spectral norm, represents a simpler problem in computing than determination of eigenvalues for arbitrary matrix. It explains the choice of the controlled scalar measure for the spectral norm of normalized innovation matrix of Kalman filter. In order to check the hypotheses $\gamma_0$ and $\gamma_1$, one-dimensional statistic for mathematical expectation of the spectral norm of the matrix $A_i$ for large values of $k$ is introduced:

$$E\left\{\|A_i\|_2\right\} \approx \overline{\|A_i\|_2} = \frac{1}{k}\sum_{j=1}^{k}\|A_j\|_2. \qquad (11.13)$$

As it is clear from (11.13), the mathematical expectation of the spectral norm of the matrix $A_i$ is substituted by its average arithmetical estimate. For determining the upper and lower limits $E\{\|A_i\|_2\}$ we use results obtained in [29], where a number of bounds have been found for the mathematical expectation of spectral norm of random matrix $A_i \in R^{n \times m}$, constituted of random Gaussian values, having zero mathematical expectation and $\sigma$ standard deviation. Let us consider some of them. Assume, $r_k^T$ and $a_j$ are rows and columns of the matrix $A$. Introduce maximum row-column norm

$$\mu \equiv \max\left[\|r_k\|_2, \|a_j\|_2\right], \qquad (11.14)$$

where $\|r_k\|_2$ and $\|a_j\|_2$ are corresponding Euclid vector norms. The following bounds for $E\{\|A_i\|_2\}$ have been obtained in [29] by means of norm $\mu$ introduced:

$$E\{\mu\} \le E\left\{\|A_i\|_2\right\} \le \left[\max(n,m)\right]^{1/2} E\{\mu\}. \qquad (11.15)$$

Using the formula (11.15) in practical calculations represents a complex problem, because of the difficulty of estimation of $E\{\mu\}$. So, the value $E\{\mu\}$ is replaced by its lower bound

$$\sigma\sqrt{\max(n,m)} = \max\left[E\left\{\|r_i\|_2\right\}, E\left\{\|a_j\|_2\right\}\right] \le E\{\mu\}. \qquad (11.16)$$

Then the Eq. (11.15) can be written as follows:

$$\sigma\sqrt{\max(n,m)} \le E\left\{\|A_i\|_2\right\} \le f\left(\max(n,m)\right)\sigma\sqrt{\max(n,m)}, \qquad (11.17)$$

where $f$ is an unknown function to be determined. It is shown in [29] by means of computer simulation, the value $\sigma\sqrt{\max(n,m)}$ is good lower bound for $E\{\|A_i\|_2\}$. It is also shown by numeric calculations that function $f$ asymptotically approaches value 2 as $n = m \to \infty$, and $f$ is always between values 1 and 2. So the value 2 is suggested to be used for estimating function $f$. Taking the above mentioned fact into consideration the following simple bounds are obtainable for $E\{\|A_i\|_2\}$:

$$\sigma\sqrt{\max(n,m)} \le E\left\{\|A\|_2\right\} \le 2\sigma\sqrt{\max(n,m)}. \qquad (11.18)$$

The expression (11.18) characterizes the connection between the standard deviation $\sigma$ of elements of the random matrix $A$ and its spectral norm.

The normalized innovation matrix $A_i$, used for finding the troubles in the estimation system consists of the Gaussian random elements with zero mathematical expectation and finite variance $a_{kj} \in N(0,1)$. The inequality (11.18) can be applied for solving the diagnostic problem formulated here. Thus it is possible to say, if elements $a_{kj}$ of the controlled normalized innovation matrix of EKF are subordinated to distribution $N(0,1)$, the inequality (11.18) is fulfilled. Nonfulfillment of the inequality (11.18) indicates a shifting zero average value of elements $a_{kj}$, changing the unitary variance or that $\{a_{kj}\}$ is other than white noise.

The algorithm offered for real system operation conditions is reduced to the following sequence of calculations to be executed at every step of filter update.

1. The EKF evaluating system state vector and vector value of the normalized innovation sequence on given step $i$ are calculated by means of expressions (11.11–11.12).
2. The normalized innovation matrix of the EKF is formed for given $n \geq 2$ and $m \geq 2$. The eigenvalues of the matrix $A^T_i A_i$ as roots of equation

$$\det\left[ A^T_i A_i - \lambda I \right] = 0 \tag{11.19}$$

and the spectral norm

$$\left\| A_i \right\|_2 \equiv \max\left\{ \left( \lambda_l \left[ A^T_i A_i \right] \right)^{1/2} \right\} \tag{11.20}$$

are determined.

3. The statistic of mathematical expectation of spectral norm of the matrix $A_i$ is calculated by means of (11.13).
4. The fulfillment of inequality (11.18) is checked and the solution is made according to the faulty operation of system.
5. The sequence of calculations is repeated for the following moment of time $i+1$.

It is necessary to note that the offered algorithm does not permit the realization of checking the non-diagonal elements of the covariance matrix of the normalized innovation sequence but permits checks only on its mathematical expectation and variance. In spite of this fact, the given approach (due to its simplicity and ease of application) can bring good results when deciding the problems of check and diagnostics under conditions of relatively limited computer memory.

## 11.2.2 INNOVATION-BASED SENSOR FAULT ISOLATION

If the sensor fault is detected, then it is necessary to determine what sensor is faulty. For this purpose, the s-dimensional sequence $\tilde{\Delta}$ is transformed into n one-dimensional sequences to isolate the faulty sensor, and for each one-dimensional sequence $\tilde{\Delta}_i \left( i = 1,2,\ldots,n \right)$ corresponding monitoring algorithm is run. The statistic of the faulty sensor is assumed to be affected much more than those of the other sensors. Let the statistics is denoted as $\xi_i(k)$. When $\max\{\xi_i(k)/i = 1,2,\ldots,n\} = \xi_p(k)$ for $i \neq j$, and $\xi_i(k) \neq \xi_j(k)$, it is judged that $p$-th control channel has failed.

Let the statistics, which is a rate of sample and theoretical variances; $\hat{\sigma}_i^2 \Big/ \sigma_i^2$ be used to verify the variances of one-dimensional innovation sequences $\tilde{\Delta}_i(k), i = 1, 2, \ldots, n$. When

$\tilde{\Delta}_i \sim N(0, \sigma_i)$ it is known that

$$\frac{v_i}{\sigma_i^2} \sim \chi_{M-1}^2, \forall i, i = 1, 2, \ldots, n, \tag{11.21}$$

where

$$v_i = (M-1)\hat{\sigma}_i^2. \tag{11.22}$$

As $\sigma_i^2 = 1$ for normalized innovation sequence, it follows that,

$$v_i \sim \chi_{M-1}^2, \forall i, i = 1, 2, \ldots, n. \tag{11.23}$$

By selecting $\alpha$ level of significance as,

$$P\left\{\chi^2 > \chi_{\alpha,M-1}^2\right\} = \alpha; \quad 0 < \alpha < 1$$

So from the equation above, the threshold value $\chi_{\alpha,M-1}^2$ will be determined.

When a fault affecting the variance of the innovation sequence occurs in the system, the statistics $v_i$ exceeds the threshold value $\chi_{\alpha,M-1}^2$ depending on the confidence probability $(1-\alpha)$, and degree of freedom $(M-1)$. Using (11.23) it can be proved that any change in the mean of the normalized innovation sequence can be detected. Let a change in the mean of the innovation sequence occur at the time $\tau$, and let $\tilde{\Delta}^*(k)$ denote the unchanged normalized innovation sequence, then the changed normalized innovation sequence is given by,

$$\tilde{\Delta}(k) = \tilde{\Delta}^*(k) \qquad k = 1, 2, \ldots, \tau - 1. \tag{11.24}$$

$$\tilde{\Delta}(k) = \tilde{\Delta}^*(k) + \mu(k - \tau) \qquad k = \tau, \tau + 1, \ldots \tag{11.25}$$

where $\mu(.)$ is an unknown change and may vary with respect to time, but there exists a quantity $L > 0$ such that $|\mu(j)| < L$, for $\forall j$. (11.24) and (11.25) yield,

$$\tilde{\Delta}(k) \sim N(0,1) \qquad k = 1, 2, \ldots, \tau - 1 \tag{11.26}$$

$$\tilde{\Delta}(k) \sim N(\mu(k-\tau),1) \qquad k = \tau, \tau + 1, \ldots \tag{11.27}$$

Let the number of shifted values from $j = k - M + 1$ to k in a window be denoted by $N$. When $k<\tau$ it can be easily shown that the mathematical expectation of investigated statistic (11.22) is $E[v_i] = M - 1$. When a fault occurs, the mathematical expectation of (11.22) can be determined by the following theorem.

**Theorem 11.1**

When $k \geq \tau$, i.e. the hypothesis $\gamma_1$ is true, the following equation is also true,

$$E\big[v(k)\big] = (M-1)\sigma^2 + E\left\{\sum_{j=k-M+1}^{k}\left[\mu(j-\tau) - \frac{\displaystyle\sum_{j=k-M+1}^{k}\mu(j-\tau)}{M}\right]^2\right\} \tag{11.28}$$

where

$$\mu(j-\tau) = \begin{cases} 0 & j < \tau \\ \mu^* = \text{constant} & j \geq \tau \end{cases}$$

The proof is given in [30].

Let the number of shifted innovation values from $j = k - M + 1$ to $k$ in a window be denoted by $N$. Two distinct cases may be considered;

(a) $N = M$, in this case,

$$E\left\{\sum_{j=k-M+1}^{k}\left[\mu(j-\tau) - \frac{\displaystyle\sum_{j=k-M+1}^{k}\mu(j-\tau)}{M}\right]^2\right\} = 0 \tag{11.29}$$

and so, $E[v(k)] = (M - 1)\sigma^2$. When the values $\tilde{\Delta}(j)$ have shifted by the same amount $\mu(j - \tau)$ in a window, it is impossible to detect the change by using (11.22).

(b) $N < M$, in this case

$$\left[\mu(j-\tau) - \frac{\displaystyle\sum_{j=k-M+1}^{k}\mu(j-\tau)}{M}\right]^2 = \left[\mu(j-\tau) - \frac{N\mu^*}{M}\right]^2 \geq 0 \tag{11.30}$$

and a shift in the innovation sequence will cause an asymptotic increase in the expected value of the statistic $v(k)$, and $v(k)$ will exceed the threshold $\chi^2_{\alpha, M-1}$. The larger $\mu^*$ the faster detection is. ∎

The sample variances $\hat{\sigma}_i$ are the diagonal components of the sample covariance matrix $S(k)$. Therefore there is no need to make heavy additional computation in the existent algorithm, but only the diagonal components of the matrix $S(k)$ are multiplied by $(M - 1)$, and compared with $\chi^2_{\alpha, M-1}$ and with one another at each iteration. The decision making for isolation is done as follows; if the hypothesis $\gamma_1$ is true and $S_{ii}(k) \neq S_{jj}(k)$, $i \neq j$ and $\max\{S_{ii}(k)/i = 1, 2, \ldots, n\} = S_{pp}(k)$ where $S_{ii}(k)$ is the $ii$th component of $S(k)$, then it is judged that there is a fault in the $p^{th}$ channel.

## 11.2.3 SIMULATION RESULTS OF FDI ALGORITHMS

To test the proposed algorithm, it is applied to the mathematical model of the small satellite's rotational motion about its center of mass [31]. It is demonstrated that the faults in a measurement channel can be detected by checking the mathematical expectation and the variance of the EKF innovation sequence. Under computer simulation of the above specified problem, as the estimation of system state vector is calculated, the values of normalized innovation sequence were determined by means of the expression (11.12). The spectral norm of matrix $A_i$ for the case $n = 6$, $m = 6$ was determined by means of expression (11.20); the mathematical expectation of spectral norm $\|A_i\|_2$ was determined by means of (11.13). Decisions on finding a system fault were made on the basis of inequality (11.18), written for the case $n = 6$, $m = 6$. If the case is $\sigma = 1, n = 6$ and $m = 6$, the inequality ((11.18)) can be written in a simpler form [31]

$$\sqrt{6} \leq E\left\{\|A\|_2\right\} \leq 2\sqrt{6} \tag{11.31}$$

The results of calculations are shown in Figures 11.1–11.3.

One can see in Figure 11.1 that the values of statistic $E\{\|A_i\|_2\}$ fall within the permissible domain (between lower and upper thresholds ) when no sensor fault occurs. The graphs of the values of statistic $E\{\|A\|_2\}$ are shown in Figure 11.2 when

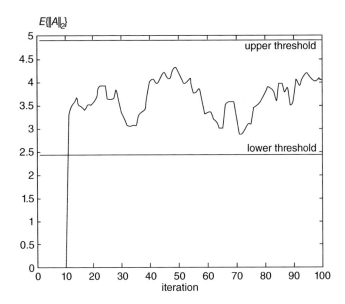

**FIGURE 11.1**   The behavior of the statistic $E\{\|A\|_2\}$ for a normal operating system.

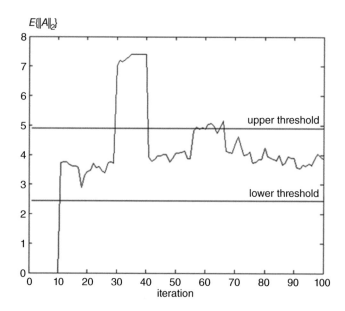

**FIGURE 11.2**    The behavior of the statistic $E\{\|A\|_2\}$ in case of shift in the pitch rate gyroscope (the moment of the shift appears at $k = 30$, the moment reveals the shift at $k = 31$).

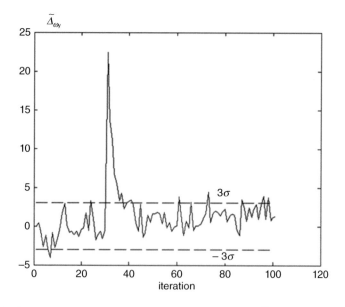

**FIGURE 11.3**    Behavior of the normalized innovation sequence $\tilde{\Delta}_{\omega_y}(k)$ in case of shift in the pitch rate gyroscope.

a shift occurs in the pitch rate gyroscope at the step 30. The behavior of the appropriate normalized innovation sequences $\tilde{\Delta}_{\omega_y}(k)$ is presented in the Figure 11.3.

This fault causes a change in the mean of the innovation sequence. As seen in Figure 11.2, when there is no sensor fault the values of statistic $E\{\|A\|_2\}$ fall within the permissible domain, and when a fault occurs in the pitch rate gyroscope $E\{\|A\|_2\}$ grows rapidly and after 1 step it exceeds the upper threshold. Hence $\gamma_1$ hypothesis is judged to be true. The Figure 11.4 shows detection of faults changing the noise variance of the pitch rate gyroscope.

In this case, the mean value of the innovation sequence does not change, but the variance changes. The graphs of the values of statistic $E\{\|A\|_2\}$ are shown in Figure 11.4 when a fault occurs in the pitch rate gyroscope at the step 30. This fault causes a change in the variance of the innovation sequence. As seen in Figure 11.4, when there is no sensor fault $E\{\|A\|_2\}$ fall between lower threshold and upper threshold lines, and when a fault occurs in the pitch rate gyroscope $E\{\|A\|_2\}$ grows rapidly and after four steeps it exceeds the threshold. Hence $\gamma_1$ hypothesis is judged to be true. The behavior of the appropriate normalized innovation sequences $\tilde{\Delta}_{\omega_y}(k)$ is presented in the Figure 11.5.

The results of computer simulation have confirmed the practical possibility of simultaneous real-time check of mathematical expectation and variance of normalized innovation sequence with the aid of the statistic introduced (11.22).

Sensor failure isolation results in case of shift in the pitch rate gyroscope are given in Figure 11.6a,b and in case of changes in noise variance of the pitch rate gyroscope in Figure 11.7a,b. As it is shown from presented figures, only the $(5,5)$ element of the

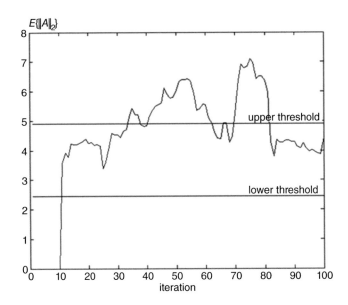

**FIGURE 11.4**   The behavior of the statistic $E\{\|A\|_2\}$ in case of changes in noise variance of the pitch rate gyroscope (the moment of variance changes at $k = 30$, the moment of revealing variance changes at $k = 34$).

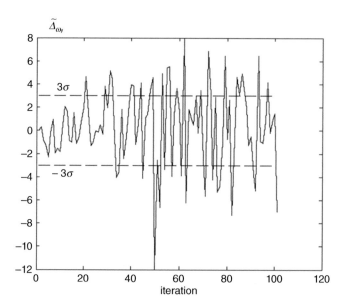

**FIGURE 11.5** Behavior of the normalized innovation sequence $\tilde{\Delta}_{\omega_y}(k)$ in case of changes in noise variance of the pitch rate gyroscope.

covariance matrix $S - S(5,5)$ exceeds the threshold $\chi^2_{\alpha,M-1}$ (for $M = 15$ and $\alpha = 0.1$ the threshold value $\chi^2_{\alpha,M-1} = 21.1$) which indicates a failure in the pitch rate gyro. $S(j,j), j \neq 5$ elements do not exceed the thresholds.

## 11.3 KALMAN FILTER RECONFIGURATION

When a sensor fault occurs in the attitude determination system, the reconfigurable EKF approach can be used for fault accommodation. In this case the faulty measurement is ignored that means it is necessary to reconfigure the filter algorithm (the update part) because of changed number for the incoming measurements.

Assume that the $i$-th measurement channel is faulty. In this case $i$-th measurement channel is ignored and the filter algorithm is reconfigured as below. The reconfigured measurement vector $z^{rec}(k)$ can be written in the following form:

$$z^{rec}(k) = H^{rec}(k)x(k) + v^{rec}(k),$$  (11.32)

where

$$z^{rec}(k) = \left[z_1(k), z_2(k), \ldots, z_{i-1}(k), z_{i+1}(k), \ldots, z_{s-2}(k), z_{s-1}(k)\right]^T$$  (11.33)

is the $(s - 1)$-dimensional reconfigured measurement vector, $v^{rec}(k)$ is the $(s - 1)$-dimensional reconfigured measurement noise vector and $H^{rec}(k)$ is the $s - 1$ by

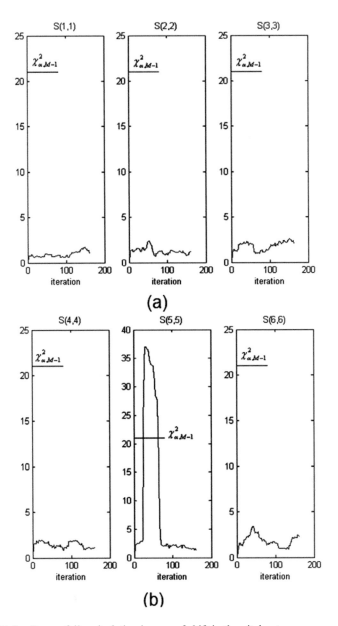

**FIGURE 11.6**  Sensor failure isolation in case of shift in the pitch rate gyroscope.

$n$ reconfigured system measurement matrix. The elements of the reconfigured measurement matrix $H_{jj}^{rec}(k)$, $j = 1, s-1$ can be determined as [32],

$$H_{jj}^{rec}(k) = 1, \quad \forall j < i, \tag{11.34}$$

$$H_{jj+1}^{rec}(k) = 1, \quad \forall j \geq i. \tag{11.35}$$

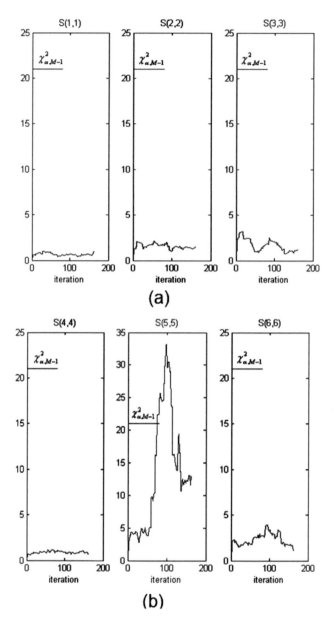

**FIGURE 11.7** Sensor failure isolation in case of changes in noise variance in the pitch rate gyroscope

The rest of the elements in the matrix $H^{rec}(k)$ are equal to zero.

The reconfigured covariance matrix of the measurement noise is

$$E\left[v^{rec}(k)\left(v^{rec}(j)\right)^{T}\right] = R^{rec}(k)\delta(kj) \qquad (11.36)$$

In the reconfigured KF instead of the parameters $z(k)$, $H(k)$, $R(k)$, the reconfigured parameters $z^{rec}(k)$, $H^{rec}(k)$, $R^{rec}(k)$ should be used. In this case, the overall computational load will increase but the estimation accuracy in comparison with the sensor fault-free EKF will decrease.

### 11.3.1 DEMONSTRATION FOR EKF RECONFIGURATION

The proposed reconfiguration scheme is demonstrated here for an EKF that is used for attitude estimation of a small satellite. The state vector of the filter is composed of the attitude parameters, gyro biases and angular velocities.

The appropriate simulation results are illustrated in Figures 11.8–11.10. Figures give EKF estimation errors for the angular velocity compared to the actual values. In Figures, blue lines indicate the estimation error and red dashed lines show the covariance bounds. Gray parts are for eclipse.

Figure 11.8 shows the angular velocity estimation error for the filter when there is no gyro error.

The angular velocity estimation results in the presence of pitch gyro bias fault are given in Figure 11.9. Starting from 400th s. gyro bias continuously changes with abrupt changes. As seen, the EKF estimation results are affected considerably by the introduced pitch rate gyro fault and diverge from the actual values.

Reconfigured EKF estimation results in the presence of pitch rate gyro fault are given in Figure 11.10. As seen, the reconfigured EKF gives the accurate estimation results without being affected from the gyro bias fault. For further comparison of the reconfigured EKF with the regular EKF in case of gyro fault, Figures 11.11 and 11.12

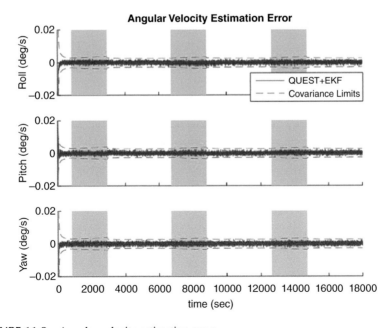

**FIGURE 11.8**   Angular velocity estimation error.

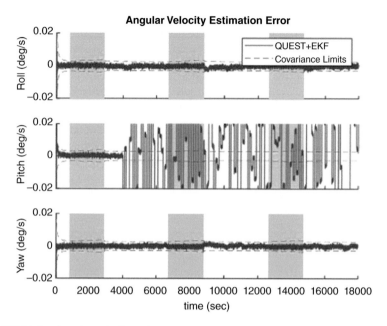

**FIGURE 11.9**    Angular velocity estimation errors in case of pitch gyro bias fault.

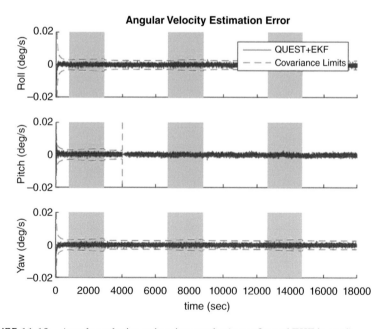

**FIGURE 11.10**    Angular velocity estimation results (reconfigured EKF is used).

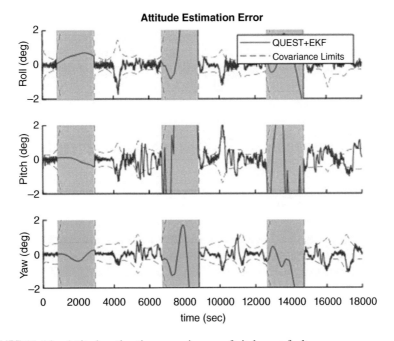

**FIGURE 11.11**    Attitude estimation errors in case of pitch gyro fault.

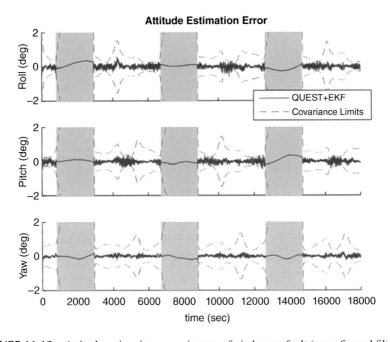

**FIGURE 11.12**    Attitude estimation errors in case of pitch gyro fault (reconfigured filter).

give attitude estimation errors. In Figure 11.11, we clearly see that the gyro fault is affecting the attitude estimation accuracy not in just the pitch angle but in all three angles. Deterioration is clearly visible especially in eclipse when there is no measurement from QUEST block and the algorithm relies mostly on the gyro measurements. Faulty pitch gyro measurements affect the propagation in all three axes. In contrast Figure 11.12 shows, the filter can keep the original accuracy limits (which we have when gyros operate normally) when it is reconfigured at 4000th s.

## 11.4 THE STRUCTURE OF THE FAULT TOLERANT ATTITUDE ESTIMATION SYSTEM

In Figure 11.13, the structural diagram of an active fault tolerant attitude estimation system is given. Kalman filter is used to detect the sensor faults, and then an innovation-based fault isolation scheme is occupied for isolation of the detected fault. If the sensor fault occurs, as seen in Figures 11.6 and 11.7, then the effect of the fault to its channel is more significant. Thus, the faulty sensor is isolated, and a new Kalman filter that ignores the feedback from the faulty sensor is designed [32]. That is carried out in the filter reconfiguration block.

As a consequence, sensor failures require modification of the estimator and the only necessity in this case is that the estimator supplies accurate estimates of the system states even after the sensor failure.

The proposed fault tolerant estimation structure can be used to isolate the faults that affect the mean and variance of the innovation process. For this purpose, the block of fault detection and isolation contains the fault detection algorithm, which is described in Section 11.2.1, based on the mathematical expectation statistics of the

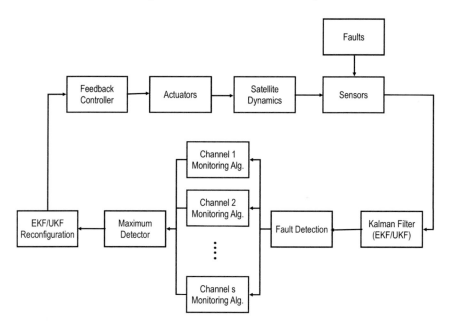

**FIGURE 11.13**    The structural diagram of an active fault tolerant attitude estimation system.

spectral norm of the normalized innovation matrix that enables simultaneous checking of the mean and variance of the innovation in real time.

The proposed innovation approach has some advantages over other structures.
Some of those are:

> This approach does not require a priori statistical characteristic of the faults and
> the computational burden is not very heavy (only one Kalman filter is used);

> The faults that change the sensor noise mean or variance are not well detected
> by the most of the present fault detection systems (For example see, Multiple
> Model Estimation Approach). Yet, the innovation-based approach is able to
> detect those faults rapidly and accurately;

> The proposed approach does not require hardware redundancy and thus the
> cost is less.

> On the other hand, a drawback of the proposed fault tolerant attitude estimation
> system structure is as follows:

If the Kalman filter diverges, the entire fault tolerant attitude estimation system
will fail catastrophically. The divergence of the filter might be caused by the following two reasons:

1. The discrepancies between the mathematical model used to derive the filter
   equations and the actual conditions under which the filter must operate.
2. One manifestation of machine-caused errors occurs in the computation of the
   state covariance matrix. After operating Kalman filter for some time, this
   matrix ceases to be positive definite and symmetric.

Then the filter weighting coefficients computed using this matrix become wrong and
consequently, motion estimates start to be incorrect.

Therefore, when the innovation approach is used, the methods for regulation and
numerical stabilization of Kalman filters, which are described in Section 7.5, have to be
applied rigorously. Another disadvantage of the approach is that it is used when only
one fault at a time occurs in the system. When more than one fault takes place (for
instance, simultaneous faults of more than one gyro), the system may not be observable.
However, by adding hardware redundancy, this scheme can be developed for two or
more faults. Furthermore, the fault detection methods, based on an innovations approach,
are not suitable to estimate the magnitude of the model parametric error due to the fault.

In spite of those drawbacks, these methods are capable of detecting the faults that
change the statistical characteristics of the innovation and hence can be used in fault
tolerant attitude estimation systems for small satellites.

## 11.5 CONCLUSION AND DISCUSSION

An active fault tolerant attitude estimation method is proposed in this Chapter. The
active category involves either an online redesign of the attitude estimator after sensor failure has occurred and has been detected. The active fault tolerant attitude

estimation system consists of two basic subsystems: sensor fault detection and isolation; attitude estimator reconfiguration.

Fault detection and isolation algorithms for small satellite attitude determination system using an approach for checking the statistical characteristics of innovation of EKF are proposed. The fault detection algorithm is based on statistic for the mathematical expectation of the spectral norm of the normalized innovation matrix of the EKF. This approach permits simultaneous real-time checking of the mathematical expectation and the variance of the innovation and does not require a priori information about the faults and statistical characteristics of the system in fault cases.

Failures in the sensors affect the characteristics of the innovation of the EKF. The failures that affect the mean and variance of the innovation have been considered. Application of the proposed fault detection algorithm to the small satellite attitude determination system has shown that sensor fault detection by the presented algorithm is possible in real time.

Assuming that the effect of the faulty sensor on its channel is more significant than the other channels, a sensor isolation method is presented by transforming $n$ dimensional innovation to one dimensional $n$ innovations. The simulations, carried out on a nonlinear dynamic model of the rotational motion of small satellite, confirm the theoretical results.

Consequently, a new structure of a fault tolerant attitude estimation system for small satellites based on innovation approach is proposed.

## NOTE

1 As in case for recent constellation missions for satellite Internet access.

## REFERENCES

1. C. Hajiyev and F. Caliskan, *Fault Diagnosis and Reconfiguration in Flight Control Systems*, vol. 2. Boston, MA: Springer US, 2003.
2. Youmin Zhang and X. Rong Li, "Detection and diagnosis of sensor and actuator failures using interacting multiple-model estimator," in *Proceedings of the 36th IEEE Conference on Decision and Control*, San Diego, USA, 1997, vol. 5, pp. 4475–4480.
3. C. Rago, R. Prasanth, R. K. Mehra, and R. Fortenbaugh, "Failure detection and identification and fault tolerant control using the IMM-KF with applications to the Eagle-Eye UAV," in *Proceedings of the 37th IEEE Conference on Decision and Control (Cat. No.98CH36171)*, Tampa, USA, 1998, vol. 4, pp. 4208–4213.
4. E. C. Larson, B. E. Parker, and B. R. Clark, "Model-based sensor and actuator fault detection and isolation," in *Proceedings of the 2002 American Control Conference (IEEE Cat. No.CH37301)*, Anchorage, USA, 2002, vol. 5, pp. 4215–4219.
5. K. K. N. Tudoroiu, "Fault detection and diagnosis for satellite's attitude control system (ACS) using an interactive multiple model (IMM) approach," pp. 1287–1292, 2005.
6. J. Lee and C. G. Park, "Cascade filter structure for sensor/actuator fault detection and isolation of satellite attitude control system," *Int. J. Control. Autom. Syst.*, vol. 10, no. 3, pp. 506–516, 2012.

7. F. N. Pirmoradi, F. Sassani, and C. W. de Silva, "Fault detection and diagnosis in a spacecraft attitude determination system," *Acta Astronaut.*, vol. 65, no. 5–6, pp. 710–729, 2009.

8. E. Lopez-Encarnacion, R. Fonod, and P. Bergner, "Model-based FDI for agile spacecraft with multiple actuators working simultaneously," *IFAC-PapersOnLine*, vol. 52, no. 12, pp. 436–441, 2019.

9. R. Fonod, D. Henry, C. Charbonnel, and E. Bornschlegl, "Position and attitude model-based thruster fault diagnosis: A comparison study," *J. Guid. Control. Dyn.*, vol. 38, no. 6, pp. 1012–1026, Jun. 2015.

10. T. Bak, R. Wisniewski, and M. Blanke, "Autonomous attitude determination and control system for the Orsted satellite," in *1996 IEEE Aerospace Applications Conference. Proceedings*, Aspen, USA, 1996, vol. 2, pp. 173–186.

11. J. N. Schreiner, "A neural network approach to fault detection in spacecraft attitude determination and control systems," ProQuest Diss. Theses, p. 141, 2015.

12. H. A. Talebi, K. Khorasani, and S. Tafazoli, "A recurrent neural-network-based sensor and actuator fault detection and isolation for nonlinear systems with application to the satellite's attitude control subsystem," *IEEE Trans. Neural Networks*, vol. 20, no. 1, pp. 45–60, 2009.

13. R. K. Mehra and J. Peschon, "An innovations approach to fault detection and diagnosis in dynamic systems," *Automatica*, vol. 7, no. 5, pp. 637–640, Sep. 1971.

14. A. S. Willsky, "A survey of design methods for failure detection in dynamic systems," *Automatica*, vol. 12, no. 6, pp. 601–611, Nov. 1976.

15. C. M. Gadzhiev, "Dynamic systems diagnosis based on Kalman filter updating sequences," *Autom. Remote Control*, vol. 1, pp. 147–150, 1992.

16. C. M. Gadzhiev, "Check of the generalized variance of the Kalman filter updating sequence in dynamic diagnosis," *Autom. Remote Control*, vol. 55, no. 8, pp. 1165–1169, 1994.

17. C. Hajiyev and F. Caliskan, "Sensor/actuator fault diagnosis based on statistical analysis of innovation sequence and Robust Kalman Filtering," *Aerosp. Sci. Technol.*, vol. 4, no. 6, pp. 415–422, Sep. 2000.

18. C. Hajiyev and F. Caliskan, "Sensor and control surface/actuator failure detection and isolation applied to F-16 flight dynamic," *Aircr. Eng. Aerosp. Technol.*, vol. 77, no. 2, pp. 152–160, Apr. 2005.

19. P. S. Pratama, A. V. Gulakari, Y. D. Setiawan, D. H. Kim, H. K. Kim, and S. B. Kim, "Trajectory tracking and fault detection algorithm for automatic guided vehicle based on multiple positioning modules," *Int. J. Control. Autom. Syst.*, vol. 14, no. 2, pp. 400–410, Apr. 2016.

20. S. Cho and J. Jiang, "Detection of sensor abnormalities in a pressurizer by means of analytical redundancy," *IEEE Trans. Nucl. Sci.*, vol. 63, no. 6, pp. 2925–2933, Dec. 2016.

21. A. Adnane, Z. Ahmed Foitih, M. A. Si Mohammed, and A. Bellar, "Real-time sensor fault detection and isolation for LEO satellite attitude estimation through magnetometer data," *Adv. Sp. Res.*, vol. 61, no. 4, pp. 1143–1157, 2018.

22. F. Caliskan and C. M. Hajiyev, "Innovation sequence application to aircraft sensor fault detection: comparison of checking covariance matrix algorithms," *ISA Trans.*, vol. 39, no. 1, pp. 47–56, Feb. 2000.

23. C. Hajiyev, "Testing the covariance matrix of the innovation sequence with sensor/actuator fault detection applications," *Int. J. Adapt. Control Signal Process.*, vol. 24, no. 9, pp. 717–730, Sep. 2010.

24. C. Hajiyev, "Generalized Rayleigh quotient based innovation covariance testing applied to sensor/actuator fault detection," *Measurement*, vol. 47, pp. 804–812, Jan. 2014.

25. C. Hajiyev, "Tracy–Widom distribution based fault detection approach: Application to aircraft sensor/actuator fault detection," *ISA Trans.*, vol. 51, no. 1, pp. 189–197, Jan. 2012.
26. T. Kailath and P. Frost, "An innovations approach to least-squares estimation--Part II: Linear smoothing in additive white noise," *IEEE Trans. Automat. Contr.*, vol. 13, no. 6, pp. 655–660, Dec. 1968.
27. C. M. Gadzhiev, "A new method of Kalman filter innovation sequence statistical characteristics checking," *Eng. Simul.*, no. 1, pp. 83–91, 1996.
28. R. A. Horn and C. R. Johnson, *Matrix Analysis*. Cambridge, UK: Cambridge University Press, 2012.
29. P. C. Hansen, "The 2-norm of random matrices," *J. Comput. Appl. Math.*, vol. 23, no. 1, pp. 117–120, Jul. 1988.
30. C. Hajiyev, "Innovation approach based measurement error self-correction in dynamic systems," *Measurement*, vol. 39, no. 7, pp. 585–593, Aug. 2006.
31. C. Hajiyev, "Innovation approach based sensor FDI in LEO satellite attitude determination and control system," in *Kalman Filter Recent Advances and Applications*, V. M. Moreno and A. Pigazo (eds.). London, UK: InTech, 2009.
32. F. Caliskan and C. Hajiyev, "Active fault-tolerant control of UAV dynamics against sensor-actuator failures," *J. Aerosp. Eng.*, vol. 29, no. 4, p. 04016012, Jul. 2016.

# 12 Fault Tolerant Attitude Estimation
## *R-Adaptation Methods*

All attitude filter designers are familiar with the covariance matrix tuning process for an attitude filter to get an estimate close to the optimal (or sub-optimal) values as much as possible. This is a difficult and time-consuming task. Selected values for the measurement and process noise covariance matrices are decision makers for the filter's performance. Composing the measurement noise covariance matrix $(R)$ is rather straightforward as we make use of the sensor specifications. Yet when there are issues such as measurement delay, the matrix components must be tuned for higher estimation accuracy and usually with an ad hoc approach. Measurement faults cause even more serious problems unless the R matrix is adapted in these cases, as will be discussed in detail in this chapter.

This chapter investigates the adaptation methods for the measurement noise covariance $(R)$ matrix of the attitude filter. As known, space is a severe environment so faults such as abnormal measurements, increase in the background noise etc. are always a potential threat for the onboard attitude sensors. The main issue is such faults degrade the estimation accuracy and may make the attitude estimator diverge in the long term. Hence the filtering algorithm that is used as the attitude estimator must be made robust using an adaptive approach. We are specifically interested in sensor fault cases but the proposed methods can be easily generalized for any other case including but not limited to:

- Experimental attitude sensors or commercial off-the-shelf (COTS) devices, for which it is difficult to get accurate noise characteristics, are used (this became very common issue together with the popularity of the low-cost small satellite missions).
- Environmental effects make the sensor noise characteristics gradually change. A common example may be thermal effects on the gyro measurements [1].
- Randomly delayed measurements are used in the filtering algorithm. This is very common case when the star trackers are used for attitude estimation. Before the sensor measurement is available to the filter, star image sampling, star position determination, star identification and attitude calculation are required. All these processes require handling time and cause a random delay in the star tracker measurements [2, 3].

Tuning the R matrix with an adaptive method has a more solid mathematical background compared to the intuitive procedures. In line, it is a very popular research topic with applications of different adaptation methods on different filtering algorithms. Tuning may be applied either all the time to increase the accuracy [4] or at certain periods to make the filter robust against measurement system uncertainties or faults [5].

Fault tolerant attitude estimation focuses on the R-adaptation methods against the measurement faults in particular. In normal operation conditions, where any kind of measurement fault is not observed, any of the attitude estimation filters, i.e. extended Kalman filter (EKF) and unscented Kalman filter (UKF), give sufficiently good estimation results, considering that the real system matches with the modeled system. If the condition of the real system departs from the modeled system as a result of the sensor faults, then the estimator accuracy degrades. In contrast, adapting the R matrix makes the filter robust against this kind of error. This is a passive fault tolerant estimation and does not require any filter reconfiguration (opposed to the active fault tolerant estimators in Chapter 11). The estimation system continues to operate with the same adaptive estimator in case of fault.

There are many different approaches to the R matrix adaptation such as innovation-based covariance scaling [5–7], residual-based scaling [8], Huber-based methods [9], entropy-based methods [10] and multiple-model-based methods [11, 12]. They have certain advantages and disadvantages especially when considered for different types of sensor faults. Specifically, for small satellite attitude estimation, as well as the capability of the adaptation algorithm to cope with different sensor faults, the algorithm complexity is important. While making the original filter robust against the sensor faults, the adaptation should not increase the computational load dramatically [13].

In this chapter we present two different approaches for adapting the R matrix of an attitude estimator. Both approaches are computationally light and do not add on significant complexity over the conventional filtering algorithms. The presented methods are applied for adaptation of both the EKF and UKF. The demonstrations for various types of sensor faults show that the despite being straightforward in nature, both algorithms are powerful R-adaptation methods for a small satellite attitude filter.

## 12.1 ROBUST UNSCENTED KALMAN FILTER

The UKF works accurately when there is no fault in the measurement system. On the contrary, in case of a fault such as abnormal measurements, step-like changes or sudden shifts in the measurement channel etc. the filter deteriorates and the estimation outputs become faulty.

Therefore, a robust algorithm must be introduced such that the filter is insensitive to the measurements in case of malfunctions and the estimation process is corrected without affecting the remaining good estimation behavior.

The robustness of the filter is secured by scaling the measurement noise covariance matrix in case of fault. In this sense two different approaches may be used: Scaling by a single scale factor (SSF) or scaling by a scale matrix built of multiple scale factors (MSFs). In general, despite its relative simplicity, using SSF is not a healthy procedure since the filter should be insensitive just to the measurements of the faulty sensor, not to the all sensors including the ones working properly [6]. In contrast, a matrix built of MSFs might be preferred since in this method the relevant terms of the measurement noise covariance are fixed individually.

The robust algorithm affects characteristic of the filter only when the condition of the measurement system does not correspond to the model used in the synthesis of the filter. Otherwise the UKF works with the regular algorithm.

It is known that the UKF innovation sequence can be determined by

$$e(k+1) = y(k+1) - \hat{y}(k+1|k),  \quad\quad (12.1)$$

where $y(k+1)$ is the measurement vector and $\hat{y}(k+1|k)$ is the predicted observation vector.

The essence of the adaptation procedure against the measurement malfunctions is to compare the real and theoretical values of the innovation covariance matrix. When there is a sensor fault in the system, the real error will exceed the theoretical one. In this case we may ensure the robustness of the filter against the sensor fault by adapting the R matrix, which is a diagonal matrix, formed of the measurement process noise covariance values. The adaptation procedure basically aims at finding an appropriate multiplier for the R, such that the real and theoretical values of the innovation covariance match. This multiplier might be either as a single factor or a matrix formed of multiple factors. In case we use a single factor, matching the real and theoretical values of the innovation covariance means that we basically increase all the terms of the R matrix and impose to the UKF that the measurements are faulty. However, we do not isolate which sensor is malfunctioning. On the contrary when we use MSFs we correct the necessary term of the R matrix (the term which corresponds to the sensor with the faulty measurement). In other words, we make UKF disregard just the measurements of this sensor which is not reliable at that sampling time [14].

### 12.1.1 ADAPTING THE R-MATRIX OF UKF USING A SINGLE SCALE FACTOR

As stated the essence of the adaptation is the covariance matching. For SSF approach we match the trace of the covariances such that

$$tr\left[e(k+1)e^T(k+1)\right] = tr\left[P_{yy}(k+1|k) + S(k)R(k+1)\right],  \quad\quad (12.2)$$

where $P_{yy}(k+1|k)$ is the observation covariance, $R(k+1)$ is the measurement noise covariance and $S(k)$ is the introduced SSF, $tr[\cdot]$ is the trace of the related matrix. We may rewrite the equation as

$$tr\left[e(k+1)e^T(k+1)\right] = tr\left[P_{yy}(k+1|k)\right] + S(k)tr\left[R(k+1)\right].  \quad\quad (12.3)$$

Then, regarding $tr[e(k + 1)e^T(k + 1)] = e^T(k + 1)e(k + 1)$ the SSF can be obtained:

$$S(k) = \frac{e^T(k+1)e(k+1) - tr\left[P_{yy}(k+1|k)\right]}{tr\left[R(k+1)\right]}. \tag{12.4}$$

The scale factor affects the Kalman gain as

$$K(k+1) = P_{xy}(k+1|k)\left[P_{yy}(k+1|k) + S(k)R(k+1)\right]^{-1}. \tag{12.5}$$

Here $K(k + 1)$ is the Kalman gain and $P_{xy}(k + 1|k)$ is the cross-correlation matrix.

In case of sensor fault, the scalar scale factor will take a larger value and that will increase all terms of the innovation covariance. Eventually the Kalman gain will decrease and the measurements will be disregarded in the state update process (or taken into consideration with lesser weight than the regular case). In such approach, the information about the faulty sensor isolation does not have any significance; all of the current information from the measurements is left out and the UKF relies mostly on the propagation information during the estimation.

## 12.1.2 ADAPTING THE R-MATRIX OF UKF USING MULTIPLE SCALE FACTORS

Considering that the fault may be only in one of the sensors and not all of them at the same time, a more reasonable way of adapting the attitude filter can be using MSFs, instead of a single one. The derivation and the essence are similar. Yet we need to calculate a matrix build of MSFs this time and ensure its mathematical properties as a scale matrix for R.

Firstly, we add a matrix built of MSFs, $S(k)$, into the algorithm in order to tune the measurement noise covariance matrix and match the real and theoretical innovation covariances,

$$\frac{1}{\xi} \sum_{j=k-\xi+1}^{k} e(j+1)e^T(j+1) = P_{yy}(k+1|k) + S(k)R(k+1). \tag{12.6}$$

Here, $\xi$ is the width of the moving window. Left-hand side of the equation represents the real innovation covariance while the right-hand side stands for the theoretical innovation covariance. Then, if we re-arrange the equation, it is clear that we can get the scale matrix by

$$S(k) = \left\{\frac{1}{\xi} \sum_{j=k-\xi+1}^{k} e(j+1)e^T(j+1) - P_{yy}(k+1|k)\right\} R^{-1}(k+1). \tag{12.7}$$

In case of a measurement fault for one of the sensors then the corresponding term of the scale matrix will be a relatively larger term and that will increase the measurement noise covariance of this sensor in the R matrix. Eventually this faulty measurement will be disregarded (or regarded with a lower gain) by the filter. On the other hand, the scale matrix affects the estimation procedure only when the measurements are faulty. Otherwise, in case of normal operation, the scale matrix will be a unit matrix as $S(k) = I_{z \times z}$, where $z$ is the size of the innovation vector.

Nonetheless, as $\xi$ is a limited number because the number of the measurements and the computations performed with the computer implies errors such as the approximation and round off errors; $S(k)$ matrix that is calculated by the use of Eq. (12.7) may not be diagonal and may have diagonal elements which are "negative" or lesser than "one." $S(k)$ matrix should be diagonal because only its diagonal terms have significance on the adaptation since each diagonal term corresponds to the noise covariance of each measurement (for the adaptation procedure $S(k)$ matrix is multiplied with the diagonal R matrix). Besides the measurement noise covariance matrix must be positive definite (that is why the multiplier $S(k)$ matrix cannot have negative terms) and also any term of this matrix cannot decrease in time for this specific problem since there is no possibility for increasing the performance of the onboard sensor (that is why the multiplier $S(k)$ matrix cannot have terms less than one).

Therefore, in order to avoid such situations, composing the scale matrix by the following rule is suggested:

$$S^* = diag\left(s_1^*, s_2^*, \ldots, s_z^*\right)$$                                              (12.8)

$$s_i^* = \max\left\{1, S_{ii}\right\} \quad i = 1, z.$$                                              (12.9)

Here, $S_{ii}$ represents the $i$th diagonal element of the matrix $S(k)$. Apart from that point, if the measurements are faulty, $S^*(k)$ will change and so affect the Kalman gain as

$$K\left(k+1\right) = P_{xy}\left(k+1|k\right)\left[P_{yy}\left(k+1|k\right) + S^*\left(k\right)R\left(k+1\right)\right]^{-1}.$$         (12.10)

In case of any kind of malfunction, the element(s) of the scale matrix, which corresponds to the faulty component(s) of the innovation vector, increases and so the terms in the related column(s) of the Kalman gain decrease. As a consequence, the effect of the faulty innovation term on the state update process reduces and accurate estimation results can be obtained even in case of measurement malfunctions.

## 12.2 ROBUST EXTENDED KALMAN FILTER

Similar to the UKF, the EKF works properly when there is no sensor fault affecting the measurement channel of the filter. However, any kind of sensor fault significantly deteriorates the filter's estimation accuracy. Similar to the UKF, we need to define single or MSFs to adapt the filter and make it robust against any kind of sensor faults.

### 12.2.1 ADAPTING THE R-MATRIX OF EKF USING A SINGLE SCALE FACTOR

The adaptation procedure of the EKF is not different from the one for the UKF. Basically this time we apply the same method by using the EKF equations.

Firstly, we match the trace of the covariances such that

$$tr\left[e(k+1)e^T(k+1)\right]=tr\left[H(k+1)P(k+1|k)H^T(k+1)+S(k)R(k+1)\right].$$

$$(12.11)$$

Considering the equality of $P_{yy}(k+1|k)$ for UKF and $H(k+1)P(k+1|k)H^T(k+1)$ for EKF, which are both standing for the observation covariance of the filters, it is straightforward to show that Eqs. (12.2) and (12.11) are same.

Departing from Eq. (12.11), the SSF for the EKF is calculated as

$$S(k)=\frac{e^T(k+1)e(k+1)-tr\left[H(k+1)P(k+1|k)H^T(k+1)\right]}{tr\left[R(k+1)\right]}.$$

$$(12.12)$$

Here $P(k+1|k)$ is the predicted covariance matrix and $H(k+1)$ is the measurement matrix constituted of the partial derivatives.

### 12.2.2 ADAPTING THE R-MATRIX OF EKF USING MULTIPLE SCALE FACTORS

To derive the equation for the MSF, similarly with the UKF, we compare the real and theoretical values of the innovation covariance matrix and add a scale matrix, $S(k)$, into the algorithm as

$$\frac{1}{\xi}\sum_{j=k-\xi+1}^{k}e(j+1)e^T(j+1)=H(k+1)P(k+1|k)H^T(k+1)+S(k)R(k+1).$$

$$(12.13)$$

Then, the definition for the scale matrix is

$$S(k)=\left\{\frac{1}{\xi}\sum_{j=k-\xi+1}^{k}e(j+1)e^T(j+1)-H(k+1)P(k+1|k)H^T(k+1)\right\}R^{-1}(k+1).$$

$$(12.14)$$

After that we use the Eqs. (12.8) and (12.9) to correct and diagonalize the scale matrix. Finally, the Kalman gain is tuned as

$$K(k+1)=P(k+1|k)H^T(k+1)\left[H(k+1)P(k+1|k)H^T(k+1)+S^*(k)R(k+1)\right]^{-1}.$$

$$(12.15)$$

### 12.3 FAULT DETECTION

As aforementioned, we shall use the robust Kalman filters only in case of the fault and in all other cases, the filters run with their conventional algorithms. The fault detection is realized via a kind of statistical information. To achieve that, the following two hypotheses may be proposed:

$\gamma_o$ – the system is normally operating.

$\gamma_1$ – there is a malfunction in the estimation system.

Then we may introduce the following statistical functions for the RUKF and REKF, respectively,

$$\beta(k) = e^T(k+1)\left[P_{yy}(k+1|k) + R(k+1)\right]^{-1} e(k+1), \qquad (12.16a)$$

$$\beta(k) = e^T(k+1)\left[H(k+1)P(k+1|k)H^T(k+1) + R(k+1)\right]^{-1} e(k+1). \qquad (12.16b)$$

These functions have $\chi^2$ distribution with $z$ degree of freedom, where $z$ is the dimension of the innovation vector.

If the level of significance, $\alpha$, is selected as

$$P\{\chi^2 > \chi^2_{\alpha,z}\} = \alpha; \quad 0 < \alpha < 1, \qquad (12.17)$$

the threshold value, $\chi^2_{\alpha,z}$ can be determined. Hence, when the hypothesis $\gamma_1$ is correct, the statistical value of $\beta(k)$ will be greater than the threshold value $\chi^2_{\alpha,s}$, i.e.:

$$\begin{aligned} \gamma_0 &: \beta(k) \le \chi^2_{\alpha,s} \quad \forall k \\ \gamma_1 &: \beta(k) > \chi^2_{\alpha,s} \quad \exists k \end{aligned} \qquad (12.18)$$

## 12.4 REMARK ON STABILITY OF THE ROBUST KALMAN FILTERS

Indeed, the proposed Robust Kalman filters are not so different from the linear Kalman filter in point of view of structure; they may be assumed as a modification. In this sense, a similar approach with the one of linear Kalman filter may be followed for the stability analysis of the Robust Kalman Filters. From Section 7.2, it is known that the following characteristic polynomial of the system can be used for stability analyses of a linear Kalman filter:

$$\text{Characteristic polynomial} = \left[zI - \{\Phi(k+1,k) - K(k+1)H(k+1)F(k+1)\}\right],$$
$$(12.19)$$

where $z$ denotes the usual z-transform variable.

The roots of this polynomial provide information about the filter stability. If all the roots lie inside the unit circle in the z-plane, the filter is stable; conversely, if any root lies on or outside the unit circle, the filter is unstable. As a matter of terminology, the roots of the characteristic polynomial are the same as the eigenvalues of

$$\left[\{\Phi(k+1,k) - K(k+1)H(k+1)F(k+1)\}\right]. \qquad (12.20)$$

Assume that the similar investigation is followed for the Robust Kalman Filters. Let's examine the case for REKF. Hence, if we substitute the necessary equality instead of Kalman filter gain, Eq. (12.20) becomes

$$
\begin{aligned}
\{F(k+1/k) - \Big[ \big( P(k+1|k) H^T(k+1) \big) \big( H(k+1) P(k+1|k) H^T(k+1) \\
+ S(k) R(k+1) \big)^{-1} \Big] H(k+1) F(k+1/k) \}.
\end{aligned}
\tag{12.21}
$$

In Eq. (12.21), $S(k)$ should be simply replaced with $S^*(k)$ if we are using MSFs instead of a single one for adapting the filter.

Now we should search for the roots of Eq. (12.21). Tests for different sensor fault cases show both versions of REKF are stable in case of a sensor fault. Nonetheless, REKF with MSFs stands more stable and usually only one root reaches to a near point of the unit cycle limit. During adaptation with MSFs, related roots (that belong to the states, innovation channel of which are corrected via scale matrix) move through the unit cycle limit but they do not approach to the marginal stability border too much. On the other hand, for REKF with SSF, where the whole innovation vector is disregarded in case of fault, all of the roots move through the stability border and usually one of them takes nearly marginal ($|z|=1$) value, so stability characteristics worsen.

In the end we can say that proposed REKFs have no risk about convergence, namely they are stable. To guarantee the accuracy of this statement, even some more harsh fault cases can be tried (e.g. far greater sensor faults in magnitude). Magnitude of the malfunction is not a factor that may break down the stability of the filter and roots of REKFs stay inside the unit cycle for every condition.

Moreover, simulation results pointed out the better stability characteristic of REKF with MSFs compared to REKF with SSFs. For every sensor fault case, roots of REKF with MSFs do not approach to the stability limit (unit cycle) as much as the ones of REKF with SSF.

Furthermore, the stability of the UKF may be examined as a special case. As discussed in [15, 16], the stability of the UKF strictly depends on the positive-definiteness of the process and measurement noise covariance matrices; as long as this certain condition is satisfied, the estimation error of the UKF remains bounded even for large initial estimation errors. Therefore, the proposed adaptation method does not have a detractive effect on the stability of the UKF used for the attitude estimation.

## 12.5  DEMONSTRATIONS OF REKF AND RUKF FOR ATTITUDE ESTIMATION OF A SMALL SATELLITE

In this section, the proposed robust Kalman filtering algorithms are tested via simulations for a small satellite model. Besides, the same simulation scenarios are repeated by using the conventional UKF or EKF algorithms and the results are compared. The attitude filters run for estimating the attitude and angular rate of the satellite having only the magnetometer measurements. It is assumed that the magnetometers are calibrated.

The simulations are realized for 7000 seconds with a sampling time of $\Delta t = 0.1$ s. This period coincides with approximately one orbit of the satellite. Nonetheless the orbit of the satellite is assumed as circular. Other orbit parameters are as follows: the magnetic dipole moment of the Earth, $M_e = 7.943 \times 10^{15}$ Wb. m; the Earth Gravitational constant, $\mu = 3.98601 \times 10^{14}$ m³/s²; the orbit inclination, $i = 31°$; the spin rate of the Earth, $\omega_e = 7.29 \times 10^{-5}$ rad/s; the magnetic dipole tilt, $\varepsilon = 11.7°$; the distance between the center of mass of the satellite and the Earth, $r_0 = 7450$ km. The inertia matrix for the used small satellite model is $J = diag(310\ 180\ 180)$kg.m².

For the magnetometer measurements, the sensor noise is characterized by zero mean Gaussian white noise with a standard deviation of $\sigma_m = 300nT$. As the filter parameters for the UKF and RUKF, $\kappa$ is selected as $\kappa = -2$, where $f = 2(a + 1)$ and $a = 1$. For the REKF and RUKF, the size of the moving window is taken as $\xi = 30$.

The initial attitude errors in the simulations are set to 50° for all three axes. Besides, the initial value of the covariance matrix is taken as $P_0 = diag\begin{bmatrix} 0.5 & 0.5 & 0.5 & 10^{-4} & 10^{-4} & 10^{-4} \end{bmatrix}$ while the process noise covariance matrix is selected as $Q = diag\begin{bmatrix} 10^{-7} & 10^{-7} & 10^{-7} & 10^{-12} & 10^{-12} & 10^{-12} \end{bmatrix}$.

Nonetheless, for the fault detection procedure, $\chi^2_{\alpha,z}$ is taken as 7.81 and this value comes from chi-square distribution when the degree of freedom is 3 and the reliability level is 95%.

Three different scenarios are taken into consideration for simulating the fault in the measurements; the continuous bias, fault of zero output and measurement-noise increment. For each scenario, a series of simulations are run by the REKF, RUKF and as well the conventional EKF and UKF algorithms.

### 12.5.1 CONTINUOUS BIAS IN MEASUREMENTS

In this scenario, a constant value is added to the measurements of the magnetometer aligned in the $x$ axis between the 3000th and 3200th s for a period of 200 s such that

$$B_x(q,t) = B_x(q,t) + 20000nT \quad t = 3000...3200 \text{ s}.$$

The constant term, selected as $20000nT$, almost doubles the magnetometer output.

In Figure 12.1, the attitude estimation error of the UKF and RUKF is given for the pitch angle. The RUKF that uses SSF is plotted with dotted line and labeled as RUKF$_s$ while the RUKF with MSFs is plotted with dashed line and labeled as RUKF$_m$. Apparently, in case of fault the estimation accuracy for the conventional UKF algorithm deteriorates. The RUKF with SSF lessens the effect of the fault but still the filter is not fully recovered and after the measurement fault ends at 3200th s. The RUKF$_s$ estimations show a fluctuating behavior. The reason for a filter that is not fully recovered is disregarding the measurements of all three magnetometers as a result of increasing all terms of the R matrix via multiplication with a single large-scale factor (see Figure 12.2 for the variation of the SSF). Instead of isolating the faulty sensor and leaving out just its measurements, the RUKF$_s$ considers all of the measurement as faulty and throughout this period it mostly relies on the

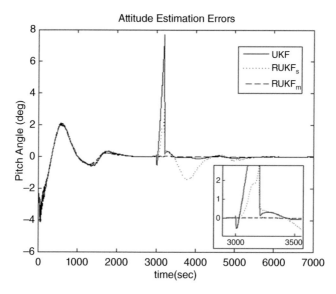

**FIGURE 12.1**    Pitch angle estimation error for the UKF and RUKF in case of continuous bias.

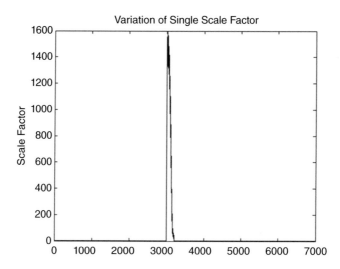

**FIGURE 12.2**    Variation of the single scale factor for the RUKF in case of continuous bias.

propagation values. In this case especially for a filter with relatively higher process noise covariance, Q, the estimation errors accumulate and the filter starts to diverge from the actual values. Table 12.1, which gives the absolute values of error at 3050th and 3150th s, supports this interpretation.

Clearly the estimation error for the RUKF$_s$ at the 3150th s is higher than the one at the 3050th s. Hence the SSF approach may be useful only for faults which last for a short period. On the contrary, the RUKF with the MSFs does not have such limitation and keeps its estimation accuracy without being affected from the fault.

**TABLE 12.1**

**Absolute Estimation Errors in Case of Continuous Bias: Regular UKF, RUKF with Single Scale Factor and RUKF with Multiple Scale Factor**

| Parameter | Abs. Err. Values for Regular UKF | | Abs. Err. Values for RUKF with SSF | | Abs. Err. Values for RUKF with MSF | |
|---|---|---|---|---|---|---|
| | 3050 s | 3150 s | 3050 s | 3150 s | 3050 s | 3150 s |
| $\varphi$(deg) | 10.580 | 10.835 | 0.1901 | 0.4936 | 0.0207 | 0.0430 |
| $\theta$(deg) | 0.6945 | 5.1327 | 0.1116 | 1.8268 | 0.0033 | 0.0178 |
| $\psi$(deg) | 0.6131 | 0.8453 | 0.0862 | 2.6225 | 0.0041 | 0.0030 |

An examination on the scale matrix at an instant between the 3000th and 3200th s of the simulation shows that the algorithm works properly; $S^* = diag(4427\ 1\ 1.59)$. Since the fault is in the measurements of the magnetometer aligned in the x axis, the correction must be applied to the first term of the R matrix as in this case. The large first diagonal term of the scale matrix decreases the terms in the first column of the Kalman gain and so the faulty innovation term (the first term of the innovation vector) is disregarded in the state update process.

In Figure 12.3, the estimation results for the REKF and EKF are given. Similar to the RUKF$_m$, the REKF$_m$ is not affected by the measurement fault and sustains reliable estimation results for the whole period. Although the REKF with single scale gives satisfactory results for a short period starting from 3000th s. it deteriorates later on because of disregarding even the healthy measurements. Table 12.2 may be seen for more detailed investigation.

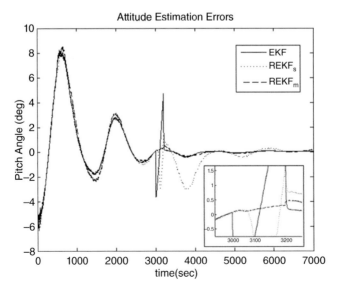

**FIGURE 12.3**  Pitch angle estimation error for the EKF and REKF in case of continuous bias.

**TABLE 12.2**

**Absolute Estimation Errors in Case of Continuous Bias: Regular EKF, REKF with Single-Scale Factor and REKF with Multiple-Scale Factor**

| Parameter | Abs. Err. Values for Regular EKF | | Abs. Err. Values for REKF with SSF | | Abs. Err. Values for REKF with MSF | |
|---|---|---|---|---|---|---|
| | 3050 s | 3150 s | 3050 s | 3150 s | 3050 s | 3150 s |
| $\varphi$(deg) | 7.1599 | 5.9440 | 3.7358 | 2.2637 | 0.5298 | 0.6140 |
| $\theta$(deg) | 2.2395 | 2.2887 | 0.1460 | 1.8108 | 0.1000 | 0.2523 |
| $\psi$(deg) | 1.9778 | 2.1547 | 1.4322 | 5.5227 | 0.1291 | 0.0056 |

Nonetheless, a comparison between the performances of the UKF and EKF (or RUKF vs. REKF) shows that the UKF algorithms outperform the EKF algorithms in terms of accuracy and the convergence speed. Obviously the EKF can satisfy attitude accuracy with less than 2 deg error approximately 1500 s later than the UKF. This is an expected result for high initial attitude error as already discussed in Chapter 9.

### 12.5.2 MEASUREMENT NOISE INCREMENT

In this scenario, the measurement noise of the magnetometer aligned in $x$ axis is multiplied with a constant between the 3000th and 3200th s for a period of 200 s. In fault case the standard deviation of this magnetometer becomes $\sigma_{m_x}^f = 300 \times 100 nT$.

In Figure 12.4, the estimation results for the UKF and RUKF are given. As can be observed, again the UKF fails about giving accurate estimation results in case of fault

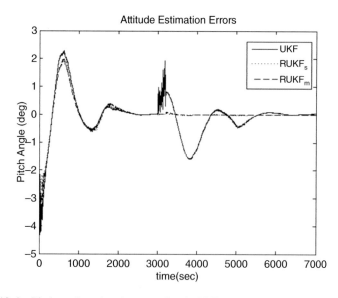

**FIGURE 12.4** Pitch angle estimation error for the UKF and RUKF in case of measurement noise increment.

and the noisy measurements make the estimations deteriorate for a longer period than the fault itself. For this simulation case it takes 3000 s for the filter to satisfy accuracy less than 0.1°. after the fault removes. On the other hand, in case of measurement noise increment both approaches for robust Kalman filtering that are scaling with single and multiple factors give accurate estimation results. Unlike the first fault case this time SSF works properly because of the fault's characteristic. Since the noise in the measurements is random, the filter does not work continuously with the robust algorithm for whole 200 s and moreover the scale factor may take values that are closer to the one. In other words, the filter tunes itself depending on the magnitude of the noise and does not disregard the measurements for the whole fault period; it only does when the $\chi^2_{\alpha,z}$ threshold is exceeded and the magnitude of the noise is high such that the scale factor takes large values. Variation of the SSF confirms that the measurements are disregarded only when the magnitude of the noise is high (Figure 12.5). Hence specifically for this fault scenario, the RUKF$_s$ and RUKF$_m$ do not have any significant difference in the sense of estimation accuracy (Table 12.3).

The performed simulations have proved that the REKF is also capable of overcoming the deteriorating effect of the measurement noise increment. For the simulation with the REKF, the REKF$_m$ gives slight more accurate estimations than the REKF$_s$ (Figure 12.6 and Table 12.4).

### 12.5.3 ZERO-OUTPUT FAILURE

The third failure case, which is fault of zero output, is simulated by simply making the measurement output of one of the magnetometers zero so it measures $0nT$ for

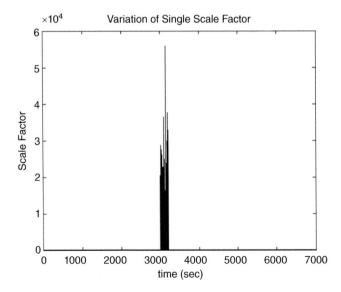

**FIGURE 12.5** Variation of the single-scale factor for the RUKF in case of measurement noise increment.

## TABLE 12.3
### Absolute Estimation Errors in Case of Measurement Noise Increment: Regular UKF, RUKF with Single-Scale Factor and RUKF with Multiple-Scale Factor

| Parameter | Abs. Err. Values for Regular UKF | | Abs. Err. Values for RUKF with SSF | | Abs. Err. Values for RUKF with MSF | |
|---|---|---|---|---|---|---|
| | 3050 s | 3150 s | 3050 s | 3150 s | 3050 s | 3150 s |
| $\varphi$(deg) | 3.4186 | 4.3199 | 0.0010 | 0.0242 | 0.0269 | 0.0412 |
| $\theta$(deg) | 0.2853 | 1.3830 | 0.0134 | 0.0048 | 0.0062 | 0.0174 |
| $\psi$(deg) | 0.9533 | 0.9003 | 0.0026 | 0.0049 | 0.0017 | 0.0015 |

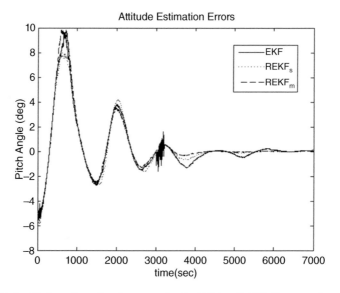

**FIGURE 12.6**  Pitch angle estimation error for the EKF and REKF in case of measurement noise increment.

## TABLE 12.4
### Absolute Estimation Errors in Case of Measurement Noise Increment: Regular EKF, REKF with Single Scale Factor and REKF with Multiple Scale Factor

| Parameter | Abs. Err. Values for Regular EKF | | Abs. Err. Values for REKF with SSF | | Abs. Err. Values for REKF with MSF | |
|---|---|---|---|---|---|---|
| | 3050 s | 3150 s | 3050 s | 3150 s. | 3050 s | 3150 s |
| $\varphi$(deg) | 0.1332 | 0.2110 | 2.7516 | 2.3680 | 0.7954 | 0.9315 |
| $\theta$(deg) | 1.1840 | 0.1027 | 0.1554 | 0.6464 | 0.1114 | 0.3182 |
| $\psi$(deg) | 1.5309 | 0.7592 | 0.6777 | 0.3859 | 0.2131 | 0.0266 |

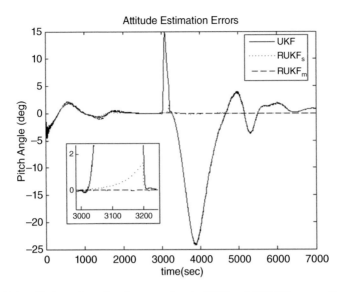

**FIGURE 12.7** Pitch angle estimation error for the UKF and RUKF in case of zero output failure.

200 s between the 3000th and 3200th s. In order to test the algorithm this time the fault is implemented to the magnetometer aligned in the z axis:

$$B_z(q,t) = 0 + \eta_{1z} \quad t = 3000...3200 \text{ s}$$

In Figure 12.7, the estimation results for the UKF and RUKF are given. Obviously, same as the first simulation scenario, the UKF cannot achieve accurate estimation, whereas the RUKF with the SSF can overcome the fault only for a short period. Because of taking none of the measurements into consideration, the RUKF$_s$ estimations get worse when the robust algorithm runs longer than 50 seconds. The SSF behaves in a similar manner with its trend for the continuous bias fault scenario (Figure 12.8). Moreover, the results show us when the filter is not robust the zero output failure has a high detractive impact on the estimation accuracy that lasts for a very long period. Even though the filter's response may vary when it is designed with different parameters (such as the Q matrix) simulations show that a fault may affect the filter for a longer period than its length. Therefore, if the magnetometer measures 0 even for just few seconds, it is not possible to compensate that with a filter other than the robust ones. In this sense the estimation results for the RUKF$_m$ clearly signify the importance of using the proposed algorithm. The RUKF$_m$ is not affected from the fault and can perform accurate estimation even when the fault lasts long by simply disregarding the measurements of the faulty magnetometer and working on the basis of the measurements from two properly operating magnetometers (Table 12.5 may be seen for further examination). The sample for the MSFs in case of fault validates that the RUKF$_m$ disregards the measurements of the magnetometer aligned in the z axis as it supposed to be $S^* = diag(1.44 \; 1 \; 4024)$.

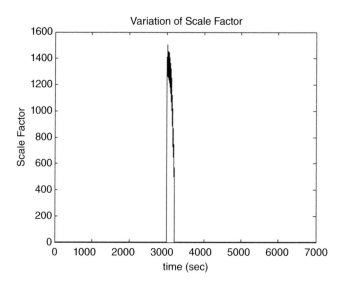

**FIGURE 12.8**    Variation of the single scale factor for the RUKF in case of zero output failure.

---

**TABLE 12.5**

**Absolute Estimation Errors in Case of Zero Output Failure: Regular UKF, RUKF with Single Scale Factor and RUKF with Multiple Scale Factor**

| Parameter | Abs. Err. Values for Regular UKF | | Abs. Err. Values for RUKF with SSF | | Abs. Err. Values for RUKF with MSF | |
|---|---|---|---|---|---|---|
| | 3050 s | 3150 s | 3050 s | 3150 s | 3050 s | 3150 s |
| $\varphi$(deg) | 9.7561 | 42.492 | 0.0321 | 0.2176 | 0.0892 | 0.0567 |
| $\theta$(deg) | 7.0059 | 8.1779 | 0.0293 | 0.3733 | 0.0521 | 0.0576 |
| $\psi$(deg) | 9.6897 | 41.673 | 0.0114 | 0.1908 | 0.0163 | 0.0191 |

In Figure 12.9 and Table 12.6, the estimation results for the EKF and REKF are given. Same as the UKF, the EKF fails at giving accurate estimation results for a longer period than the fault itself, whereas the $REKF_m$ is superior to $REKF_s$ regarding the estimation accuracy.

## 12.5.4  Discussion on the Implementation of Robust Attitude Filters on Real Small Satellite Missions

The simulations show that the robust Kalman filtering is more reasonable in case of measurement faults. For small satellites, the risk of being affected from the external and internal disturbances is high. The interaction between the tightly placed subsystems and the external disturbances such as the ionospheric charges may change the characteristics of the magnetometers and that is reflected to the measurements as

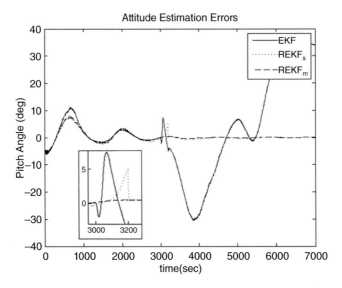

**FIGURE 12.9**   Pitch angle estimation error for the EKF and REKF in case of zero output failure.

---

## TABLE 12.6
## Absolute Estimation Errors in Case of Zero Output Failure: Regular EKF, REKF with Single Scale Factor and REKF with Multiple Scale Factor

| Parameter | Abs. Err. Values for Regular EKF | | Abs. Err. Values for REKF with SSF | | Abs. Err. Values for REKF with MSF | |
|---|---|---|---|---|---|---|
| | 3050 s | 3150 s | 3050 s | 3150 s | 3050 s | 3150 s |
| $\varphi$(deg) | 53.686 | 148.50 | 2.8558 | 1.8026 | 1.5884 | 1.2380 |
| $\theta$(deg) | 5.2576 | 0.8919 | 0.0760 | 2.2197 | 0.2049 | 0.4120 |
| $\psi$(deg) | 30.024 | 45.173 | 0.9518 | 3.6684 | 0.3210 | 0.0242 |

additional bias, increase in the noise etc. For small satellite applications, the magnetometers are usually preferred as the primary sensors since they are light and small, appropriately with the concept of the satellite.

Hence the attitude estimation accuracy relies on the magnetometer measurements and as shown with the simulations even a short-term fault affects the estimation accuracy in a significant amount. That is why the proposed robust Kalman filters are especially important for small satellite applications.

The existing R-adaptation techniques in the literature mostly depend on either direct estimation of the R matrix itself or scaling the R matrix using a complex algorithm. The benefits of our approach are being simple and giving more stable results than the existing estimation methods. The required extra computational load compared to the regular filter algorithms is insignificant [13] and a small satellite processor that is running the original filter algorithm can also easily run the robust version.

When implementing the MSF version of the adaptive algorithms, there is no need for an additional fault isolation scheme; the isolation is autonomous as only the measurement noise covariance term corresponding to the fault sensor is scaled. This saves considerable amount of computational load especially when compared with fault isolation processes that require extra filters to be run (see the discussions in Chapter 11). Last but not least when our method is used there is no chance of obtaining negative definite R matrix, which is a high risk when R is estimated rather than being scaled [5]. This ensures stability of the filter.

### 12.5.5 COMPARISON OF RECONFIGURED UKF AND RUKF IN THE PRESENCE OF MEASUREMENT FAULTS

In the application phase, as an alternative to the presented Robust Kalman filtering methods, ignoring the faulty measurement after the fault identification might be another option. However, the REKF and RUKF algorithms, especially with the MSFs, have multiple advantages over this approach:

1. When the faulty measurement is ignored that means it is necessary to reconfigure the filter algorithm (the update part) because of changed number of incoming measurements. In this case, especially if we regard that we need to have more than one reconfiguration options depending on which sensor is faulty, the overall computational load will increase. However, in the REKF/RUKF algorithms with MSFs, the only extra thing that we need to calculate compared to the regular UKF/EKF is the fault detection statistics and the scale matrix when necessary. So in this robust Kalman filtering method, the increment in the computational load is insignificant. The fault detection, faulty sensor isolation and fault compensation are achieved by just using a simple modification over the traditional EKF/UKF algorithm.

2. Another problem, when the faulty sensor is ignored, is the process of recovery. After the fault we still need to check faulty sensor to understand whether it did recover or not. If it recovered, then we will start using its measurements again. On the other hand, if we consider a scenario where the faulty sensor is ignored for the rest of the estimation procedure then that will decrease the estimation accuracy remarkably and eventually the filter will diverge. In Table 12.7,

### TABLE 12.7
### Absolute Error Values for the Reconfigured UKF and the RUKF

| Parameter | Abs. Err. Values for the Reconfigured UKF | | Abs. Err. Values for RUKF | |
|---|---|---|---|---|
| | 3050 s. | 6500 s. | 3050 s. | 6500 s. |
| $\varphi$(deg) | 0.03380 | 0.32211 | 0.02803 | 0.01733 |
| $\theta$(deg) | 0.00331 | 0.03330 | 0.00524 | 0.00900 |
| $\psi$(deg) | 0.00850 | 0.02028 | 0.00591 | 0.01153 |

absolute error values are tabulated for such scenario. After the fault that occurs at 3000th s (the fault lasts for 100 s.) the UKF is reconfigured and the measurement of this sensor is ignored for the rest of the procedure. In contrast, the RUKF continues to use the measurements of this sensor. The degradation in the estimation accuracy for the reconfigured UKF is apparent especially at 6500th s where reconfigured UKF is working with the measurements of two sensors but the RUKF is using all sensors including the one that recovered at 3100th s. It should be again stressed that the reconfigured UKF in this scenario will eventually diverge.

3. Suppose that we are experiencing measurement noise increment failure for one of the magnetometers. In this case for each sampling the noise will have random noise characteristics with a high standard deviation as discussed. A measurement sample with high noise may be followed by another sample with low noise because of the randomness of the process. That means the measurement may still contain some healthy information. The REKF and RUKF algorithms detect such variations automatically and reflect to the estimation process by changing the calculated scale matrix. However, when we completely ignore the sensor we will also loose possible healthy information.

## 12.6 CONCLUSION AND DISCUSSION

In this chapter, various robust Kalman filters for small satellite attitude estimation are presented in scope of passive fault tolerant attitude estimators. EKF and UKF are taken into consideration as different Kalman filtering algorithms for nonlinear satellite dynamics and kinematics and robust versions of these filters are presented.

Two simple methods for the measurement noise covariance matrix scaling are introduced. In order to show the clear effects of disregarding only the data of the faulty sensor, we performed the adaptation by using both single and MSFs which are two different approaches for the same problem. It is possible to adapt filter by using single adaptive factor as a corrective term on the filter gain, but that is not a healthy procedure as long as the filter performance differs for each state for the complex systems with multivariable. The preferred method is to use an adaptive matrix built of multiple measurement noise scale factors to fix the relevant term of the Kalman gain matrix, individually. The MSFs based adaptation scheme has advantages because in this case the filters are modified such that the unnecessary information loss is prevented by disregarding only the data of the faulty sensor.

Using only magnetometers is a common preference for the small satellite applications (especially for the cubesats); and having only a limited number of sensors onboard increases the significance of the given robust Kalman filtering methods. Throughout the chapter, Robust Kalman filter algorithms with single and MSFs are tested for attitude estimation problem of a small satellite which has only three magnetometers onboard as the attitude sensors. Results of these proposed algorithms are compared and discussed for different types of magnetometer sensor faults. The simulation results show that both the RUKF and REKF perform well in case of various different sensor fault types. On the other hand, the conventional filters (the UKF and EKF) fail at giving accurate estimation results for the period of the fault and as well for some additional time that is necessary for filter to converge again. Moreover, the RUKF outperforms all other filters including the REKF for all simulation cases when the initial attitude error is high.

The proposed robust algorithms are free of computational burden and they can be easily run by simple microprocessors. These algorithms do not bring an extra requirement and they can be used with the systems suitable for general EKF or UKF processing.

## REFERENCES

1. J. Halverson, R. Harman, and R. Carpenter, "AAS 17-591 Tuning the solar dynamics observatory onboard kalman filter," *Proceedings of 2017 AAS/AIAA Astrodynamics Conference*, Stevenson, USA, 2017, pp. 1–14.
2. S. Wang, S. Zhang, B. Zhou, and X. Wang, "A factor graph approach for attitude estimation of spacecraft using a stellar gyroscope," *IEEE Access*, vol. 7, pp. 1–1, 2019.
3. E. A. Hogan and B. A. Woo, "AAS 17-554 treatment of measurement variance for star tracker-based attitude estimation," *Proceedings of 2017 AAS/AIAA Astrodynamics Conference*, Stevenson, USA, 2017, pp. 1–19
4. L. Cao, D. Ran, X. Chen, X. Li, and B. Xiao, "Huber second-order variable structure predictive filter for satellites attitude estimation," *Int. J. Control. Autom. Syst.*, vol. 17, no. 7, pp. 1781–1792, 2019.
5. H. E. Soken, C. Hajiyev, and S.-I. Sakai, "Robust Kalman filtering for small satellite attitude estimation in the presence of measurement faults," *Eur. J. Control*, vol. 20, no. 2, 2014.
6. H. E. Soken and C. Hajiyev, "Pico satellite attitude estimation via Robust Unscented Kalman Filter in the presence of measurement faults," *ISA Trans.*, vol. 49, no. 3, pp. 249–256, 2010.
7. A. Adnane, Z. Ahmed Foitih, M. A. Si Mohammed, and A. Bellar, "Real-time sensor fault detection and isolation for LEO satellite attitude estimation through magnetometer data," *Adv. Sp. Res.*, vol. 61, no. 4, pp. 1143–1157, 2018.
8. H. X. Le and S. Matunaga, "A residual based adaptive unscented Kalman filter for fault recovery in attitude determination system of microsatellites," *Acta Astronaut.*, vol. 105, no. 1, pp. 30–39, 2014.
9. L. Cao, D. Qiao, and X. Chen, "Laplace ℓ1 Huber based cubature Kalman filter for attitude estimation of small satellite," *Acta Astronaut.*, vol. 148, no. February, pp. 48–56, 2018.
10. M. Kiani, A. Barzegar, and S. H. Pourtakdoust, "Entropy-based adaptive attitude estimation," *Acta Astronaut.*, vol. 144, no. July 2017, pp. 271–282, 2018.
11. Q. M. Lam and J. L. Crassidis, "Precision attitude determination using a multiple model," *Proceedings of the IEEE Aerospace Conference*, Big Sky, USA, 2007, pp. 1–20.
12. H. Stearns and M. Tomizuka, "Multiple model adaptive estimation of satellite attitude using MEMS gyros," *Proc. 2011 Am. Control Conf*, San Francisco, USA, 2014 pp. 3490–3495.
13. D. Cilden-Guler, M. Raitoharju, R. Piche, and C. Hajiyev, "Nanosatellite attitude estimation using Kalman-type filters with non-Gaussian noise," *Aerosp. Sci. Technol.*, vol. 92, pp. 66–76, 2019.
14. H. E. Soken, C. Hajiyev, and S. Sakai, "Robust Kalman filtering with single and multiple scale factors for small satellite attitude estimation," in *Advances in Estimation, Navigation, and Spacecraft Control*, J. Bordeneuve-Guibe, A. Drouin and C. Roos (eds.). Berlin, Heidelberg: Springer Berlin Heidelberg, 2015, pp. 391–411.

15. K. Xiong, L. D. Liu, and H. Y. Zhang, "Modified unscented Kalman filtering and its application in autonomous satellite navigation," *Aerosp. Sci. Technol.*, vol. 13, no. 4–5, pp. 238–246, Jun. 2009.

16. G. Dymirkovksy, Y. Jing, and J. Xu, "Discrete-time unscented Kalman filter: Comprehensive study of stochastic stability," in *Itzhack Y. Bar-Itzhack Memorial Symposium on Estimation, Navigation, and Spacecraft Control*, Haifa, Israel, 2012, pp. 782–789.

# 13 Fault Tolerant Attitude Estimation

## Q-Adaptation Methods

The high possibility of any kind of unexpected events in space environment makes satellites' attitude determination system vulnerable to faults. These faults do not happen only in the sensors but also in the actuators. Actually, any uncertainty in the satellite dynamics should be accounted in this category. In this case it is not possible to get precise estimation results by optimal regular attitude filters since the model for the process noise covariance mismatches with the real value.

In any case determining the process noise covariance (Q matrix) of an attitude filter is a tough procedure. For a gyro-based filter this is rather easier, since we have an analytical approximation for the process noise covariance based on the gyro characteristics as discussed in Chapter 9. However, especially when a dynamics-based filter is used, there are various uncertainties and faults which may be affecting the dynamics of the satellite. Filter's performance highly depends on our knowledge for the process noise.

In specific, there is need for a Q-adaptive attitude filter in the below cases.

1. In a gyro-based filter, the gyro error characteristics may change in-orbit, especially due the thermal variations [1].
2. In a gyro-based filter, there may be a need for estimating several other additional parameters such as magnetometer biases. This arises difficulties in tuning the process noise covariance since the analytical approximation mostly fail when the state dimension increases [2].
3. In a dynamics-based filter, any uncertainty in the dynamics parameters (e.g. the disturbance torque, inertia parameters, actuator alignments) will affect the filter's accuracy and must be compensated by adapting the process noise covariance. Uncertainty may be permanent (e.g. due to disturbance torques) or temporary (e.g. due to thermal variations).
4. In a dynamics-based filter, any actuator fault will cause mismatch between the modeled and real dynamics. The filter must be robust against these faults by means of adapting the process noise covariance.

In this chapter, the Q-adaptive Kalman filtering methods are proposed for the small satellite attitude estimation problem. The presented methods are applied for both the EKF and UKF. For different cases that are listed above, we give different Q-adaptive attitude filters and evaluate their performance with demonstrations and comparisons with the regular filtering algorithms.

## 13.1 ADAPTATION OF A GYRO-BASED ATTITUDE FILTER

The gyro-based attitude filter is well described in Chapter 9. Here our concern is to tune this filter's process noise covariance matrix for increased estimation accuracy and compensating any uncertainties, mostly due to the gyro performance.

The first case that we investigate in this chapter for the gyro-based attitude filter is a regular filter estimating the quaternion vector $q$ and gyro biases $b_g$. Thus we have the state vector as

$$x = \begin{bmatrix} q \\ b_g \end{bmatrix}. \tag{13.1}$$

As discussed in Chapter 9, when a gyro-based attitude filtering algorithm is used for estimating the above state vector of Eq. (13.1) the process noise covariance matrix has an analytical expression.

$$Q = \begin{bmatrix} \sigma_v^2 I_{3\times3} & 0_{3\times3} \\ 0_{3\times3} & \sigma_u^2 I_{3\times3} \end{bmatrix}, \tag{13.2}$$

which is approximated as

$$Q_k = \begin{bmatrix} \left( \sigma_v^2 \Delta t + \frac{1}{3}\sigma_u^2 \Delta t^3 \right) I_{3\times3} & -\left( \frac{1}{2}\sigma_u^2 \Delta t^2 \right) I_{3\times3} \\ -\left( \frac{1}{2}\sigma_u^2 \Delta t^2 \right) I_{3\times3} & \left( \sigma_u^2 \Delta t \right) I_{3\times3} \end{bmatrix} \tag{13.3}$$

in discrete form [3]. This is a fair approximation especially if the gyro sampling rate is high. Nonetheless, it may not give the optimal result in practice and we may require tuning this Q matrix.

### 13.1.1 INTUITIVE TUNING OF AN ATTITUDE FILTER

In line with the detailed description given in [4] the covariance tuning steps can be summarized as[1]

1. Check the $\pm 3\sigma$ bounds calculated from the state covariance for each state and compare with the estimation errors.
2. If the error for any state is exceeding the corresponding $\pm 3\sigma$ bound, increase its process noise covariance term.
3. If the $\pm 3\sigma$ bound for any specific state is far above the estimation error, decrease its process noise covariance term.
4. While increasing or decreasing the Q terms, keep the Q/R ratio fixed as much as possible.

Here we demonstrate this tuning approach for a Multiplicative Extended Kalman Filter for estimating the attitude of the satellite and the gyro biases.

This satellite which is approximately 90 kg is assumed to be orbiting at a circular orbit with an altitude of 685 km.

The angular random walk and rate random walk values for the used gyros are $\sigma_v = 1.1975 \times 10^{-5}$ rad $/\sqrt{s}$ and $\sigma_u = 3.0834 \times 10^{-9}$ rad $/\sqrt{s^3}$, respectively. The initial value for the gyro bias is $b_{g,0} = 0.05°/s$ in all three axes.

A type of star tracker designed for small satellite missions is used for measuring the attitude. Considering the maximum noise level among the star trackers for specifically nanosatellite missions, the standard deviation for the measurements both around the boresight and cross-boresight is assumed as $\sigma_{sst,xy} = \sigma_{sst,z} = 0.04°$. No other error sources (e.g. sensor misalignment) are affecting the measurements in this investigation case.

The proposed algorithm is run for 14000 s, which is more than two orbit periods for the satellite. The sampling time for all measurements is assumed to be $\Delta t = 1$ s.

Initial values for the MEKF part are set as $\hat{q}_0 = \begin{bmatrix} 0 & 0 & 0 & 1 \end{bmatrix}^T$ for quaternions and $\hat{b}_{g,0} = 0$ for gyro biases.

In Figure 13.1 the attitude estimation error for the MEKF is given when the Q matrix is set using the analytical approximation in Eq. (13.3). As clearly seen the estimation error in all three axes exceeds the associated $\pm 3\sigma$ bounds and this is periodically happening in roll and yaw angles. Hence, we may deduce that the filter is not robust enough and any unexpected error, which is very likely in practice, may disturb the filter's stability.

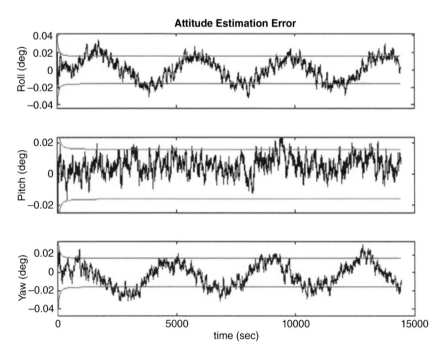

**FIGURE 13.1**　Attitude estimation error when Q matrix is set using Eq. (13.3). Waveform is for the estimation error and the straight lines are for the associated $\pm 3\sigma$ bounds.

Figure 13.2 presents the attitude estimation results, this time for the MEKF that is intuitively tuned by following the discussed guidelines. Filter is behaving more robust with errors bounded with the $\pm 3\sigma$ bounds. This shows us that even for a straightforward scenario, we may need to follow an intuitive tuning scheme for proper filter performance. Such procedure becomes crucial for filter's successful operation in practice when there are several possible factors to affect the filter's estimation performance (e.g. unknown disturbances, measurement delay etc.).

### 13.1.2 PROCESS NOISE COVARIANCE MATRIX ESTIMATION FOR MEKF

One method to estimate the process noise covariance matrix is to use the innovation vector for covariance matching [5]. In this method Q matrix is updated at every recursive step as

$$Q^* = \Delta x_k \Delta x_k^T + P_k^- - P_k - Q_{k-1}. \tag{13.4}$$

Here, $\Delta x_k$ is the residual vector, which is the difference in between the estimated and predicted states as

$$\Delta x_k = \hat{x}_k - \hat{x}_k^-. \tag{13.5}$$

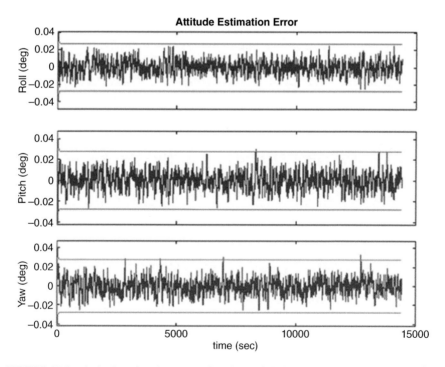

**FIGURE 13.2**    Attitude estimation error when Q matrix is intuitively tuned. Waveform is for the estimation error and the straight lines are for the associated $\pm 3\sigma$ bounds.

In Eq. (13.4), $P_k^-$ is the predicted covariance matrix and $P_k$ is the estimated covariance matrix.

Furthermore, as known, the attitude estimator is sensitive against any kind of variations in the process noise covariance matrix. Therefore, directly forcing the estimated process noise covariance matrix into the estimator at the next step may cause stability issues especially if the initial process noise covariance matrix is not appropriately selected. One possible method is to consider the $Q^*$ within a low pass filter such as [6]:

$$Q_k = \lambda Q^* + \left(1 - \lambda\right)Q_{k-1}, \tag{13.6}$$

where $\lambda$ is the scale factor for low-pass effect.

To see how this covariance matching based $Q$-adaptation algorithm improves the filter's accuracy, it is used to tune the $Q$ matrix for the MEKF for attitude estimation. Other than the $Q$-adaptation, the estimation scenario is exactly same with the one presented in previous section. Low-pass scale factor is selected as $\lambda = 0.8$.

Figure 13.3 shows the attitude estimation error for the filter together with the associated $\pm 3\sigma$ bounds for estimation error covariance. The filter tries to minimize the estimation error by tuning the $Q$ matrix. Especially around the end of the simulation, the error bounds are tightened (compared to Figure 13.2 which presents results for an intuitively tuned version) and the estimation error considerably decreases. Yet root mean square error values for two filters (intuitively and adaptively tuned) are of the same order ($\begin{bmatrix} 0.0078 & 0.0083 & 0.0081 \end{bmatrix}^\circ$ for the intuitively tuned MEKF and

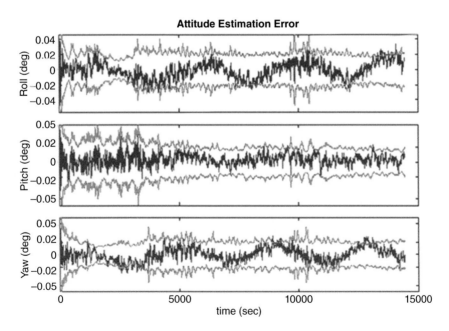

**FIGURE 13.3** Attitude estimation error when Q matrix is adaptively tuned. Waveform is for the estimation error and the straight lines are for the associated $\pm 3\sigma$ bounds.

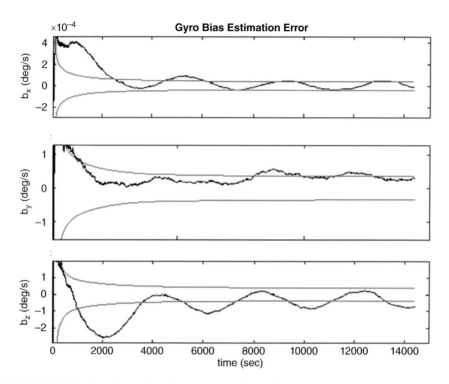

**FIGURE 13.4**  Gyro bias estimation error when Q matrix is adaptively tuned. Waveform is for the estimation error and the straight lines are for the associated ±3σ bounds.

$\begin{bmatrix} 0.0105 & 0.0085 & 0.0109 \end{bmatrix}^\circ$ for the adaptively tuned MEKF). Regarding this result, we can conclude that the adaptive tuning may be an alternate tuning method for the attitude filter. However, the gyro bias errors (Figure 13.4) that are exceeding their associated ±3σ bounds imply that the filter is not performing in an optimal fashion [7].

### 13.1.3 Process Noise Covariance Matrix Estimation for UKF

Adaptive estimation of the process noise covariance for the UKF can be also performed using the covariance matching method on the basis of a maximum likelihood estimator, which is well discussed in [5, 6, 8] and applied to MEKF for attitude estimation in the previous section. Here we show that Eq. (13.4) is also valid for UKF.

The adaptive Kalman filtering problem deals with satisfying the necessary condition for the following equation [8]:

$$\sum_{j=j_0}^{k+1} \left[ tr\left\{ C_e^{-1}(j) \frac{\partial C_e(j)}{\partial \alpha(k+1)} \right\} - e^T(k+1) C_e^{-1}(j) \frac{\partial C_e(j)}{\partial \alpha(k+1)} C_e^{-1}(j) e(k+1) \right] = 0,$$

$$(13.7)$$

where $\alpha(k + 1)$ is the adaptive parameters, $tr(\cdot)$ is the trace operator, $j_0$ is $j_0 = k - N + 2$ for a $N$ size of moving window, $e(k + 1)$ is the innovation sequence and $C_e$ is the innovation covariance which may be written for the UKF as

$$C_e(k+1) = R(k+1) + P_{yy}(k+1|k).$$
(13.8)

Here, $R(k + 1)$ and $P_{yy}(k + 1|k)$ are the measurement noise and observation covariance matrices, respectively. Nonetheless the basic equation used for the process noise covariance adaptation is given as

$$\sum_{j=j_0}^{k+1} tr\left\{H^T(j)\left[C_e^{-1}(j) - C_e^{-1}(j)e(k+1)e^T(k+1)C_e^{-1}(j)\right]H(j)\right\} = 0. \quad (13.9)$$

$H(k + 1)$ is the measurement matrix and $H(k + 1|k) = P_{xy}(k + 1|k)^T P(k + 1|k)^{-1}$ for the UKF, where $P_{xy}(k + 1|k)$ is the cross correlation matrix and $P(k + 1|k)$ is the predicted covariance matrix. We need to show Eq. (13.9) holds true for the UKF and it can be derived from Eq. (13.7) using the UKF equations. First let us rewrite the Eq. (13.8) in the light of formulations from [9]

$$C_e(k+1) = R(k+1) + P_{xy}^T(k+1|k)P^{-1}(k+1|k)P_{xy}(k+1|k).$$
(13.10)

Then the derivative of the innovation covariance with respect to the adaptive factors becomes

$$\frac{\partial C_e(k+1)}{\partial \alpha(k+1)} = \frac{\partial R(k+1)}{\partial \alpha(k+1)} - P_{xy}^T(k+1|k)P^{-1}(k+1|k)$$
$$\times \frac{\partial Q(k+1)}{\partial \alpha(k+1)}P^{-1}(k+1|k)P_{xy}(k+1|k)$$
(13.11)

Here, while taking the derivative, we assume that $P_{xy}(k + 1|k)$ and $P^*(k + 1|k)$ are not a function of $\alpha(k + 1)$, where $P^*(k + 1|k)$ is the propagated covariance without the additive noise as $P(k + 1|k) = P^*(k + 1|k) + Q(k + 1)$. Then regarding the equality of

$$P_{xy}(k+1|k) = P(k+1|k)H^T(k+1|k).$$
(13.12)

Eq. (13.11) can be rewritten as

$$\frac{\partial C_e(k+1)}{\partial \alpha(k+1)} = \frac{\partial R(k+1)}{\partial \alpha(k+1)} - H(k+1|k)\frac{\partial Q(k+1)}{\partial \alpha(k+1)}H^T(k+1|k). \quad (13.13)$$

If we replace Eq. (13.13) in Eq. (13.7), we get

$$\sum_{j=j_0}^{k+1} tr\left\{\left[C_e^{-1}(j) - C_e^{-1}(j)e(k+1)e^T(k+1)C_e^{-1}(j)\right]\right.$$
$$\left.\left[\frac{\partial R(j)}{\partial \alpha(k+1)} - H(j)\frac{\partial Q(j-1)}{\partial \alpha(k+1)}H(j)^T\right]\right\} = 0.$$
(13.14)

Finally, if we consider that the measurement noise covariance $R$ is completely known and independent of $\alpha$ and takes $\alpha_i = Q_{ii}$, then the Eq. (13.14) reduces to the same equation as Eq. (13.9).

After that, next step is to use this basic equation (Eq. (13.9)) in order to find out a definition for the covariance matrix of the system noise. This procedure is same as it is presented in [5]. Consequently, the estimation of the $Q(k)$ can be written as [2]

$$Q^* = \Delta x\left(k+1\right)\Delta x^T\left(k+1\right) + P\left(k+1|k\right) - P\left(k+1|k+1\right) - Q(k) \quad (13.15)$$

which is same as the Q-adaptation equation for EKF in Eq. (13.4).

### 13.1.4 DEMONSTRATION OF ADAPTIVE UKF FOR AUGMENTED STATES

When the UKF is used for estimating the state vector of Eq. (13.1) the analytical approximation for the process noise covariance matrix is same as the one given in Eq. (13.3). Suppose now we include the magnetometer biases into the state vector [10, 11], which is one favorable method for calibrating the magnetometers in real-time for small satellite missions[2]. So, the state vector is $x = \begin{bmatrix} q & b_g & b_m \end{bmatrix}^T$. Then according to the theory, the $9 \times 9$ process noise covariance matrix must be [12]:

$$Q = \begin{bmatrix} \left(\sigma_v^2\Delta t + \dfrac{1}{3}\sigma_u^2\Delta t^3\right)I_{3\times3} & -\left(\dfrac{1}{2}\sigma_u^2\Delta t^2\right)I_{3\times3} & 0_{3\times3} \\[2ex] -\left(\dfrac{1}{2}\sigma_u^2\Delta t^2\right)I_{3\times3} & \left(\sigma_u^2\Delta t\right)I_{3\times3} & 0_{3\times3} \\[2ex] 0_{3\times3} & 0_{3\times3} & 0_{3\times3} \end{bmatrix} \quad (13.16)$$

However, as aforementioned, this approximation fails in practice for an UKF that is used for the attitude, gyro bias and magnetometer bias estimation. Hence, we use the adaptive estimation algorithm for the noise covariance given by Eqs. (13.15) and (13.6).

The process noise covariance matrix estimated by Eq. (13.15) is a non-diagonal matrix because of the state residual term. So specifically, for the attitude estimation problem, we modify it such that it fits in the form given as

$$Q = \begin{bmatrix} Q_q I_{3\times3} & Q_{q\_gb} I_{3\times3} & 0_{3\times3} \\ Q_{q\_gb} I_{3\times3} & Q_{gb} I_{3\times3} & 0_{3\times3} \\ 0_{3\times3} & 0_{3\times3} & Q_{mb} I_{3\times3} \end{bmatrix}, \quad (13.17)$$

when the attitude, gyro bias and magnetometer bias are estimated. Here $Q_q$, $Q_{gb}$ and $Q_{mb}$ are the terms of the process noise covariance matrix which correspond to the attitude quaternion, gyro bias and magnetometer bias, respectively, and $Q_{q\_gb}$ are the terms for the noise covariance in between the quaternion and gyro bias states.

Adaptive UKF algorithm in this case is tested with simulations for a small satellite. For the magnetometer measurements, the sensor noise is characterized by zero mean Gaussian white noise with a standard deviation of $\sigma_m = 300nT$ and the constant magnetometer bias terms are accepted as $\bar{b}_m = \begin{bmatrix} 0.14 & 0.019 & 0.37 \end{bmatrix}^T \times 10^4 nT$.

Besides, the gyro random error is taken as $\sigma_v = 1 \times 10^{-3} \left[ \circ / \sqrt{h} \right]$, whereas the standard deviation of the gyro biases is $\sigma_u = 2 \times 10^{-3} \left[ \circ / \sqrt{h^3} \right]$. In this case, the analytically approximated process noise covariance matrix becomes

$$Q = \begin{bmatrix} (8.46E-14)I_{3\times3} & -(1.305E-21)I_{3\times3} & 0_{3\times3} \\ -(1.305E-21)I_{3\times3} & (2.61E-20)I_{3\times3} & 0_{3\times3} \\ 0_{3\times3} & 0_{3\times3} & 0_{3\times3} \end{bmatrix}. \qquad (13.18)$$

As the UKF parameters, $\kappa$ is selected as $\kappa = 1$ for the first scenario and $\kappa = -2$ for the second scenario where $f = 2(a + 1)$ and $a = 1$. Initial attitude errors for both filters are set to 30, 25 and 25° for pitch, yaw and roll axes, respectively. The initial estimation values for the gyro and magnetometer biases are taken as 0. Besides, the initial value of the covariance matrix is $P_0 = diag \begin{bmatrix} 0.1 & 0.1 & 0.1 & 10^{-7} & 10^{-7} & 10^{-7} & 10^{-3} & 10^{-3} & 10^{-3} \end{bmatrix}$ where $diag(\cdot)$ refers to diagonal matrix. The window size for the adaptive covariance estimation is taken as $\gamma = 0.01$ for both scenarios.

Last but not least, the initial value of the process noise covariance matrix for the AUKF is set to

$$Q = \begin{bmatrix} (1E-5)I_{3\times3} & -(1E-15)I_{3\times3} & 0_{3\times3} \\ -(1E-15)I_{3\times3} & (1E-12)I_{3\times3} & 0_{3\times3} \\ 0_{3\times3} & 0_{3\times3} & (1E-15)I_{3\times3} \end{bmatrix} \qquad (13.19)$$

Figure 13.5 gives the norm of the attitude estimation errors for UKF and AUKF, respectively. The mean of 10 different runs for each filter are presented. As seen, the AUKF gives better estimation results than the UKF with the analytically approximated process noise covariance matrix (UKF$_a$) and actually the results obtained by UKF$_a$ are far from being accurate. As discussed, the main reason for such estimation characteristics is assuming that the process noise covariance for the magnetometer biases is zero. That means we have perfect initial guess for the magnetometer biases and the filter shall trust the propagation model more than the incoming measurements. The magnetometer bias estimation results also support this statement and clearly show that the UKF$_a$ is incapable of estimating these parameters (Figure 13.6).

The converged values for the terms of the $Q$ matrix (Eq. (13.17)) show us the values that correspond to the attitude and gyro bias states differ from the analytically

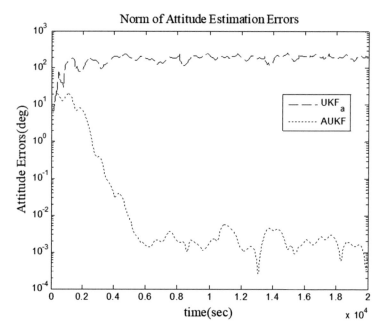

**FIGURE 13.5**    Norm of attitude estimation errors for UKF and AUKF.

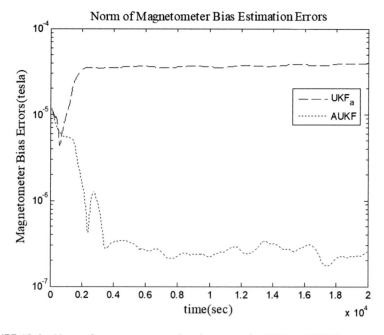

**FIGURE 13.6**    Norm of magnetometer estimation errors for UKF and AUKF.

approximated values. Besides, obviously the elements that correspond to the magnetometer biases are not zero in contrast with the analytical approximation.

$$
\hat{Q}_{adap} = \begin{bmatrix}
2.015E-7 & 0 & 0 & -1.047E-10 & 0 & 0 & 0 & 0 & 0 \\
0 & 4.956E-7 & 0 & 0 & -1.757E-10 & 0 & 0 & 0 & 0 \\
0 & 0 & 2.385E-7 & 0 & 0 & -9.987E-11 & 0 & 0 & 0 \\
-1.047E-10 & 0 & 0 & 1.178E-13 & 0 & 0 & 0 & 0 & 0 \\
0 & -1.757E-10 & 0 & 0 & 8.214E-14 & 0 & 0 & 0 & 0 \\
0 & 0 & -9.987E-11 & 0 & 0 & 6.573E-14 & 0 & 0 & 0 \\
0 & 0 & 0 & 0 & 0 & 0 & 7.515E-18 & 0 & 0 \\
0 & 0 & 0 & 0 & 0 & 0 & 0 & 1.000E-17 & 0 \\
0 & 0 & 0 & 0 & 0 & 0 & 0 & 0 & 6.568E-18
\end{bmatrix}
$$

(13.20)

Further investigations show that the $UKF_a$ may perform well only if we have the perfect initial guess for all the estimated parameters (such as zero initial attitude, gyro and magnetometer bias error). However, this is simply contradictory with the nature of the filtering concept. Although we may have rather better initial guesses for the attitude, the parameters like sensor bias are not known at all.

In conclusion, although analytical approximation for the process noise covariance matrix may give sufficiently accurate estimation results for the case, we use the UKF for estimation of attitude and gyro biases, such method fails when we include terms constant in time into the state vector, such as magnetometer biases. It is always possible to tune the filter by trial--error method as we showed in previous sections but it is a time-consuming and difficult process for large size state vectors. The given adaptive tuning method and so the AUKF algorithm is considerably simpler. If we compare the AUKF with the filter that uses analytically approximated covariance, when such approximation is possible, we see that both filters give similar results. Therefore, we may state that the given adaptation method is an easy way of tuning the filter especially in the absence of any analytical approximation for the calculation of $Q$ matrix and by using the AUKF it is possible to get accurate estimation results close to the optimality.

## 13.2 ADAPTATION OF A DYNAMICS-BASED ATTITUDE FILTER

UKF is a filtering algorithm which gives sufficiently good estimation results for estimation problems of nonlinear systems even when high nonlinearity is in question. However, in case of system uncertainty, UKF becomes to be inaccurate and diverges by time. In other words, if any change occurs in the system model, which is known as a priori, the filter fails. For an attitude filter, this kind of changes is more likely if we are using the satellite dynamics in the filter, thus a dynamics-based filter is being used. As discussed, changes in the system model may be due to the environmental changes (e.g. the disturbance torques), variations in the dynamics parameters such as spacecraft inertias and the actuator faults.

This section introduces an Adaptive Fading Unscented Kalman Filter (AFUKF) algorithm based on the correction of process noise covariance (Q-adaptation) for the case of mismatches with the system model. By the use of the presented adaptation scheme for the conventional UKF algorithm, change in the noise covariance is detected and corrected. Different than the adaptive filtering algorithm discussed previously in this chapter, the process noise covariance is not updated at each step; it is only corrected when the change occurs. This is advantageous for an agile filter design that is responding to the faults instantaneously and adapting itself. The investigated algorithm is tested as a part of the attitude estimation algorithm of a pico satellite. In this sense, different change scenarios for the process noise covariance are taken into consideration.

### 13.2.1 ADAPTIVE FADING UKF

In case of normal operation, where the model for the process noise covariance matches with the real values, UKF works well. However, when a change occurs in the system model, the filter fails and the estimation outputs become faulty [13].

Hence, an adaptive algorithm must be introduced so as an adaptation on process noise covariance (Q-adaptation) is performed and the estimations of the filter are corrected without affecting good estimation characteristic of the remaining process.

Adaptive algorithm performs the correction only when the real values of the process noise covariance does not match with the model used in the synthesis of the filter. Otherwise the filter works with regular algorithm in an optimal way. Adaptation occurs as a change in the predicted covariance. First, let us rewrite the predicted covariance as

$$P\left(k+1|k\right) = P^*\left(k+1|k\right) + Q\left(k\right) \tag{13.21}$$

So $P^*(k + 1|k)$ is the predicted covariance without the additive process noise. In order to adapt the covariance an adaptive fading factor is put into the procedure [13];

$$P\left(k+1|k\right) = P^*\left(k+1|k\right) + \Lambda\left(k\right)Q\left(k\right), \tag{13.22}$$

where $\Lambda(k)$ is the adaptive fading factor calculated in the base of residual for the state vector estimation, $\tilde{e}\left(k+1\right)$, which may be defined as

$$\tilde{e}\left(k+1\right) = y\left(k+1\right) - H\left(k+1\right)\hat{x}\left(k+1|k+1\right), \tag{13.23}$$

where $H(k + 1)$ is the measurement matrix constituted of the partial derivatives as

$$H\left(k+1\right) = \frac{\partial y}{\partial x}\bigg|_{\hat{x}\left(k+1|k+1\right)}. \tag{13.24}$$

Nonetheless the covariance matrix of the residue in Eq. (13.23) can be written as

$$P_{\tilde{e}}\left(k+1\right) = R\left(k+1\right) - H\left(k+1\right)P\left(k+1|k+1\right)H\left(k+1\right)^T. \tag{13.25}$$

The gain matrix is changed when the condition of

$$tr\left[\tilde{e}\left(k+1\right)\tilde{e}^T\left(k+1\right)\right] \geq tr\left[R\left(k+1\right)\right] - tr\left[H\left(k+1\right)P\left(k+1|k+1\right)H\left(k+1\right)^T\right]$$

$$\tag{13.26}$$

is the point at issue. Here $tr(\cdot)$ is the trace of the related matrix. Left-hand side of (13.26) represents the real filtration error while the right-hand side is the accuracy of the residue known as a result of a priori information. When the estimated observation vector $H(k+1)\hat{x}(k+1|k+1)$ is reasonably different from measurement vector, $y(k + 1)$, because of any change that occurs in $Q(k)$ real filtration error exceeds the theoretical one. Hence, the process noise covariance must be fixed hereafter by the use of defined adaptive fading factor $\Lambda(k)$. In order to calculate the adaptive fading factor, equality of

$$tr\left[\tilde{e}(k+1)\tilde{e}^T(k+1)\right] = tr\left[R(k+1)\right]$$
$$-tr\left[H(k+1)P(k+1|k+1)H(k+1)^T\right] \quad (13.27)$$

is used. If we replace $P(k + 1|k + 1)$ with its definition, then

$$tr\left[\tilde{e}(k+1)\tilde{e}^T(k+1)\right] = tr\left[R(k+1)\right]$$
$$-tr\left\{H(k+1)\left[P(k+1|k) - K(k+1)\right.\right. \quad (13.28)$$
$$\left.\left.\times P_{vv}(k+1|k)K^T(k+1)\right]H(k+1)^T\right\}$$

After that we should put Eq. (13.22) into Eq. (13.28) and

$$tr\left[\tilde{e}(k+1)\tilde{e}^T(k+1)\right] = tr\left[R(k+1)\right] - tr\left[H(k+1)P^*(k+1|k)H(k+1)^T\right]$$
$$-\Lambda(k)tr\left[H(k+1)Q(k)H(k+1)^T\right]$$
$$+tr\left[H(k+1)K(k+1)P_{vv}(k+1|k)K^T(k+1)H(k+1)^T\right]$$
$$(13.39)$$

Finally, if the knowledge of

$$tr\left[\tilde{e}(k+1)\tilde{e}^T(k+1)\right] = \tilde{e}^T(k+1)\tilde{e}(k+1) \quad (13.30)$$

is taken into consideration then the adaptive factor can be found as (note that discretization indices are not written for sake of readability) [13]

$$\Lambda = \frac{tr\left[R\right] - tr\left[HP^*H^T\right] + tr\left[HKP_{vv}K^TH^T\right] - \tilde{e}^T\tilde{e}}{tr\left[HQH^T\right]}. \quad (13.31)$$

On the other hand, as a main difference from the existing AFUKF algorithms process noise covariance matrix is not updated for whole the estimation procedure; adaptive algorithm is used only in case of changes and in all other cases procedure is run optimally with regular UKF. Fault detection is satisfied via a kind of statistical information. In order to achieve that, following two hypotheses may be introduced:

$\gamma_o$ – the system is normally operating;

$\gamma_1$ – there is a malfunction in the estimation system.

Failure detection is realized by the use of following statistical function [14]:

$$\beta(k) = \tilde{e}^T (k+1) \left[ R(k+1) - H(k+1) P(k+1|k+1) H(k+1)^T \right]^{-1} \tilde{e}(k+1) \quad (13.32)$$

This statistical function has $\chi^2$ distribution with $s$ degree of freedom, where $s$ is the dimension of the residual vector $\tilde{e}(k+1)$.

If the level of significance, $\alpha$, is selected as

$$P\{\chi^2 > \chi_{\alpha,s}^2\} = \alpha; \quad 0 < \alpha < 1, \quad (13.33)$$

the threshold value, $\chi_{\alpha,s}^2$ can be determined. Hence, when the hypothesis $\gamma_1$ is correct, the statistical value of $\beta(k)$ will be greater than the threshold value $\chi_{\alpha,s}^2$, i.e.:

$$\begin{aligned} \gamma_0 &: \beta(k) \le \chi_{\alpha,s}^2 \quad \forall k \\ \gamma_1 &: \beta(k) > \chi_{\alpha,s}^2 \quad \exists k. \end{aligned} \quad (13.34)$$

### 13.2.2 DEMONSTRATIONS FOR AN ADAPTIVE FADING UKF

Simulations for the Adaptive Fading UKF algorithm as a dynamics-based attitude estimator are realized for 2000 s with a sampling time of $\Delta t = 0.1$ s. For the used pico satellite model the inertia matrix is taken as

$$J = \begin{bmatrix} 2.1 \times 10^{-3} & 0 & 0 \\ 0 & 2.0 \times 10^{-3} & 0 \\ 0 & 0 & 1.9 \times 10^{-3} \end{bmatrix} \text{kg.m}^2. \quad (13.35)$$

Nonetheless the orbit of the satellite is a circular orbit with an altitude of $r = 550$ km.

For magnetometer measurements, sensor noise is characterized by zero mean Gaussian white noise with a standard deviation of $\sigma_m = 200nT$.

During simulations, for testing AFUKF algorithm, two different scenarios are taken into consideration. First the process noise covariance matrix is changed temporarily for a period. Second, a continuous change is implemented to the matrix; change is implemented at a specific time step and it is left as it is. Both changes are simply formed by multiplying covariance with a constant.

Besides, in case of a change in the process noise covariance, simulation is also achieved for regular UKF in order to compare results with AFUKF and understand efficiency of the adaptive algorithm in a better way. For AFUKF $\chi^2_{\alpha,s}$ is taken as 12.592, and this value comes from chi-square distribution when the degree of freedom is 6 and the reliability level is 95%.

First part of figures gives UKF and AFUKF parameter estimation results and the actual values in a comparing way. Second part of the figures shows the error of the estimation process based on the actual estimation values of the satellite. The last part indicates the variance of the estimation.

### 13.2.2.1 Temporary Uncertainty in Dynamics

In this scenario, the real process noise covariance matrix is changed in between 1700th and 1800th s. This change is simply formed by multiplying covariance with a constant and increasing the covariance for this time interval.

As it is apparent form Figure 13.7, regular UKF fails at estimating the attitude parameters when the change occurs. On the other hand, as given in Figure 13.8, AFUKF is not affected from the change in the process noise covariance and it satisfies its good estimation characteristic for the whole process (less than ±1° of estimation error for this case).

In Figure 13.9, an example for the estimation of an angular velocity via UKF is given. Besides Figure 13.10 shows same estimation procedure via AFUKF. It is obvious that AUFKF also corrects the error in the estimation of angular rates. Another fact that should be regarded, even though the estimations of UKF

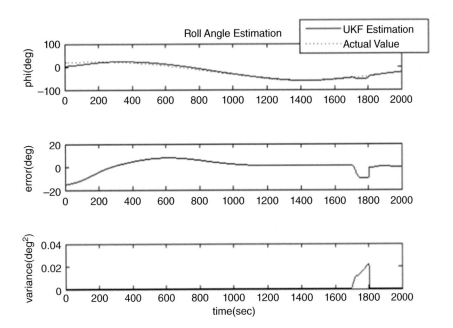

**FIGURE 13.7** Roll angle estimation via UKF in case of temporary change in system model.

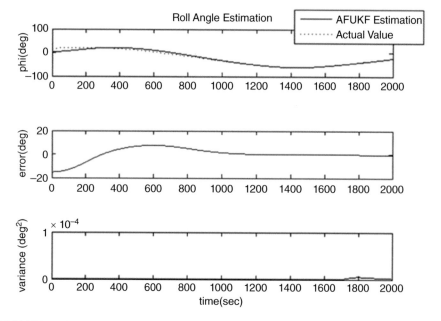

**FIGURE 13.8**  Roll angle estimation via AFUKF in case of temporary change in system model.

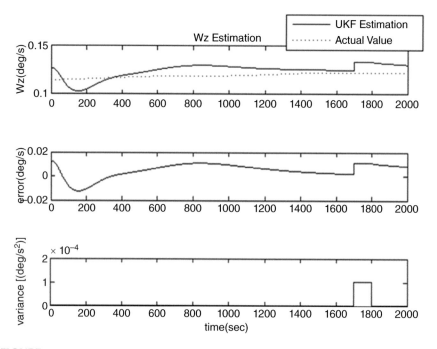

**FIGURE 13.9**  Wz estimation via UKF in case of temporary change in system model.

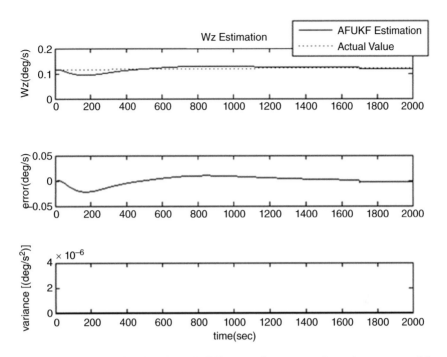

**FIGURE 13.10**   Wz estimation via AFUKF in case of temporary change in system model.

deteriorate after the 1700th s, the filter starts to converge again to the actual value when the change disappears and the process noise covariance returns to its original value. However as seen from Figure 13.9, for some of the estimated parameters it takes some time for filter to satisfy estimation accuracy within the desired limits. Hence, a change in the process noise covariance for 100 s affects the accuracy of the filter for a longer time. Similar results have been obtained for all other state parameters.

Last but not least, for the AFUKF even though we have additional matrix calculations for adaptive fading factor calculation, the computational load of the adaptive algorithm is not high compared to the regular UKF. Using the adaptive algorithm only when the fault is detected, further saves the computational load.

### 13.2.2.2  Permanent Uncertainty in Dynamics

A second scenario is formed by multiplying the process noise covariance matrix at 1700th s so as to increase it and leaving it as it is for the remaining process. Hence for the second scenario a permanent change is the point at issue.

Figure 13.11 shows that implemented change makes the regular UKF diverge by time. In other words, estimation error increases as the estimation goes on. On the other hand, AFUKF compensates this change by adapting itself and it gives accurate estimation results (Figure 13.12). Note that similar results can be obtained for all other parameters.

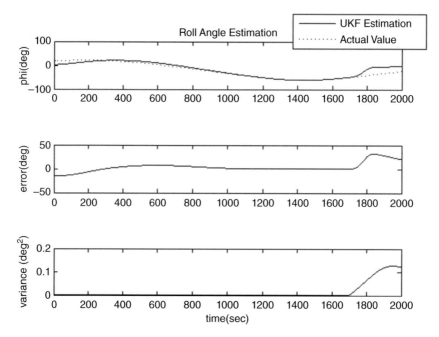

**FIGURE 13.11** Roll angle estimation via UKF in case of permanent change in system model.

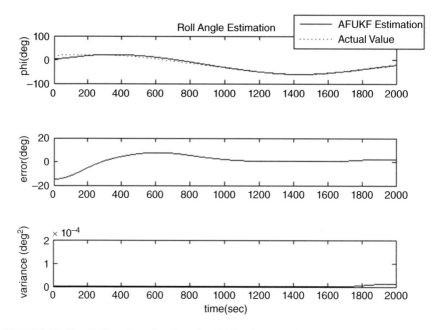

**FIGURE 13.12** Roll angle estimation via AFUKF in case of permanent change in system model.

## 13.3 CONCLUSION AND DISCUSSION

In this chapter, we discussed the details of adapting the process noise covariance (Q) matrix of an attitude filter. We first investigated the applicability of common adaptive methods for gyro-based attitude filtering and provided a performance comparison in between intuitively and adaptively tuned attitude filters. We showed that adaptively tuning the attitude filter can be specifically advantageous if additional states are being estimated together with the attitude and gyro biases. We demonstrated the performance of adaptive attitude filter when both the EKF and UKF are used as the main filtering algorithm. We specifically analyzed the attitude filter's performance when the process noise covariance matrix is tuned for optimal performance or adapted in real-time.

The high possibility of any kind of unexpected events in the space environment makes satellites vulnerable vehicles in point of view of attitude determination system. This is especially troublesome when a dynamics-based attitude filter is being used. In the second part of this chapter, we showed that it is not possible to get precise estimation results by regular UKF if the model for the process noise covariance mismatches with the real value as a result of change occurred somehow. On the other hand, the presented AFUKF algorithm for dynamics-based attitude estimation is not affected from those changes and secures its good estimation characteristic all the time.

Depending on the used attitude sensors, the preferred method of attitude filtering (e.g. gyro-based or dynamics-based) and the preferred algorithm for nonlinear filtering there are several options for an attitude determination algorithm design for a small satellite. Different adaptation methods presented in this chapter are suitable to be applied in these different scenarios and designers/engineers can easily adapt one of these methods to their algorithms.

## NOTES

1 Here we present the intuitive tuning process as a part of our discussion for adaptation of the gyro-based attitude filter. However, it should be noted that same procedure can be applied for any of the designed attitude filters.
2 Details will be discussed in Chapter 16.

## REFERENCES

1. J. Halverson, R. Harman, and R. Carpenter, "Tuning the solar dynamics observatory Kalman filter," *Proc. 2017 AAS/AIAA Astrodynamics Specialist Conference*, Stevenson, USA, 2017, pp. 1–14.
2. H. E. Soken and S.-I. Sakai, "Adaptive tuning of the unscented kalman filter for satellite attitude estimation," *J. Aerosp. Eng.*, vol. 28, no. 3, 2015.
3. R. L. Farrenkoph, "Analytic steady-state accuracy solutions for two common spacecraft attitude estimators," *J. Guid. Control. Dyn.*, vol. 1, no. 4, pp. 282–284, 1978.
4. Q. M. Lam and M. Jacox, "Can tuning really get your design back into the acceptable pointing knowledge requirement region?," *Proc. AIAA Guidance, Navigation, and Control (GNC) Conference*, Boston, USA, 2013.
5. A. H. Mohamed and K. P. Schwarz, "Adaptive Kalman filtering for INS/GPS," *J. Geod.*, vol. 73, no. 4, pp. 193–203, 1999.

6. D.-J. Lee and K. Alfriend, "Adaptive sigma point filtering for state and parameter estimation," in *AIAA/AAS Astrodynamics Specialist Conference and Exhibit*, Rhode Island, USA, 2004.

7. H. E. Soken and C. Hacizade, "Tuning the attitude filter: A comparison of intuitive and adaptive approaches," in *2019 9th International Conference on Recent Advances in Space Technologies (RAST)*, Istanbul, Turkey, 2019, pp. 747–752.

8. P. S. Maybeck, *Stochastic Models, Estimation, and Control*. New York, USA: Academic Press, 1982.

9. S. J. Julier, J. K. Uhlmann, and H. F. Durrant-Whyte, "A new approach for filtering nonlinear systems," *Proc. 1995 Am. Control Conf. - ACC'95*, Seattle, USA, pp. 0–4, 1995.

10. O. Khurshid, J. Selkäinaho, H. E. Soken, E. Kallio, and A. Visala, "Small satellite attitude determination during plasma brake deorbiting experiment," *Acta Astronaut.*, vol. 129, pp. 52–58, 2016.

11. H. E. Soken, "UKF Adaptation and Filter Integration for Attitude Determination and Control of Nanosatellites with Magnetic Sensors and Actuators," The Graduate University for Advanced Studies, 2013.

12. A. Fosbury, "Steady-state accuracy solutions of more spacecraft attitude estimators," in *AIAA Guidance, Navigation, and Control Conference*, Portland, USA, 2011.

13. H. E. Soken and C. Hajiyev, "Adaptive fading UKF with Q-adaptation: Application to picosatellite attitude estimation," *J. Aerosp. Eng.*, vol. 26, no. 3, 2013.

14. C. Hajiyev and H. E. Soken, "Robust adaptive unscented Kalman filter for attitude estimation of pico satellites," *Int. J. Adapt. Control Signal Process.*, vol. 28, no. 2, 2014.

# 14 Integration of R- and Q-Adaptation Methods

In Chapters 12 and 13, we presented different methods for adapting the measurement (R) and process (Q) noise covariance matrices of the attitude filter. The Q- and R-adaptation algorithms are solutions for different problems. As discussed, the Q-adaptation may be used as a tuning algorithm for the process noise covariance of the filter in order to ease the difficult tuning procedure and make the filter more efficient in the sense of estimation accuracy. Q-adaptation is also necessary when the system model used in the filter mismatches with the real one, as in case for environmental changes and/or actuator faults. On the other hand, the R-adaptation is performed as a measure against the possible measurement faults in the harsh space environment. In this section, we integrate these two filters.

The integration of the Q- and R-adaptation techniques is an open topic and there are numerous researches in the literature [1–3]. Indeed, there is not any stable integration method when both the R and Q matrices are estimated based on the innovation covariance throughout the whole estimation duration. In this case, the Q must be estimated assuming full knowledge of the R and vice versa. Nonetheless, we consider three cases for integrated Q and R adaptation in an attitude filter:

1. Making use of a fault isolation block the first the fault type is identified and then the necessary adaptation (either R or Q) is applied. This is a useful scheme specifically for a dynamics-based attitude filter when we assume the faults (or changes in the system) are not continuous and last only for a period of time. In this case, it is simply assumed that the faults in the system (dynamics) and measurements are not concurrent. The R and Q adaptation takes place in different times.

2. The Q-adaptation method estimates the Q matrix based on the residual covariance and the adaptation method for the R matrix is an innovation covariance-based scaling method, not the direct estimation of the matrix itself. The adaptation methods use different information sources, the R-adaptation uses the innovation and the Q adaptation uses the residual. Moreover, the Q-adaptation is performed by directly estimating the matrix whereas the R is adapted by scaling. This integrated scheme is especially useful when we adaptively tune the attitude filter (presumably a gyro-based one) and still make it robust against the measurement faults. In this case, Q-adaptation runs all the time, whereas R-adaptation is applied when a measurement fault is detected.

3. We use the nontraditional approach for attitude filtering that is discussed in Chapter 10. In this case, the attitude filter is inherently robust against the

measurement faults. We adapt only the Q matrix for satisfying the system fault tolerance in the filter.

Following subsections introduce the design of the integrated adaptive algorithms with these three approaches (by referring necessary equations from Chapters 12 and 13) and demonstrate the algorithms for small satellite attitude estimation.

## 14.1 INTEGRATION OF R- AND Q-ADAPTATION METHODS BY FAULT ISOLATION

### 14.1.1 INTEGRATION OF R- AND Q-ADAPTATION METHODS

In this section, we build a robust adaptive attitude filter using one of the R-adaptation method presented in Chapter 2 and the Q-adaptation method specifically presented in Chapter 13, Section 13.2.1. Any of the R-adaptation methods (adaptation using single scale factor or multiple scale factors) can be used for measurement fault tolerance. Yet, as discussed and demonstrated via simulations, most of the time R-adaptation with multiple scale factors is more advantageous. Thus, here, we also use multiple scale factors for adapting the R-matrix. As mentioned, the Q matrix is adapted using the adaptive fading factor given in Eq. (13.31).

Considering the nature of the problem, and necessary conditions to adapt the Q-matrix using Eq. (13.31) in particular, the integrated scheme here is more appropriate for using with a dynamics-based attitude filter. It can be used for any of the nonlinear filtering algorithms. Here we present it based on the Unscented Kalman Filter (UKF).

The fault, irrelevant of its type, is detected by the use of the following statistical function, which is same as the function we used in previous chapters for both detecting the faults in system and measurement:

$$\beta(k) = e^T(k+1)\left[P_{yy}(k+1|k) + R(k+1)\right]^{-1} e(k+1). \tag{14.1}$$

This statistical function has $\chi^2$ distribution with $s$ degree of freedom, where $s$ is the dimension of the innovation vector $e(k + 1)$.

The threshold value, $\chi^2_{\alpha,s}$ corresponding to the level of significance $\alpha$, can be determined. Hence, when the hypothesis $\gamma_1$ is correct, the statistical value of $\beta(k)$ will be greater than the threshold value $\chi^2_{\alpha,s}$, i.e.:

$$\begin{aligned} \gamma_0 &: \beta(k) \le \chi^2_{\alpha,s} \quad \forall k \\ \gamma_1 &: \beta(k) > \chi^2_{\alpha,s} \quad \exists k. \end{aligned} \tag{14.2}$$

The key point for integrated R and Q adaptive scheme is detecting the type of the fault (either a measurement malfunction or system-dynamics model fault). After that, the appropriate adaptation (R-adaptation or Q-adaptation) may be applied. The fault isolation can be realized by an algorithm similar to the one proposed for aircrafts in [4]. A Kalman filter (KF) that satisfies the Doyle-Stein condition is referred in [4] as the Robust KF insensitive to the actuator failures and may be used for this purpose.

A similar Robust UKF can be used in order to isolate the detected measurement and system failures as it is insensitive to the latter. Overall scheme for the integration of Q and R adaptations and so Robust Adaptive UKF (RAUKF) for attitude estimation can be seen in Figure 14.1.

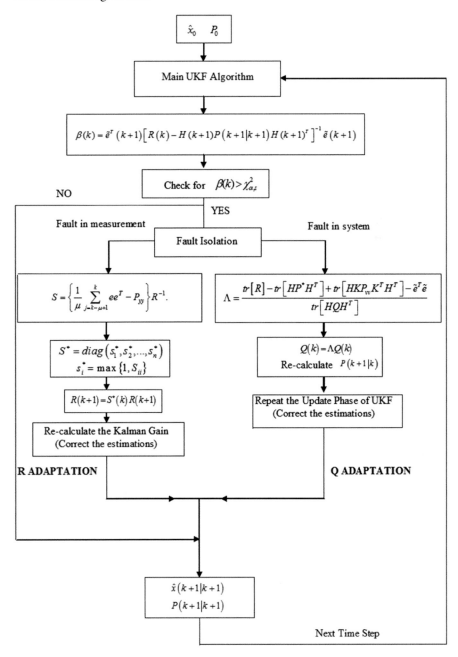

**FIGURE 14.1**   The overall estimation scheme for RAUKF

## 14.1.2 DEMONSTRATION OF R- AND Q-ADAPTATION METHODS

As a numerical example for the proposed RAUKF, the algorithm is tested via simulations for the attitude estimation of a pico satellite. The simulations are realized for 3000 seconds with a sampling time of $\Delta t = 0.1$ sec. For the used pico satellite model the inertia matrix is taken as

$$J = diag\left(2.1\times10^{-3}, 2.0\times10^{-3}, 1.9\times10^{-3}\right) \text{kg.m}^2.$$

Nonetheless the orbit of the satellite is a circular orbit with an altitude of $r = 550$ km.

For the magnetometer measurements, sensor noise is characterized by zero mean Gaussian white noise with a standard deviation of $\sigma_m = 200nT$.

During the simulations, for testing the fault detection and isolation procedures and also the RAUKF algorithm, two different faults are implemented to the systems. First, the process noise covariance matrix is changed temporarily for a period (in between 1700th and 1800th s). Second, a continuous bias is implemented to one of the magnetometers as a measurement fault (in between 2400th and 2500th s). For fault detection $\chi^2_{a,s}$ is taken as 12.592 and this value comes from chi-square distribution when the degree of freedom is 6 and the reliability level is 95%.

The graph for the statistical values of $\beta(k)$ in case of regular UKF utilization is shown in Figure 14.2. As seen, except the periods, where the measurement and system failures occur, $\beta(k)$ is lower than the threshold. On the contrary, when one of

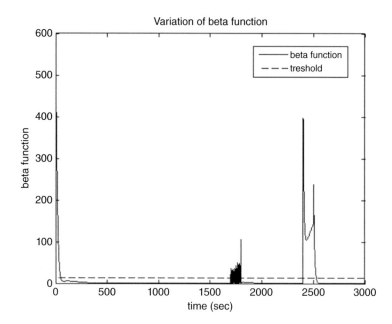

**FIGURE 14.2** Behavior of the statistics $\beta(k)$ in the case of system failure (in between 1700th and 1800th s.) and measurement failure (in between 2400th and 2500th s.) when the regular UKF is used.

these failures occurs $\beta(k)$ grows rapidly and exceeds the threshold value. Hence, $\gamma_1$ hypothesis in Eq. (14.2) is judged to be true and it is shown that by using such procedure the fault can be detected.

The results for the fault isolation process are given in Figure 14.3. As mentioned, the Robust UKF insensitive to system failures is used for this isolation process. The behavior of the statistics $\beta(k)$ in the case of measurement/system failures shows that detecting the system failure is not possible when the Robust UKF is used; on the other hand the sensor failure is detected immediately. Hence, it is shown that such procedure can be used for isolating the fault.

The results of the simulations achieved for testing the RAUKF algorithm in the presence of measurement/system failures are given in Figures 14.4 and 14.5. The simulations for the systems with faults are both performed via the regular UKF and RAUKF in order to compare results. The first part of Figures 14.4 and 14.5 gives the UKF and RAUKF parameter estimation results and the actual values by comparison. The second part of the figures shows the error of the estimation process based on the actual estimation values of the satellite. The last part indicates the variance of the estimation.

Figure 14.4 shows the estimation characteristic of the regular UKF in case of faults. As seen, in the cases of both system and measurement fault, the regular UKF algorithm fails and the estimations for these periods become inaccurate. On the other hand, Figure 14.6 shows that the RAUKF stands robust against both faults and satisfies accurate attitude estimation for the whole estimation procedure. Similar results achieved for the rest of the estimated parameters.

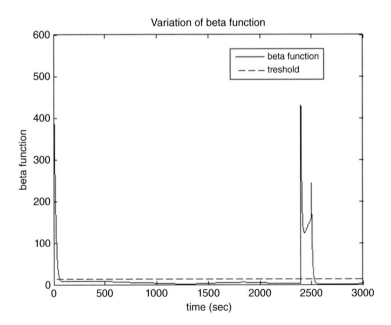

**FIGURE 14.3** Behavior of the statistics $\beta(k)$ in the case of system failure (in between 1700th and 1800th s) and measurement failure (in between 2400th and 2500th s) when the Robust UKF insensitive to the system failures is used.

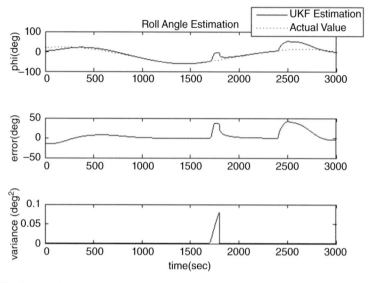

**FIGURE 14.4**    Estimation of roll angle via UKF.

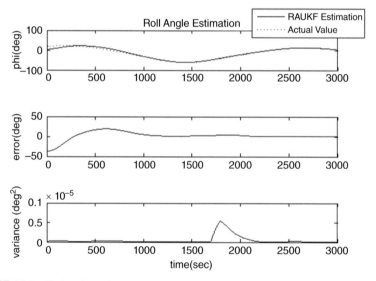

**FIGURE 14.5**    Estimation of roll angle via RAUKF.

In order to clarify the superiority of the RAUKF, the absolute values of estimation errors are tabulated for both regular and RAUKFs (Table 14.1). Columns, given for 1750th s, show the estimation errors for the time steps where system fault is the point at issue. Other columns are for the time steps where measurement fault occurs. Table supports the main idea that is stated by the figures and proves that in case of fault the RAUKF gives accurate estimation results even though the regular UKF fails.

Thus, the presented RAUKF algorithm performs correction for process noise covariance (Q-adaptation) or measurement noise covariance (R-adaptation) depending on

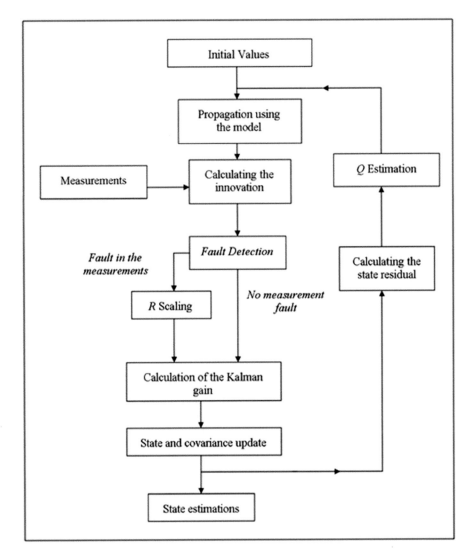

**FIGURE 14.6**   The Block-Scheme of RAUKF.

the type of the fault. Simulation results prove that proposed algorithm works well and provides accurate estimation results without being affected from both the system and measurement fault.

## 14.2  SIMULTANEOUS Q AND R ADAPTATION

The Q-adaptation method presented in this section estimates the Q matrix based on the residual covariance as presented in Section 13.1. The adaptation method for the R matrix is the innovation covariance-based scaling method given in Chapter 12. Thus, the adaptation methods use different information sources, the R-adaptation

**TABLE 14.1**

**Comparison of Absolute Values of Error for Regular and Robust Adaptive UKFs**

| Par. | Abs. Values of Error for Regular UKF | | Abs. Values of Error for RAUKF | |
|---|---|---|---|---|
| | 1750 s | 2450 s | 1750 s | 2450 s |
| $\varphi$ (deg) | 35.867 | 32.136 | 1.3067 | 0.2821 |
| $\theta$ (deg) | 5.9056 | 16.350 | 0.6813 | 0.5480 |
| $\psi$ (deg) | 16.780 | 46.254 | 0.9942 | 0.7500 |
| $\omega_x$ (deg/s) | 0.0137 | 0.1102 | 0.0075 | 0.0022 |
| $\omega_y$ (deg/s) | 0.0137 | 0.0694 | 0.0075 | 0.0017 |
| $\omega_z$ (deg/s) | 0.0137 | 0.0215 | 0.0075 | 0.0005 |

uses the innovation and the Q adaptation uses the residual. Moreover, the Q-adaptation is performed by directly estimating the matrix, whereas the R is adapted by scaling. Hence, these two methods can be run at the same time. Figure 14.6 shows the integration method with the key steps of the RAUKF.

There are two important points that should be regarded while designing the RAUKF [5]:

- The R scaling is performed only when a fault is detected in the measurements as given with Eq. (14.1). In all other cases, the filter runs with the regular algorithm only with the Q estimation (when there is no fault the algorithm is same as the UKF with Q estimation).
- The scale factor for the Q-adaptation in Eq. (13.6) should be selected carefully. If more aggressive adaptation is performed (such that $\gamma \approx 1$), the stability of the RAUKF might be affected in case of a measurement fault when both R and Q adaptations are necessary.

In order to test the RAUKF, a series of simulations are performed for a hypothetical nanosatellite. The attitude filter is a gyro-based filter that is running for an augmented state of $x = [q \; b_g \; b_m]^T$, including the magnetometer bias terms.

First, we show the effects of the Q-adaptation by comparing a regular UKF with the RAUKF. Then the results for the UKF and RAUKF are compared in case of sensor malfunction to clarify the necessity of the R-adaptation. The advantages of using the RAUKF as the attitude estimation algorithm are clearly shown by the discussing the estimation results.

The satellite is assumed to be a 3U cubesat with dimensions of 10 cm × 10 cm × 30 cm, an approximate mass of 3 kg and an approximate inertia matrix of $J = diag(0.055 \; 0.055 \; 0.017)$kg.m$^2$. A real mission example for a similar 3U nanosatellite might be seen in [6].

For the magnetometer measurements, the sensor noise is characterized by zero mean Gaussian white noise with a standard deviation of $\sigma_m = 300nT$ and the constant magnetometer bias terms are accepted as $\bar{b}_m = \begin{bmatrix} 0.14 & 0.019 & 0.37 \end{bmatrix}^T \times 10^4 nT$, which is reasonable when compared to the values in [7]. Moreover, the gyro random error is taken as $\sigma_v = 2.47 \left[ arcsec / \sqrt{s} \right]$, whereas the standard deviation of the gyro biases is $\sigma_u = 6.36 \times 10^{-4} \left[ arcsec / \sqrt{s^3} \right]$. Initial attitude errors are set to 30°, 25° and 25° for pitch, yaw and roll axes, respectively. The initial estimation values for the gyro and magnetometer biases are all taken as 0.

In Figure 14.7, the pitch angle estimation results that are obtained when we use the regular UKF or the proposed RAUKF are given in the same plot. As clearly seen, especially from the zoomed subplot, the results obtained by the RAUKF are far more accurate. This is mainly because of the nearly optimal values of the $Q$ matrix for the RAUKF that we cannot easily obtain by the trial–error method.

For the simulations the process noise covariance matrix for the UKF is

$$Q = \begin{bmatrix} \left(1 \times 10^{-3}\right) I_{3\times3} & -\left(1.5 \times 10^{-7}\right) I_{3\times3} & 0_{3\times3} \\ -\left(1.5 \times 10^{-7}\right) I_{3\times3} & \left(1 \times 10^{-10}\right) I_{3\times3} & 0_{3\times3} \\ 0_{3\times3} & 0_{3\times3} & \left(1 \times 10^{-12}\right) I_{3\times3} \end{bmatrix}.$$

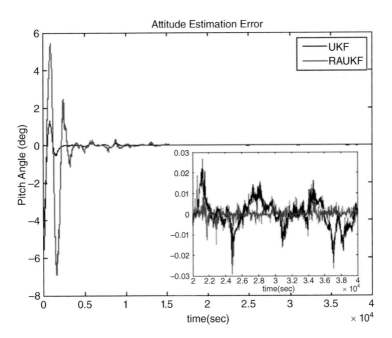

**FIGURE 14.7**  Estimation of the pitch angle via the RAUKF (red line) and UKF (black line).

That is also the same value with which the RAUKF is initialized. As to the values that the RAUKF converged are

$$
\bar{Q} = \begin{bmatrix} \bar{Q}_q & \bar{Q}_{q\_gb} & 0_{3\times3} \\ \bar{Q}_{q\_gb} & \bar{Q}_{gb} & 0_{3\times3} \\ 0_{3\times3} & 0_{3\times3} & \bar{Q}_{mb} \end{bmatrix},
$$

where

$$
\bar{Q}_q = \begin{bmatrix} 1.9372\times10^{-6} & 0 & 0 \\ 0 & 3.1132\times10^{-6} & 0 \\ 0 & 0 & 3.6348\times10^{-6} \end{bmatrix},
$$

$$
\bar{Q}_{gb} = \begin{bmatrix} 1.9528\times10^{-13} & 0 & 0 \\ 0 & 2.7443\times10^{-13} & 0 \\ 0 & 0 & 2.5297\times10^{-13} \end{bmatrix},
$$

$$
\bar{Q}_{mb} = \begin{bmatrix} 1.694\times10^{-15} & 0 & 0 \\ 0 & 1.4116\times10^{-15} & 0 \\ 0 & 0 & 2.3095\times10^{-15} \end{bmatrix},
$$

$$
\bar{Q}_{q\_gb} = \begin{bmatrix} -5.6722\times10^{-10} & 0 & 0 \\ 0 & -8.6334\times10^{-10} & 0 \\ 0 & 0 & -8.7782\times10^{-10} \end{bmatrix}.
$$

As seen the estimated $Q$ values are absolutely different from the initial values and it is not easy to guess such values by trial–error method.

In case we use the RAUKF the sensor calibration performance is also highly increased as may be seen in Figure 14.8, which presents the magnetometer bias estimation error for the magnetometer aligned in the $z$ axis. Indeed, such increment is tightly related to the increased attitude estimation accuracy. The RAUKF itself increases both attitude estimation and sensor calibration performance as the estimated Q values match with the real values. Besides, when the magnetometer biases are estimated more precisely that brings about better attitude estimation results since the accuracy of the incoming measurements is increased.

Next the RAUKF is tested for the continuous bias failure. A constant value is added to the measurements of the magnetometer aligned in the $x$ axis between the 30000th and 30200th s for a period of 200 s such that

$$
B_x(\boldsymbol{q},t) = B_x(\boldsymbol{q},t) + 20000nT \quad t = 30000...30200 \text{ s.}
$$

A deviation in the bias with this amount is reasonable when the values given in [7] are taken into consideration.

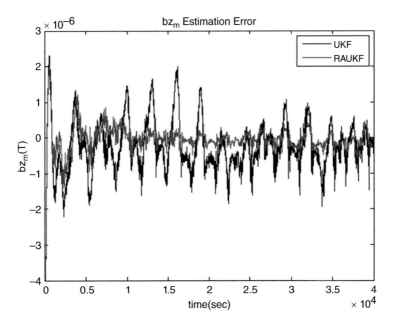

**FIGURE 14.8**    Bias estimation error for the magnetometer aligned in the z axis: the RAUKF (red line) and UKF (black line).

Figure 14.9 gives the roll angle estimation result comparing the cases that the estimator is the RAUKF or just the regular UKF. As expected, the UKF estimations deteriorate in case of measurement fault and it takes an additional 1000 s for the filter to converge again and satisfy estimation results with error less than 0.1°. On the other hand, the RAUKF maintains its good estimation performance even in case of the fault.

Similar behavior can be seen for the magnetometer bias estimation results (Figure 14.10). An additional bias that is experienced because of the sensor fault is considered as a variation in the estimated bias terms by the regular UKF. Therefore, UKF bias estimations, specifically for the magnetometer with the fault, worsen. However, the RAUKF keep providing the accurate bias estimation results.

The simulation results show that the RAUKF performs well under all conditions including the sensor fault case and give better estimation results than the regular UKF algorithm. Besides, the demonstrations prove that the proposed integration scheme for two different adaptation methods works properly. The presented filtering algorithm considerably increases the estimation performance and is fault tolerant against the sensor malfunctions.

## 14.3  NONTRADITIONAL ATTITUDE FILTERING WITH Q-ADAPTATION

The main advantages of the nontraditional approach for attitude filtering, which is integrating one of the single-frame attitude estimation methods with the filtering algorithm, were discussed in Chapter 10. In this chapter, we will discuss another

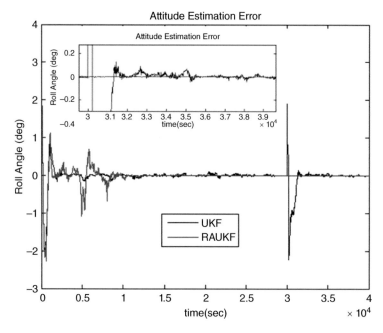

**FIGURE 14.9**  Estimation of the roll angle via the RAUKF (red line) and UKF (black line) in case of measurement malfunction.

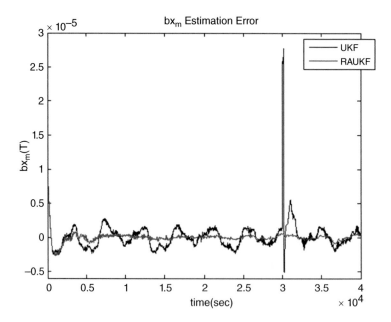

**FIGURE 14.10**  Bias estimation error for the magnetometer aligned in the x axis in case of measurement malfunction: the RAUKF (red line) and UKF (black line).

advantage of the nontraditional attitude filtering that appears when a fault tolerant attitude estimator is being designed.

The nontraditional approach has the ability to be robust against measurement faults and has a better accuracy performance compared to a filter that adapts the measurement noise covariance matrix (R matrix) based on the innovation sequence [8]. In [8], a nontraditional filter that integrates Singular Value Decomposition (SVD) method with the UKF and its R-adaptive version are compared for different fault scenarios. The SVD/UKF is used without any further modification for sensor faults. For the R-adaptive version (SVD/RUKF), the filter's measurement noise covariance matrix is tuned with a scale factor. It is concluded that the nontraditional approach is inherently robust against the sensor faults as the vector measurements are first preprocessed in the SVD before running the UKF and the R matrix for the UKF is autonomously tuned depending on the covariance for the SVD estimates. Although the fault cannot be compensated completely, its effect on the attitude estimation accuracy is minimized. Results show that it is more advantageous to use the SVD/UKF algorithm in both faulty and fault-free modes. In [9], the authors analyzed the filter's agility (how quickly it responds to the changes in the environment), especially while going into/out of the eclipse, by testing with three pairs of process noises without any adaptation rule. However, they did not consider any type of faults and suggest any adaptation rule.

Here, we use the SVD/UKF algorithm and adapt the UKF's Q matrix with the covariance scaling algorithm similar to one given in Section 13.2. It should be noted here for Q adaptation in SVD/UKF algorithm any of the methods in Chapter 13 can be used depending on the type of the attitude filter (e.g. dynamics-based or gyro-based). Our intention is to adapt the dynamics-based filter against any changes in the system model. Moreover, since the measurement model becomes linear, when we first process the measurements using the SVD, we can apply a straightforward covariance matching-based adaptation for the Q matrix, which is slightly different than the one given in Section 13.2. As the nontraditional attitude filtering algorithm is inherently adaptive against especially noise-increment type of measurement faults, adaptation of the Q matrix, in addition, provides us an algorithm which is capable of adapting the process and measurement noise covariance matrices simultaneously. Thus, the attitude and attitude rates of the satellite can be estimated accurately in spite of any change in the process and measurement noise covariance [10].

### 14.3.1 ADAPTING THE Q MATRIX IN NONTRADITIONAL FILTER

To adapt the SVD/UKF to the changing conditions the Q matrix of the UKF is tuned with an algorithm similar to the one presented in Section 13.2. However, since the measurement model becomes linear in the nontraditional filtering approach, the adaptation rule is similar to that of a linear KF and easier to apply.

Here we adapt the algorithm for possible faults in the system (dynamics) model. Thus, adaptive algorithm performs the correction only when the real values of the process noise covariance does not match with the model used in the synthesis of the filter. Otherwise, the filter works with regular algorithm [11]. Adaptation occurs as a change in the predicted covariance. First, let us rewrite the predicted covariance of UKF as [12]

$$P(k+1|k) = P^*(k+1|k) + Q(k).$$  (14.3)

So $P^*(k + 1|k)$ is the predicted covariance without the additive process noise. In order to adapt the covariance an adaptive scaling factor is put into the procedure;

$$P(k+1|k) = P^*(k+1|k) + \Lambda(k)Q(k),$$  (14.4)

where $\Lambda(k)$ is the adaptive scaling factor calculated in the base of innovation sequence of UKF[1]. The innovation $\nu(k + 1)$ is found as the difference between the actual observation and the predicted observation:

$$v(k+1) = y(k+1) - \hat{y}(k+1|k),$$  (14.5)

The innovation covariance is

$$P_{vv}(k+1|k) = H(k+1)P(k+1/k)H^T(k+1) + R(k+1).$$  (14.6)

Here $H(k)$ is the measurement matrix, $R(k + 1)$ is the covariance matrix of measurement noise.

The innovation covariance Eq. (14.6) can be expressed in the form below by substituting Eq. (14.4) into Eq. (14.6):

$$\begin{aligned} P_{vv}(k+1|k) = H(k+1)P^*(k+1/k)H^T(k+1) \\ + \Lambda(k)\left[H(k+1)Q(k)H^T(k+1)\right] + R(k+1). \end{aligned}$$  (14.7)

The scaling factor $\Lambda(k)$ is defined from the expression (14.7) as [13]

$$\Lambda(k) = \frac{P_{vv}(k+1|k) - H(k+1)P^*(k+1/k)H^T(k+1) - R(k+1)}{H(k+1)Q(k)H^T(k+1)}$$  (14.8)

or

$$\Lambda(k) = \frac{\dfrac{1}{\mu}\sum_{j=k-\mu+1}^{k} v(k+1)v^T(k+1) - H(k+1)P^*(k+1/k)H^T(k+1) - R(k+1)}{H(k+1)Q(k)H^T(k+1)}.$$  (14.9)

Furthermore, as discussed, this Q-adaptation method for the SVD/UKF approach is proposed for coping with possible changes in the system model. Thus, the UKF is not adapted all the time and the procedure given above is applied whenever a change is detected in the system. For change detection, Eqs. (14.1) and (14.2) can be used.

The structural scheme of the presented SVD-aided adaptive UKF algorithm is given in Figure 14.11.

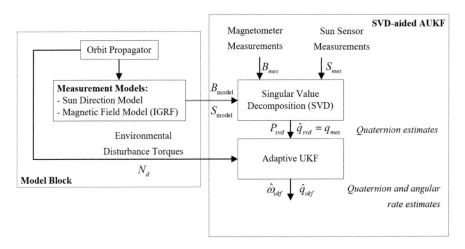

**FIGURE 14.11**   The proposed SVD/AUKF algorithm.

## 14.3.2 DEMONSTRATIONS OF SVD/AUKF

The proposed nontraditional attitude filter is demonstrated with simulations for a nanosatellite. We consider the nanosatellite with principal moments of inertia $J = diag\begin{bmatrix} 0.055 & 0.055 & 0.017 \end{bmatrix} kg \cdot m^2$. The orbit is almost circular with inclination $i = 74.1°$, eccentricity $e = 0.0007$, and average altitude 600 km. Algorithm runs for 6000 s and with 1 Hz sampling rate for the sensors. The filters are also propagated with a sampling time of $\Delta t = 1$ s. It is considered that the satellite has magnetometers and Sun sensors providing vector measurements. The sensor noise is characterized using the standard deviations of $\sigma_B = 300$ nT for magnetometers and $\sigma_S = 0.002$ for Sun sensors. The initial state of the filters is taken as $x_0 = \begin{bmatrix} 0 & 0 & 0 & 1 & 0 & 0 & 0 \end{bmatrix}$. The initial dynamic system's noise covariance is selected as

$$Q = \begin{bmatrix} 10^{-4} I_{3\times 3} & 0 \\ 0 & 10^{-9} I_{3\times 3} \end{bmatrix}.$$

The single-frame attitude estimation methods can only work when at least two measurements are available at the same time and vectors are not parallel to each other (see Chapters 5 and 6 for more details). Thus, the single-frame method fails when the satellite is in eclipse and two vectors are parallel. In order to show the behavior of the filter in these intervals, an eclipse period is introduced between 2000th and 3000th s. In the first section, two different sensor faults are considered for both magnetometer and Sun sensor: continuous bias and noise increment. The measurement faults occur between 4500th and 5500th s. Then, in the second section, process noise increment fault is considered for the simultaneous adaptation case for the same interval of the measurement faults.

### 14.3.2.1 SVD/AUKF with Continuous Bias at Measurements

In the first simulation scenario, the continuous bias term is simulated by adding a constant vector $\begin{bmatrix} 1000 & 2000 & 3000 \end{bmatrix}^T nT$ to the measurements of the magnetometers between 4500th and 5500th s as

$$B_b(k) = A(k)B_o(k) + \eta_1(k) + \begin{bmatrix} 1000 & 2000 & 3000 \end{bmatrix}^T, (4500 \le k \le 5500),$$

$$(14.10)$$

where $B_m(k)$ is the magnetometer measurements, $v_{Bm}(k)$ is the zero mean magnetometer measurement noises. In this interval, variation in the measurement variance values provided by the SVD algorithm can be seen in Figure 14.12.

As discussed in Chapter 12, when we use a traditional filtering algorithm, the filter should be tuned using an adaptation method to make it robust against the additional bias that is experienced for a period of time. In the SVD/AUKF case, the filter is R adaptive in nature as the nontraditional approach is used. Together with SVD algorithm's estimation covariance the R matrix terms for the filter should autonomously increase in case of a bias in the measurements. Thus, we expect that the filter can compensate any additional bias. However, the attitude estimation results in Figure 14.13 show that the filtering results are highly dependent on the SVD estimations. As the Figure 14.12 shows, R terms are increasing very slightly and not at a sufficient level to compensate the bias. The bias in the magnetometer measurements is reflected to the results of both the SVD algorithm and the proposed SVD/AUKF algorithm. SVD/AUKF algorithm's capability to cope with the bias type measurement faults, which is supposed to be an inherent characteristic, is limited.

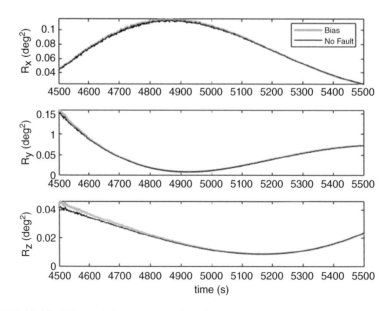

**FIGURE 14.12** Diagonal elements of $R$ when there is bias type of fault in the magnetometer measurements.

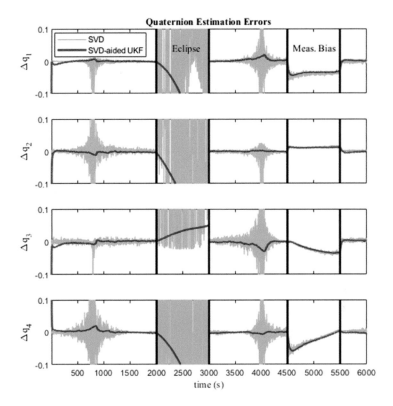

**FIGURE 14.13** Quaternion estimation errors when there is bias type of fault in the magnetometer measurements.

In Figure 14.13, there are two increases for SVD estimation results seen at about 700 and 4000 s. In those periods, Sun direction and magnetic field vectors are getting closer to being parallel. Therefore, attitude estimation error for the SVD algorithm, which needs at least three independent parameters for attitude estimation, increases. Yet the SVD/AUKF algorithm can provide much accurate attitude results compared to the SVD. This shows in any case the SVD/AUKF approach is more advantageous to use compared to a standalone SVD algorithm.

### 14.3.2.2 SVD/AUKF with Measurement Noise Increment

In the second measurement malfunction scenario, measurement fault is characterized by multiplying the standard deviation of the noise of the magnetometer measurements with a constant term in between 4500th and 5500th s. For the test case, the constant term is selected as 10,

$$B_b(k) = A(k)B_o(k) + \eta_1(k) \times 10, \; (4500 \leq k \leq 5500). \tag{14.11}$$

In Figure 14.14, we see that the R matrix elements are varying depending on the intensity of the noise to compensate the noise increment in between 4500th and 5500th s. Here the R matrix that is directly coming from the SVD algorithm (as the

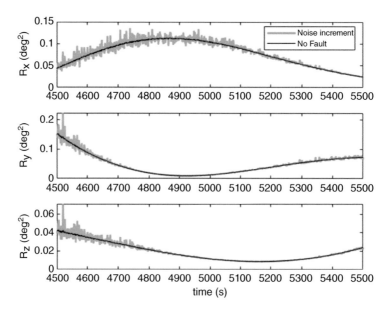

**FIGURE 14.14** Diagonal elements of when there is noise increment type of fault in the magnetometer measurements.

estimation covariance of the SVD) is shown. What we would like to emphasize here, in case we adapt R matrix using an adaptation rule (one of the methods in Chapter 12), the behavior of the adapted R matrix would be similar to this one. Yet this time, in the proposed SVD/AUKF algorithm, it is happening inherently. The attitude estimation errors by the SVD and the SVD/AUKF are presented in Figure 14.15. Even if the level of the measurement noise increment is large, the filter is capable of reducing the effect of the noise on the estimated attitude. The results clearly show that the filter works better for the noise increment type of measurement fault than the continuous bias case (Figures 14.13 and 14.15). It should be noted that in the figures, noise increment is abbreviated as N.I.

The nontraditional filter has the ability to update the measurement covariance at each step. The variations of $R$ because of the noise increment and continuous bias type of faults in the magnetometer measurements are directly sensed via the measurement error covariance matrix calculated by the SVD. The proposed algorithms are tested for only the magnetometer faults. It cannot remove the fault completely, but it is advantageous to use a nontraditional filtering algorithm in case of any malfunction in order to have a reasonably well attitude estimate.

### 14.3.2.3 SVD/AUKF with System Change

So far, the inherent R adaptation of the SVD/AUKF has been investigated. Here, we show that the nontraditional attitude filter can be made simultaneously $R$ and $Q$ adaptive. Similarly, with the previous section, eclipse period is from 2000th to 3000th s. The process noise is increased between 4500th and 5500th s to simulate the

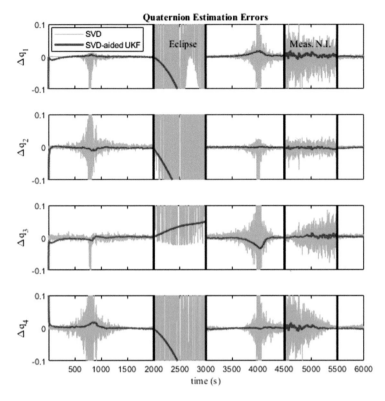

**FIGURE 14.15** Quaternion estimation errors when there is noise increment type of fault in the magnetometer measurements.

environment change and evaluate the performance of the proposed $Q$-adaptation algorithm for the attitude filter.

As seen in Figure 14.16, the proposed nontraditional filter can estimate all the quaternion terms accurately except two periods, which are eclipse and the process noise increment. We are especially interested in the deterioration around the process noise increment and the results show that the filter's output may become even worse than the SVD estimates in this period. Also, in general we can get angular velocity estimates close to the actual values (Figure 14.17). Yet the accuracy is not as high as one may expect. Low process noise values for the angular velocity estimation disturb the settling of the filter. Furthermore, eclipse and environmental changes prevent the filter to settle to an accurate angular velocity estimate. Tuning of the SVD/AUKF algorithm for more accurate angular rate estimation should be investigated. However, what is important here is the fact that the process noise increment deteriorates the angular rate estimates even more. Obviously longer, the process noise increment lasts higher the oscillations for angular rate estimations (specifically for $\omega_x$ and $\omega_y$) become. This may cause filter's divergence if the process noise increment lasts for a long duration.

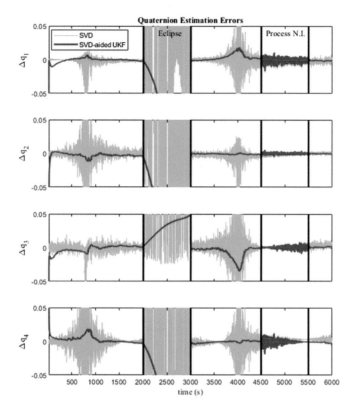

**FIGURE 14.16**   Quaternion estimation errors using the SVD and SVD/UKF in case of process noise increment.

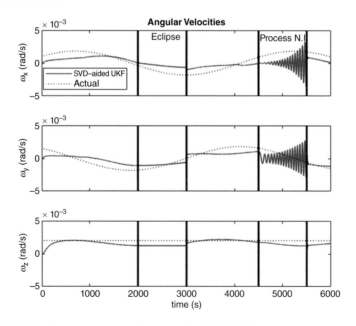

**FIGURE 14.17**   Angular rate estimation errors for SVD/UKF.

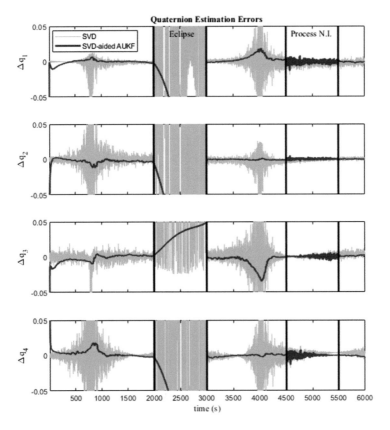

**FIGURE 14.18**  Quaternion estimation errors using the SVD and SVD/AUKF in case of process noise increment.

Figure 14.18 shows the SVD and the SVD/AUKF quaternion estimation errors. We see that compared to the SVD-aided UKF provides more accurate results with lesser noise in the fault region. The RMSE values presented in Table 14.2 give us opportunity to make a clearer accuracy evaluation for these two filters including only the parts where sensor fault and solar eclipse occurred. As expected, the angular rate estimation

**TABLE 14.2**
**RMSE for SVD/UKF and SVD/AUKF**

| RMSE | SVD/UKF | SVD/AUKF |
|---|---|---|
| $\Delta q_1$ | 0.0035 | 0.0024 |
| $\Delta q_2$ | 0.0014 | 0.0011 |
| $\Delta q_3$ | 0.0023 | 0.0016 |
| $\Delta q_4$ | 0.0032 | 0.0021 |
| $\omega_x$ (rad/s) | 0.0019 | 0.0014 |
| $\omega_y$ (rad/s) | 0.0013 | 0.0006 |
| $\omega_z$ (rad/s) | 0.0006 | 0.0006 |

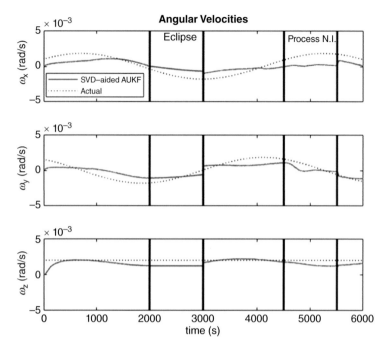

**FIGURE 14.19**  Angular rate estimation errors for SVD/AUKF.

by the SVD/AUKF algorithm is more accurate in the period of process noise increment as well (Figure 14.19) and we do not observe any increasing oscillation type
error in the estimation results. In fact, any change in the dynamics or environment
affects the angular rate estimations more than the attitude estimations. Thus, we have
higher improvement in the attitude rates estimates, when the Q-adaptive filter is used.

## 14.4  CONCLUSION AND DISCUSSION

In this chapter, simultaneous adaptation of the process and measurement noise covariance matrices for the traditional and nontraditional attitude filtering algorithms is
discussed. By making use of the R- and Q-adaptation methods, three different integration schemes are considered.

The first of these integration schemes is making use of a fault isolation block. In
this scheme, first the fault type is identified and then the necessary adaptation (either
R or Q) is applied. This is a useful scheme specifically for a dynamics-based attitude
filter when we assume the faults (or changes in the system) are not continuous and
last only for a period of time. In this case, it is simply assumed that the faults in the
system (dynamics) and measurements are not concurrent. The R and Q adaptation
take place in different times.

The second adaptation scheme is adaptively tuning the Q matrix all the time and
adapting the R matrix whenever a measurement malfunction is detected. The
Q-adaptation method estimates the Q matrix based on the residual covariance and the

adaptation method for the R matrix is an innovation covariance-based scaling method, not the direct estimation of the matrix itself. The adaptation methods use different information sources, the R-adaptation uses the innovation and the Q adaptation uses the residual. Moreover, the Q-adaptation is performed by directly estimating the matrix whereas the R is adapted by scaling. This integrated scheme is especially useful when we adaptively tune the attitude filter (presumably a gyro-based one) and still make it robust against the measurement faults.

In the last integration method, we used the nontraditional approach for attitude filtering that is discussed in Chapter 10. In this case the attitude filter is inherently robust against the measurement faults. We adapt only the Q matrix for satisfying the system fault tolerance in the filter. We showed that the Q-adaptation method is rather simpler for a nontraditional attitude filter since the measurement model becomes linear. A comprehensive discussion on the fault tolerance of the filter to different types of fault is provided.

## NOTE

1 This is the main difference compared to the adaptive fading UKF algorithm presented in Section 13.2. Here we adapt the filter based on the innovation sequence rather than the residual sequence.

## REFERENCES

1. A. Almagbile, J. Wang, and W. Ding, "Evaluating the performances of adaptive Kalman filter methods in GPS/INS integration," *J. Glob. Position. Syst.*, vol. 9, no. 1, pp. 33–40, 2010.
2. C. Hajiyev and H. E. Soken, "Robust adaptive unscented Kalman filter for attitude estimation of pico satellites," *Int. J. Adapt. Control Signal Process.*, vol. 28, no. 2, 2014.
3. H. E. Soken, *UKF adaptation and filter integration for attitude determination and control of nanosatellites with magnetic sensors and actuators*. Hayama, Japan: The Graduate University for Advanced Studies, 2013.
4. C. Hajiyev and F. Caliskan, *Fault Diagnosis and Reconfiguration in Flight Control Systems*, vol. 2. Boston, MA: Springer US, 2003.
5. H. E. Soken, C. Hacizade, and S. Sakai, "Simultaneous adaptation of the process and measurement noise covariances for the UKF applied to nanosatellite attitude estimation," *IFAC Proc. Vol.*, vol. 47, no. 3, pp. 5921–5926, 2014.
6. J. Reijneveld and D. Choukroun, "Attitude control of the Delfi-n3Xt satellite," in *AIAA Guidance, Navigation, and Control Conference*, Minneapolis, USA, 2012.
7. S. Sakai, N. Bando, S. Shimizu, Y. Maru, and H. Fuke, "Real-time estimation of the bias error of the magnetometer only with the gyro sensors," in *28th International Symposium on Space Technology and Science*, Okinawa, Japan, 2011, pp. 1–6.
8. D. Cilden-Guler, H. E. Soken, and C. Hajiyev, "Non-Traditional Robust UKF against Attitude Sensors Faults," in *31st International Symposium on Space Technology and Science (ISTS)*, Matsuyama, Japan, 2017, p. d-077.
9. D. Cilden, H. E. Soken, and C. Hajiyev, "Nanosatellite attitude estimation from vector measurements using SVD-AIDED UKF algorithm," *Metrol. Meas. Syst.*, vol. 24, no. 1, 2017.
10. C. Hajiyev, H. E. Soken, and D. Cilden-Guler, "Nontraditional attitude filtering with simultaneous process and measurement covariance adaptation," *J. Aerosp. Eng.*, vol. 32, no. 5, p. 04019054, Sep. 2019.

11. C. Hajiyev, D. C. Guler, and H. E. Soken, "Nontraditional UKF based nanosatellite attitude estimation with the process and measurement noise covariances adaptation," in *Proceedings of 8th International Conference on Recent Advances in Space Technologies, RAST 2017*, Istanbul, Turkey, 2017.

12. H. E. Soken and C. Hajiyev, "Adaptive fading UKF with Q-adaptation: Application to picosatellite attitude estimation," *J. Aerosp. Eng.*, vol. 26, no. 3, 2013.

13. C. Hajiyev, H. E. Soken, and D. C. Guler, "Q-Adaptation of SVD-aided UKF algorithm for nanosatellite attitude estimation," *IFAC-PapersOnLine*, vol. 50, no. 1, pp. 8273–8278, Jul. 2017.

# 15 In-Orbit Calibration of Small Satellite Magnetometers
## Batch Calibration Algorithms

Magnetometers are an integral part of attitude determination system for the low-Earth orbiting small satellites as they are lightweight, inexpensive and reliable. Yet using magnetometers for attitude determination is not straightforward because of the sensor errors. These errors limit the overall achievable attitude determination accuracy. Thus far different methods to cope with magnetometer errors and calibrate the magnetometers have been proposed. A new research field is the specific errors for magnetometers onboard the small satellites and their time-variation characteristics. In accordance, algorithms which consider also the time-varying error terms are proposed.

In Chapters 15 and 16, we review recent calibration algorithms for small satellite magnetometers. We emphasize the algorithms that consider time-varying magnetometer errors or algorithms that are responsive to the time-variation in the errors. In this sense, in this chapter, we first present the batch methods which are capable of calibrating the magnetometers in different circumstances against the time-varying errors. In particular, we discuss the optimization algorithms for batch magnetometer calibration methods. We review new optimization algorithms as a part of the search for ensuring algorithm's convergence to the global minima.

This and the following chapter discuss the underlying ideas and assumptions of the magnetometer calibration methods and present examples for selected ones together with their results. Chapters mainly follow the survey articles given in [1, 2]. The reader should note that, although the algorithms in some of the referenced papers are implemented on platforms different than satellites, we present these algorithms considering their applicability for calibration of small satellite magnetometers against time-varying errors. Our aim is to present the algorithms in a broad perspective such that the reader easily understands how they work and what their pros and cons are. The relationship between different algorithms is examined to show the reader how the algorithms can be integrated to improve the accuracy of the calibration. Further research directions for the small satellite magnetometer calibration are discussed as well.

## 15.1 REQUIREMENT FOR MAGNETOMETER CALIBRATION

Three-axis magnetometers (TAMs) are part of the attitude sensor package for almost all of the low-Earth orbiting small satellites [3, 4]. Being lightweight, small, reliable and having low-power requirements make them ideal for small satellite applications.

In fact, in addition to their well-known implementation for attitude determination, TAMs can be used also for orbit determination [5].

The main challenge for using the TAM for attitude estimation is sensor errors. In fact, TAMs are the most prone ones to having errors among the attitude sensors presented in Chapter 3. Calibration of all the attitude sensors is a necessity to improve the attitude estimation accuracy; however, one-time calibration might be sufficient for all of them except the magnetometers. The magnetometer errors may change throughout the operation as will be discussed in this chapter. These errors limit the overall achievable attitude estimation accuracy, unless they are taken care of. Until now researchers suggested various solutions for dealing with the magnetometer errors. These include straightforward satellite design issues such as keeping the TAMs far from the electromagnetic interference. Best example is locating the sensors on platforms separated from the satellite main body, such as the tip of a boom [6]. However, for especially nanosatellite missions, this is not an option as the size of the satellite should be kept to a minimum. In this case, the magnetometers must be in-flight calibrated.

A new research area is the time-varying errors for small satellite magnetometers [5–9].Onboard the small satellites the magnetometers are located close to the other subsystems because of size limitations. Thus, nearby electronics and satellite magnetic torquers (MTQs) cause time-varying magnetometer errors. Various algorithms exist to estimate both time-invariant and time-varying errors, which are covered by three general error types: bias, scale factors and nonorthogonality [9]. Researchers usually address both time-invariant and time-varying parameter estimation together and propose an overall calibration algorithm for magnetometers. This algorithm must be in-flight applicable and capable of calibrating the magnetometers against both constant and time-varying error terms.

## 15.2 MAGNETOMETER ERRORS IN DETAIL

Errors for a small satellite magnetometer are summarized in Figure 15.1. In Figure 15.1, errors that are originated from the internal/external disturbances are given in white boxes whereas the inherent sensor errors are given in dark gray boxes. Errors that may be either inherent to sensor or originated from the disturbances are given with light gray boxes.

In the following subsections, we briefly present different error types [9] and discuss how they affect the magnetometer measurements.

### 15.2.1 SOFT IRON ERROR

Soft irons are materials that generate magnetic fields in response to an externally applied field. The field generated by these materials can vary over a wide range depending on both the magnitude and direction of the applied external magnetic field. In result, soft irons impose error on the magnetometer measurements depending on the specifications of the generated magnetic field. In this book, we represent the soft iron error with a $3 \times 3$ matrix, $D_{si}$.

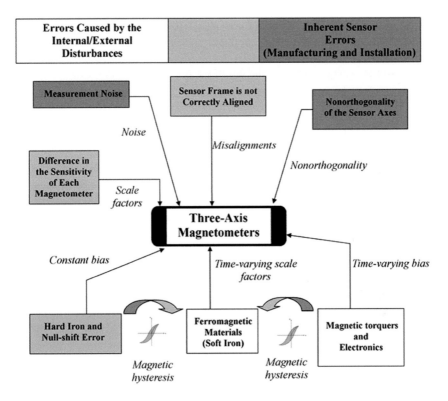

**FIGURE 15.1** Magnetometer errors.

In most applications, $D_{si}$ is assumed to be a matrix with constant terms [10]. In this case, $D_{si}$ is given as

$$D_{si} = \begin{bmatrix} \alpha_{xx} & \alpha_{xy} & \alpha_{xz} \\ \alpha_{yx} & \alpha_{yy} & \alpha_{yz} \\ \alpha_{zx} & \alpha_{zy} & \alpha_{zz} \end{bmatrix}, \tag{15.1}$$

where $\alpha_{ij}$ are the effective soft iron coefficients that show the proportionality between the magnetic field applied to a soft iron and resulting induced magnetic field.

Assumption of constant soft iron coefficients is valid as long as the objects that generate large magnetic field (e.g. MTQs, current carrying wires) are not located close to the magnetometers. For small satellites, where all the subsystems are closely located, an accurate soft iron error model that considers the error's time-variation is needed.

## 15.2.2 HARD IRON AND NULL-SHIFT ERROR

Any unwanted magnetic field, generated by materials with permanent magnetic field (hard irons), impose bias on the magnetometer measurements. This constant bias term, $b_{hi,}$ must be in-flight estimated for small satellite applications since it may vary after the launch [3]. It is assumed that the hard iron bias remains constant once the satellite is in orbit and fully functional.

Null-shift error, $b_{ns}$, which is inherent to the sensor, has also constant bias effect on the magnetometer measurements.

### 15.2.3 TIME-VARYING BIAS

Items inside the satellite such as MTQs may generate unwanted magnetic fields that are time-varying [7]. Depending on the satellite configuration, the strength of the time-varying bias, $b_{tv}$, may be either small or large compared to the permanent hard iron bias. All bias terms – hard iron, null-shift and time-varying biases – may be treated together and in-orbit estimated as a single time-varying bias vector $b = b_{hi} + b_{ns} + b_{tv}$.

### 15.2.4 SCALING

Ideally, three sensors of a magnetometer triad are identical so their output will be the same when they are subjected to an identical magnetic field. However, in practice sensitivity of each magnetometer is different and to represent this difference the output of each sensor must be multiplied with a scale factor. Scaling error is represented with a 3 × 3 matrix, $D_{sf}$ built of scale factors,

$$D_{sf} = \begin{bmatrix} 1+\xi_x & 0 & 0 \\ 0 & 1+\xi_y & 0 \\ 0 & 0 & 1+\xi_z \end{bmatrix}. \tag{15.2}$$

Here, $\xi_x$, $\xi_y$ and $\xi_z$ are scale factors representing the input-to-output sensitivity of each magnetometer.

Scale factor terms may vary over time due to the environmental influences, e.g. temperature [11].

### 15.2.5 NONORTHOGONALITY

In case the sensors are not orthogonal to each other, this error should be reflected to the measurements as a transformation of vector space basis. Nonorthogonality is parameterized by $D_{no}$ matrix:

$$D_{no} = \begin{bmatrix} 1 & 0 & 0 \\ \sin(\rho) & \cos(\rho) & 0 \\ \sin(\phi)\cos(\lambda) & \sin(\lambda) & \cos(\phi)\cos(\lambda) \end{bmatrix}, \tag{15.3}$$

where $\rho$, $\phi$ and $\lambda$ are, respectively, the angles between the y-sensor and y-axis, the z-sensor and y–z plane, and the z-sensor and y–z plane (Figure 15.2) [12].

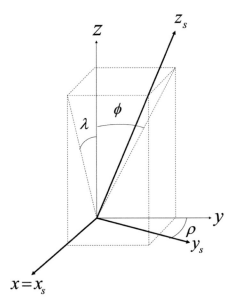

**FIGURE 15.2**  Representation of nonorthogonality for TAM. $x$, $y$ and $z$ are axes of orthogonal frame and $x_s$, $y_s$ and $z_s$ are axes of sensor frame. $x_s$ and $x$ coincide and $y_s$ is in $x$–$y$ plane in this representation. Figure is adapted from [3].

### 15.2.6 MISALIGNMENT

In ideal case the alignment of the TAM frame with respect to the spacecraft body frame would be known accurately. However, in practice a perfect alignment cannot be achieved. As a result of different factors like mounting errors or vibrations during the launch misalignments exist between the ideal and real sensor frames. The misalignment error can be represented with a skew-symmetric matrix formed of three small rotation angles:

$$
D_m = \begin{bmatrix} 1 & -\varepsilon_z & \varepsilon_y \\ \varepsilon_z & 1 & -\varepsilon_x \\ -\varepsilon_y & \varepsilon_x & 1 \end{bmatrix}.
\tag{15.4}
$$

$D_m$ may vary over time, specifically due to the thermal stress.

## 15.3 MAGNETOMETER MEASUREMENT MODELS

### 15.3.1 GENERAL MODEL

Although we know the source of different errors and how they affect the magnetometer measurements, all these errors cannot be distinguished mathematically. The magnetometer model captures the total bias, scaling and nonorthogonality errors [13].

Thus the magnetometer measurement model is exactly same as the one presented in Chapter 4. For sake of completeness we present here again:

$$B_k = (I_{3\times3} + D)^{-1} (A_k H_k + b_k + v_k), \qquad (15.5)$$

where, $B_k$ is the magnetometer measurement vector, $H_k$ is the Earth's magnetic field in the inertial frame, $D$ is the scaling matrix which reflects the scaling, nonorthogonality and soft iron errors, $b_k$ is the bias vector and $v_k$ is the the Gaussian zero-mean measurement noise with covariance $\Sigma_k$ as $v_k \sim \mathcal{N}(0, \Sigma_k)$. $A_k$ is the attitude matrix for the spacecraft which defines the attitude of the spacecraft body frame with respect to the inertial frame.

For a properly calibrated magnetometer the magnitude of the measured magnetic field is not dependent on the orientation of the platform; data collected by rotating the magnetometer forms a sphere [14–16] (Figure 15.3). All bias terms shift the center of this magnetometer measurement locus. Diagonal terms of $D$ matrix scale the sphere into an ellipsoid. The non-diagonal terms of $D$ matrix skew and rotate the ellipsoid.

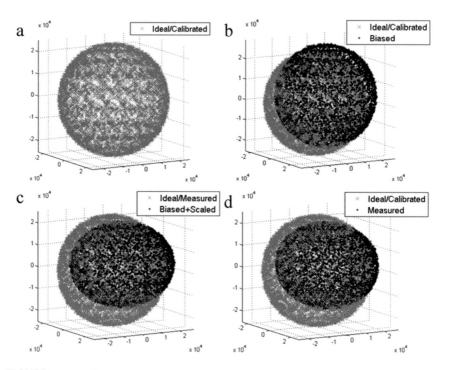

**FIGURE 15.3** Effects of magnetometer errors on the measurements. a) Calibrated magnetometer measurements form a sphere when the senor is rotated; b) Bias shifts the center of sphere; c) Scaling terms scale the sphere into an ellipsoid; d) Measured magnetometer measurements that include effects of misalignment, nonorthogonality and soft iron error, form a skewed and rotated ellipsoid.

For attitude-independent estimation methods, the $D$ matrix parameters that only affect the measured magnetic field direction rather than its magnitude are not observable and cannot be estimated [13]. A fully-populated $D$ matrix with nine independent parameters can be only estimated if the spacecraft's attitude is known or if there is an additional direction information [16, 17].

The assumptions for $D$ matrix vary in different researches. In [18], a diagonal $D$matrix is assumed disregarding the nonorthogonality and non-diagonal soft-iron coefficients. In [7, 13, 19], the antisymmetric part of $D$, which may be caused by sensor misalignments and non-diagonal soft-iron coefficients, is neglected and asymmetric $D$ matrix with six parameters is estimated. In [3, 20], it is assumed that one of the sensor axis is perfectly aligned with the reference axis (spacecraft body axis in our case) and the soft-iron error is split into a combination of a scale error and a misalignment error. As a result, a lower-triangular $D$ matrix is estimated.

### 15.3.2 MODELS CONSIDERING TIME-VARYING ERRORS

Specifically, for batch methods for magnetometer error estimation, the researchers suggested improvements in the model given with Eq. (15.5) to consider time-variation of the errors.

In [7], the MTQ coupling effect is considered and the model in Eq. (15.5) is modified as

$$B_k = (I_{3\times3} + D)^{-1}\left(A_kH_k + b_k + GM_k + v_k\right), \qquad (15.6)$$

where $M_k$ is the magnetic dipole moment vector by the MTQ and $G$ is a $3 \times 3$ constant matrix representing the coupling between the MTQs and TAM. Thus the MTQs corrupt the magnetometers with an additional bias whose quantity depends on the magnetic dipole moment vector.

In [3], the bias caused by the nearby electronics is included in the model as

$$B_k = \left(I_{3\times3} + D\right)^{-1}\left(A_kH_k + b_k + SI_k + v_k\right), \qquad (15.7)$$

where $I_k$ is $c \times 1$ current measurement vector and $S$ is the $3 \times c$ coefficient matrix representing each current measurement's contribution to the bias vector. In [21], a similar model is used to estimate the bias caused by the solar panel currents as a part of the pre-launch calibration process.

## 15.4 BATCH MAGNETOMETER CALIBRATION ALGORITHMS

Batch magnetometer calibration methods use an entire set of measurements to estimate the unknown calibration parameters. Number of the parameters to be estimated depends on the assumed magnetometer measurement model.

In general sense, the batch magnetometer calibration methods are not capable of capturing the time variation in the magnetometer error terms and produce a single set of estimated parameters for the entire available measurements. Yet by applying different calibration procedures or building a model, which can represent the time variation in the parameters (as in Eqs. 15.6 and 15.7), batch methods can be made capable of estimating the time-varying error terms.

Surely the most straightforward method to capture the time-variation in the magnetometer error terms, by using the batch methods, is to repeat the estimation for multiple times with different data sets. For example in [22], the estimation routine is repeating the batch estimation method for each MTQ state and estimating a different set of bias vector. The main drawback, in this case, is the number of bias vector sets that increases in proportion with the number of MTQ states. Besides, this procedure requires an extended amount of magnetometer measurements for each MTQ state. To collect the required amount of measurement, data may take several orbits when the method is applied in-flight.

As discussed, the $D$ matrix parameters that only affect the measured magnetic field direction rather than its magnitude cannot be estimated unless the spacecraft's attitude is known or there is additional direction information. The batch calibration methods address this issue by using measurements of an additional sensor. In [14, 17], a fully populated $D$ matrix is estimated by using the measurements from an inertial sensor. As a result, the magnetometers are also calibrated against the misalignment with respect to the inertial sensor. In [16], the magnetometer misalignment with respect to a rotation axis, whose information is obtained from gyro or an attitude sensor (e.g. star tracker) is estimated. In [23], it is shown that the misalignment between two three-axis sensors can be estimated using a batch method. The algorithm is verified by applying for calibration of small satellite magnetometers; misalignment between two TAMs is estimated.

## 15.4.1 THE COST FUNCTION

The essence of batch magnetometer calibration methods is using an attitude-independent observation for estimating the magnetometer error terms. This observation is derived based on the fact that the magnitude of the magnetic field vector must be same in the spacecraft body frame and inertial frame [13, 19]. It is given as

$$Z_k = \left|\boldsymbol{B}_k\right|^2 - \left|\boldsymbol{H}_k\right|^2 = f(\boldsymbol{B}_k, \boldsymbol{\theta}) + \eta_k, \tag{15.8}$$

where $\boldsymbol{\theta}$ is the vector of unknown magnetometer error terms and $\eta_k$ is the effective measurement noise, which is approximately Gaussian as $\eta_k \sim \mathcal{N}\left(\mu_k, \sigma_k^2\right)$. Here,

$$\mu_k = E\{\eta_k\} = -\mathrm{tr}\left(\Sigma_k\right), \tag{15.9a}$$

$$\sigma_k^2 = E\{\eta_k^2\} - \mu_k^2. \tag{15.9b}$$

The estimation of error parameters vector, $\boldsymbol{\theta}$, according to the maximum likelihood criterion leads us minimizing the following cost function:

$$J(\boldsymbol{\theta}) = \frac{1}{2}\sum_{k=1}^{N}\left[\frac{1}{\sigma_k^2}\left(Z_k - f(\boldsymbol{B}_k, \boldsymbol{\theta}) - \mu_k\right)^2 + \log\sigma_k^2 + \log 2\pi\right]. \tag{15.10}$$

The Eqs. (15.8, 15.10) must be modified in accordance when we include the time-varying terms into the model as in Eqs. (15.6, 15.7). In these cases, $f$ becomes also a function of $M_k$ and $I_k$.

## 15.4.2 THE MINIMIZATION ALGORITHM

There are different methods to minimize Eq. (15.10) and estimate the magnetometer error terms. The two-step method first applies centering to the measurements to obtain a coarse estimate and then uses nonlinear least squares (LS) to solve for global minima [13]. In [3], the authors use directly the iterative nonlinear LS method without any centering approximation and initialization. In [24], an adaptive LS method is used. In [25], authors extend this survey and use hyper LS method to solve for error parameters.

In [17], the authors re-formulate the calibration algorithm as a solution to the maximum likelihood problem. The formulation requires running an Extended Kalman Filter (EKF) to estimate the sensor orientation before evaluating the cost function. In addition to the scaling, nonorthogonality and bias errors, the misalignment between the TAM and inertial sensors is estimated.

In [26], the batch calibration method is applied only for estimating the magnetometer biases. The genetic algorithm (GA) is used to obtain a coarse estimate for bias vector, instead of the centering approximation, which constitutes the first phase of the two-step algorithm.

In [7, 16], the particle swarm optimization (PSO) is used to solve for the magnetometer error terms. To solve for the TAM misalignment error, an additional cost function is included in the optimization algorithm in [16]. This second cost function minimizes the angular deviation between the rotation axis of the TAM and a known rotation vector.

Among batch magnetometer calibration methods, the algorithms that consider time-varying errors have large number of parameters to estimate. For example in [3], 24 parameters are estimated. In this case, using a variant of LS algorithm to estimate the error parameters may be inefficient due to Jacobian and Hessian matrix calculations. The classical gradient-based optimizations suffer a weakness, as they do not guarantee a global minimum in multimodal optimization problems [7].

The GA and PSO algorithms have a better theoretical capability in finding the global solution without a good initial estimate compared to the LS algorithm variants [27]. Thus using these methods to perform batch magnetometer calibration with time-varying error parameters is advantageous in terms of the quality of the solutions. Accuracy of the estimates is similar for these two algorithms [28].

Main disadvantage of GAS and PSO algorithms is their computational load. Although the PSO is computationally lighter compared to the GA [28], its load may be still too heavy to be applied in-flight. The computational load of the PSO algorithm is more than 100 times higher than that of the two-step algorithm [27].

## 15.4.3 DISCUSSION ON BATCH MAGNETOMETER CALIBRATION

The main advantage of batch magnetometer calibration algorithms is not having a requirement to run the algorithm all the time throughout the mission. Magnetometer

**TABLE 15.1**

**Results for Batch Magnetometer Calibration using Two-Step Algorithm for a Spinning Small Satellite**

| Parameter | Actual Value | Two-Step Estimated Value Mean ± Standard Dev. |
|---|---|---|
| $b_x$ | 5000 $nT$ | 4998.50 ± 15.56 $nT$ |
| $b_y$ | 3000 $nT$ | 2999.14 ± 15.83 $nT$ |
| $b_z$ | 4000 $nT$ | 4009.55 ± 11.83 $nT$ |
| $D_{11}$ | 0.05 | 0.0496 ± 0.0007 |
| $D_{22}$ | 0.1 | 0.0996 ± 0.0008 |
| $D_{33}$ | 0.05 | 0.0495 ± 0.0005 |
| $D_{12}$ | 0.05 | 0.0499 ± 0.0006 |
| $D_{13}$ | 0.05 | 0.0499 ± 0.0005 |
| $D_{23}$ | 0.05 | 0.0499 ± 0.0005 |

measurements can be calibrated using a single set of parameter estimates as long as the estimation results are consistent. In [3], the authors show that the estimated calibration parameters do not change drastically over time when they apply their batch magnetometer calibration algorithm using real data. In this case, if the calibration parameters are constant, they can provide very accurate estimation results. Table 15.1 presents the results for 1000 Monte Carlo runs of the two-step algorithm for estimating the standard set of calibration parameter for a small satellite magnetometer as an example. Note here the standard deviation of the magnetometer noise is $\Sigma_k = 300nT$, and the satellite is a spin-stabilized one with $Z_b$ as the spin axis.

The batch magnetometer calibration methods are suitable for estimating the time-varying magnetometer errors only if the proposed measurement model is accurate. This is the main disadvantage of batch algorithms. There are also limitations that make the algorithms mission oriented. For example in [3], it is suggested that the magnetometer and current measurements are sampled simultaneously. In [21], it is discussed that the algorithm's accuracy is not sufficient because of the erroneous solar panel current measurements.

Lastly, it is trivial that most of the batch magnetometer calibration algorithms require high computational load. This makes their in-flight applicability to small satellite missions with limited computational resources questionable.

## 15.5 CONCLUSION

In this chapter, we first reviewed the magnetometer errors and presented possible extensions to the general magnetometer model for considering the time-varying errors. Then we discussed the batch methods for magnetometer calibration and their capability for calibration of a small satellite magnetometer. A special attention was given to the optimization algorithms, which are required for cost-function minimization in batch calibration.

The batch magnetometer calibration algorithms consider only the time-variation of the bias vector. None of the studies examine the variation of the scaling matrix terms which is very likely for small satellites. The presence of ferromagnetic materials onboard the spacecraft and/or the thermal effects may cause such variations in the scaling matrix terms. Lacking information about the characteristics of variations is the main difficulty. More information, which is specifically gathered with actual in-flight data, is needed before an accurate model is built.

The models used for batch magnetometer calibration methods are accurate as long as the assumptions hold. The error variation for a magnetometer onboard the small satellites must be investigated in depth before validating these algorithms' generalizability.

One another difficulty is the computational load of the algorithm, especially when multiple parameters are included to the estimation state vector to account for the time-variation of the errors. Optimization algorithms, which can ensure global minima in these cases, are computationally very costly, and these algorithms' in-orbit application is questionable.

## REFERENCES

1. H. E. Soken, "A survey of calibration algorithms for small satellite magnetometers," *Meas. J. Int. Meas. Confed.*, vol. 122, no. September 2017, pp. 417–423, 2018.
2. H. E. Soken, "A survey of calibration algorithms for small satellite magnetometers," in *2017 IEEE International Workshop on Metrology for AeroSpace (MetroAeroSpace)*, Padua, Italy, 2017, pp. 62–67.
3. J. C. Springmann and J. W. Cutler, "Attitude-independent magnetometer calibration with time-varying bias," *J. Guid. Control. Dyn.*, vol. 35, no. 4, pp. 1080–1088, 2012.
4. H. Ma and S. Xu, "Magnetometer-only attitude and angular velocity filtering estimation for attitude changing spacecraft," *Acta Astronaut.*, vol. 102, pp. 89–102, 2014.
5. K.-M. Roh, S.-Y. Park, and K.-H. Choi, "Orbit determination using the geomagnetic field measurement via the unscented Kalman filter," *J. Spacecr. Rockets*, vol. 44, no. 1, pp. 246–253, 2007.
6. T. Bak, "Onboard attitude determination for a small satellite," in *3rd ESA Int. Conf. Spacecr. Guid. Navig. andControl Syst.*, Noordwijk, the Netherlands, 1996.
7. E. Kim, H. Bang, and S.-H. Lee, "Attitude independent magnetometer calibration considering magnetic torquer coupling effect," *J. Spacecr. Rockets*, vol. 48, no. 4, pp. 691–694, 2011.
8. T. Inamori, R. Hamaguchi, K. Ozawa, P. Saisutjarit, N. Sako, and S. Nakasuka, "Online magnetometer calibration in consideration of geomagnetic anomalies using Kalman filters in nanosatellites and microsatellites," *J. Aerosp. Eng.*, vol. 29, no. 6, p. 04016046, 2016.
9. H. E. Soken and S. Sakai, "Magnetometer calibration for advanced small satellite missions," in *30th International Symposium on Space Technology and Science*, Kobe, Japan, July 2015, pp. 10–12.
10. D. Gebre-Egziabher, G. H. Elkaim, J. David Powell, and B. W. Parkinson, "Calibration of strapdown magnetometers in magnetic field domain," *J. Aerosp. Eng.*, vol. 19, no. April, pp. 87–102, 2006.
11. F. Yin and H. Lühr, "Recalibration of the CHAMP satellite magnetic field measurements," *Meas. Sci. Technol.*, vol. 22, no. 5, p. 055101, 2011.
12. J. F. Vasconcelos, G. Elkaim, C. Silvestre, P. Oliveira, and B. Cardeira, "Geometric approach to strapdown magnetometer calibration in sensor frame," *IEEE Trans. Aerosp. Electron. Syst.*, vol. 47, no. 2, pp. 1293–1306, 2011.
13. R. Alonso and M. D. Shuster, "Complete linear attitude-independent magnetometer calibration," *J. Astronaut. Sci.*, vol. 50, no. 4, pp. 477–490, 2002.
14. M. Kok, J. Hol, T. Schön, F. Gustafsson, and H. Luinge, "Calibration of a magnetometer in combination with inertial sensors," *Proc. 15th Int. Conf. Inf. Fusion*, Singapore, 2012, pp. 787–793.

15. J. De Amorim and L. S. Martins-Filho, "Experimental magnetometer calibration for nanosatellites' navigation system," *J. Aerosp. Technol. Manag.*, vol. 8, no. 1, pp. 103–112, 2016.

16. B. A. Riwanto, T. Tikka, A. Kestila, and J. Praks, "Particle swarm optimization with rotation axis fitting for magnetometer calibration," *IEEE Trans. Aerosp. Electron. Syst.*, vol. 9251, no. c, pp. 1–1, 2017.

17. M. Kok and T. B. Schön, "Magnetometer calibration using inertial sensors," *IEEE Sens. J.*, vol. 16, no. 14, pp. 5679–5689, 2016.

18. H. E. Soken and C. Hajiyev, "UKF-based reconfigurable attitude parameters estimation and magnetometer calibration," *IEEE Trans. Aerosp. Electron. Syst.*, vol. 48, no. 3, pp. 2614–2627, Jul. 2012.

19. J. L. Crassidis, K.-L. Lai, and R. R. Harman, "Real-time attitude-independent three-axis magnetometer calibration," *J. Guid. Control. Dyn.*, vol. 28, no. 1, pp. 115–120, 2005.

20. C. C. Foster and G. H. Elkaim, "Extension of a two-step calibration methodology to include nonorthogonal sensor axes," *IEEE Trans. Aerosp. Electron. Syst.*, vol. 44, no. 3, pp. 1070–1078, 2008.

21. K. Han, H. Wang, T. Xiang, and Z. Jin, "Magnetometer compensation scheme and experimental results on ZDPS-1A pico-satellite," *Chinese J. Aeronaut.*, vol. 25, no. 3, pp. 430–436, 2012.

22. J. Frey, J. Hawkins, and D. Thorsen, "Magnetometer calibration in the presence of hard magnetic torquers," *IEEE Aerosp. Conf. Proc.*, Big Sky, USA, 2014.

23. G. H. Elkaim, "Misalignment calibration using body frame measurements," *Am. Control Conf. ACC13*, Washington, DC, vol. 1, no. 2, 2013.

24. V. Renaudin, M. H. Afzal, and G. Lachapelle, "Complete triaxis magnetometer calibration in the magnetic domain," *J. Sensors*, vol. 2010, 2010.

25. M. Kiani, S. H. Pourtakdoust, and A. A. Sheikhy, "Consistent calibration of magnetometers for nonlinear attitude determination," *Measurement*, vol. 73, no. May, pp. 180–190, 2015.

26. Eunghyun Kim and Hyo-Choong Bang, "Bias estimation of magnetometer using genetic algorithm," *Int. Conf. Control. Autom. Syst*, Seoul, Korea, 2007, pp. 195–198.

27. Z. Wu, Y. Wu, X. Hu, and M. Wu, "Calibration of three-axis magnetometer using stretching particle swarm optimization algorithm," *IEEE Trans. Instrum. Meas.*, vol. 62, no. 2, pp. 281–292, 2013.

28. R. Hassan, B. Cohanim, and O. de Weck, "A comparison of particle swarm optimization and the genetic algorithm," in *46th AIAA/ASME/ASCE/AHS/ASC Structures, Structural Dynamics and Materials Conference*, Austin, USA, 2005, pp. 1–13.

# 16 In-Orbit Calibration of Small Satellite Magnetometers

## *Recursive Calibration Algorithms*

In this chapter, we continue our discussion on the in-orbit calibration algorithms for small satellite magnetometers. Chapter 15 was dedicated to the batch calibration algorithms. This chapter, on the other hand, presents the recursive estimation algorithms which are capable of sensing the time-variation in the estimated error terms. We categorize the recursive magnetometer calibration algorithms in both the following: attitude-dependent and attitude-independent algorithms. For each we provide several algorithm examples together with the demonstration results.

Recursive estimation algorithms calibrate the magnetometer errors in real-time. In these algorithms the magnetometer error parameters are modeled as constant terms,

$$\dot{\theta} = 0. \tag{16.1}$$

Yet the algorithm can be made capable of capturing the time-variation in the parameters with a proper tuning [1, 2]. Examples for such estimation results will be presented for associated recursive magnetometer calibration algorithms.

Attitude-dependent magnetometer calibration algorithms require spacecraft attitude knowledge. There may be two sources for the attitude information: 1) the algorithm itself may estimate also the attitude (simultaneous attitude estimation and magnetometer calibration); 2) the attitude may be obtained from a sensor (e.g. star tracker) or a separate attitude estimator (calibration with known attitude). In contrast, the attitude-independent algorithms do not need the absolute attitude knowledge so they are specifically advantageous in lost-in-space conditions and can be implemented independently from the attitude estimation algorithms. As will be discussed, attitude-independent algorithms rely on the magnitude difference for measured and modeled magnetic field and other additional measurement such as the angular rates of the spacecraft.

For each type of recursive calibration algorithm, presented in this chapter, we first briefly introduce the preliminaries and provide a literature survey. Then we give example algorithms with demonstration results in the following sections.

## 16.1 SIMULTANEOUS ATTITUDE ESTIMATION AND MAGNETOMETER CALIBRATION

Recursive algorithms for simultaneous attitude estimation and magnetometer calibration are the first type of attitude-dependent algorithms. They run for a combined state of

$$X = \begin{bmatrix} q^T & b_g^T & \theta^T \end{bmatrix}^T, \tag{16.2}$$

where $q$ is the attitude vector for the spacecraft (in terms of quaternions, Euler angles etc.) and $b_g$ is the gyro bias vector. Note that for a dynamics-based attitude filter the gyro bias vector should be replaced with the angular velocity vector of $\omega_{bi}$.

Due to the nonlinearity of the estimation problem, a nonlinear filtering algorithm must be used in this case. In [3], the authors propose using the EKF to estimate the magnetometer bias terms and scale factors in addition to the attitude and attitude rate of the spacecraft. In [4, 5], this time, an Unscented Kalman Filter (UKF) is used to estimate the attitude of the spacecraft together with the magnetometer bias vector and the terms for a diagonal scale matrix. The scale matrix terms are not estimated as a part of the state vector but by using a covariance matching method. In [6], a two-stage method to estimate the magnetometer bias terms together with spacecraft's attitude and gyro biases is introduced. The method runs an UKF with different states at different stages to estimate all the parameters.

In [7] and [9], performances of both the EKF and UKF are investigated for simultaneous attitude estimation and magnetometer calibration. In addition to the magnetometer bias terms, the magnetometer error vector includes terms corresponding to the magnetic anomalies in the Earth's magnetic field.

In [10], an UKF is used to estimate the attitude of the spacecraft together with bias vector and the nine independent terms of a full-populated scale matrix. For the same estimation scenario, a marginal modified UKF is used to reduce the computational load in [35].

## 16.2 UKF FOR SIMULTANEOUS ATTITUDE ESTIMATION AND MAGNETOMETER CALIBRATION

This section presents our first example for simultaneous attitude estimation and magnetometer calibration. Using an UKF both the attitude of the satellite and the magnetometer error terms are estimated. For scale factor estimation, a covariance matching-based procedure is used and the scale factors are not included in the state vector.

If the kinematics of the pico satellite is derived in the base of the Euler angles, then the mathematical model can be expressed with a nine-dimensional system vector which is formed of the attitude Euler angles ($\varphi$ is the roll angle about $x$ axis; $\theta$ is the pitch angle about $y$ axis; $\psi$ is the yaw angle about $z$ axis) vector, body angular rate vector with respect to the inertial axis frame and vector formed of bias terms of magnetometers. Hence,

$$\bar{x} = \begin{bmatrix} \varphi & \theta & \psi & \omega_x & \omega_y & \omega_z & b_{m_x} & b_{m_y} & b_{m_z} \end{bmatrix}^T, \tag{16.3}$$

Here, $\varphi$, $\theta$ and $\psi$ are roll, pitch and yaw angles defining the attitude of the space-craft with respect to the orbit frame, $\omega_{bi} = \begin{bmatrix} \omega_x & \omega_y & \omega_z \end{bmatrix}^T$ is the body angular rate vector with respect to the inertial axis frame and $b_m = \begin{bmatrix} b_x & b_y & b_z \end{bmatrix}^T$ is the magnetometer bias vector.

## 16.2.1 SCALE FACTOR ESTIMATION

When the diagonal terms of the scale matrix $S$ are not equal to unity, that indicates an error in the corresponding axis' scale factor and shows up as a mismatch between the real condition of the measurement system and the model used in the synthesis of the filter. Hence, scale factors should be estimated and the filter's measurement model must be corrected.

A diagonal scaling and misalignment matrix is assumed accounting for only the scale factors as

$$\left(I_{3\times3} + D\right) = S = \begin{bmatrix} s_{11} & 0 & 0 \\ 0 & s_{22} & 0 \\ 0 & 0 & s_{33} \end{bmatrix} = diag\left(s_{11}, s_{22}, s_{33}\right). \tag{16.4}$$

The scale factor terms are not included in the state vector but are estimated using an innovation-based calculation process that checks the mismatches between the real and theoretical innovation covariance matrices [11]:

$$tr\left\{e(k+1)e^T(k+1)\right\} \le tr\left\{P_{yy}\left(k+1\middle|k\right) + R(k+1)\right\}, \tag{16.5}$$

Here $tr\{\cdot\}$ denotes the trace of the related matrix. Left-hand side of (16.5) represents the real filtration error while the right-hand side is the accuracy of the innovation sequence known as a result of a priori information [12]. However, when there is a need for scale factor estimation (diagonal terms of scale matrix (16.4) are not equal to unity) the real error will exceed the theoretical one. In this case a single-scale factor, $S(k)$, may be added into the algorithm as

$$tr\left\{e(k+1)e^T(k+1)\right\} = tr\left\{P_{yy}\left(k+1\middle|k\right)\right\} + S(k)tr\left\{R(k+1)\right\}, \tag{16.6}$$

and it can be obtained by simply solving the Eq. (16.6). However single-scale factor estimation for three different magnetometers is not a healthy approach and it may not reflect corrective effects of the scale correction [11]. Technique, which can be implemented to overcome this problem, is to use distinct scale factors for each magnetometer as usual. In this case innovation should be taken into account within a moving window and directly the matrix for scale factors should be estimated [13],

$$\frac{1}{\zeta}\sum_{j=k-\zeta+1}^{k} e(k+1)e^T(k+1) = P_{yy}\left(k+1\middle|k\right) + S(k)R(k+1)S^T(k). \tag{16.7}$$

Here, $\zeta$ is the width of the moving window.

Then, since it is known that $S(k)$ is diagonal (there in no misalignment/nonorthag-onality for magnetometers) and also $R(k + 1)$ is diagonal by its nature, it is possible to rewrite (16.7),

$$\frac{1}{\varsigma}\sum_{j=k-\varsigma+1}^{k} e(k+1)e^{T}(k+1) = P_{yy}(k+1|k) + [S(k)]^{2}R(k+1). \qquad (16.8)$$

Finally, scale factor matrix can be found as given below,

$$S(k) = \sqrt{\left\{\frac{1}{\varsigma}\sum_{j=k-\varsigma+1}^{k} e(k+1)e^{T}(k+1) - P_{yy}(k+1|k)\right\}R^{-1}(k+1)}. \qquad (16.9)$$

## 16.2.2 DEMONSTRATION OF UKF FOR SIMULTANEOUS ATTITUDE ESTIMATION AND MAGNETOMETER CALIBRATION

In order to test the proposed algorithm, simulations for attitude estimation and mag-netometer calibration procedure of a pico satellite are performed. Chosen pico satel-lite is a 1U cubesat which means its mass is about 1 kg and it is 10 cm in each size.

Simulations are realized for 500 s with a sampling time of $\Delta t = 0.1$ s. The inertia matrix of the cubesat is taken as

$$J = \begin{bmatrix} 2.1\times10^{-3} & 0 & 0 \\ 0 & 2.0\times10^{-3} & 0 \\ 0 & 0 & 1.9\times10^{-3} \end{bmatrix}.$$

Nonetheless the orbit of the satellite is a circular orbit with an altitude of $r = 550$ km. For magnetometer measurements, sensor noise is characterized by zero mean Gaussian white noise with a standard deviation of $\sigma_m = 200nT$. Besides constant magnetometer bias terms are accepted as $\bar{b}_m = [0.14 \quad 0.019 \quad 0.37]^T \times 10^4 nT$.

Bias estimation example is given in Figure 16.1. As it is apparent, bias terms are estimated accurately. Besides, in simulations, scale factors that are implemented to the magnetometer measurements were

$$S = \begin{bmatrix} 1.2 & 0 & 0 \\ 0 & 1.3 & 0 \\ 0 & 0 & 1.5 \end{bmatrix}$$

These are the values that are implemented to the magnetometer measurements in order to simulate estimation process. In return, following scale factors are estimated via the proposed method:

$$\hat{S} = \begin{bmatrix} 1.1953 & 0 & 0 \\ 0 & 1.2955 & 0 \\ 0 & 0 & 1.4933 \end{bmatrix}$$

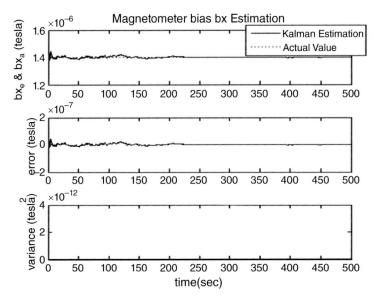

**FIGURE 16.1** Estimation of the bias for magnetometer aligned through "x" axis.

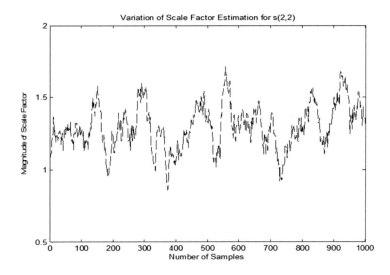

**FIGURE 16.2** Estimation of the scale factor $\hat{s}(2,2)$ for 1000 samples.

As seen, estimation error for the scale factors is not high. On the other hand, Figure 16.2, which gives the variation of the one of the scale factor estimation $((\hat{s}(2,2))$ for 1000 samples, sets an example for the noisy scale factor estimation process. So as to understand the reason of such process, a group of simulations have been performed mainly on the sensitivity of the scale factor estimations to the changes in the size of the moving window $\zeta$.

**TABLE 16.1**

**Estimation of Scale Factors for Different Moving Window Sizes**

|  | $\zeta = 10$ | $\zeta = 20$ | $\zeta = 30$ | $\zeta = 50$ | $\zeta = 80$ | Mean | Variance |
|---|---|---|---|---|---|---|---|
| $\hat{s}(1,1)$ | 1.1986 | 1.2100 | 1.1989 | 1.2080 | 1.2185 | 1.2068 | 6.963E-05 |
| $\hat{s}(2,2)$ | 1.2995 | 1.3110 | 1.2991 | 1.3114 | 1.3209 | 1.3084 | 8.418E-05 |
| $\hat{s}(3,3)$ | 1.4978 | 1.5129 | 1.4974 | 1.5105 | 1.5240 | 1.5085 | 1.254E-04 |

In this demonstration, window size of $\zeta = 30$ is chosen and that choice is empirical as usual. In general, a reasonable size for moving window may vary between 10 and 30 [14]. If the size of the moving window is too small that will bring about a noisy estimation of the innovation covariance and so scale factor. Per contra, when the size is larger estimation process will be smoother but at the expense of long transient time. Performed simulations proved that general statement. In Table 16.1, scale factor estimation results are given for different sizes of moving window as well as the mean and variance of the given estimation values (given results are the mean of scale factor estimation samples gained till the bias estimation stopping instant). In Table 16.1, although some deviation about the real value is apparent, it is not possible to comment on the effect of varying window size in real terms. However, it is seen that for all sizes of the moving window, estimation accuracy is not bad.

As a consequence, another investigation has been realized and this time estimation of one of the scale factors, $s(2,2)$, has been plotted for 100 consecutive samples starting from a random time instant and the window size has been changed at each simulation (Figure 16.3). Simulation proves the aforementioned theoretical statement; when the size of the moving window is small-scale factor estimation becomes noisy and in contrast when the size is larger it is possible to obtain a smother estimation process but at the cost of a longer transient time. That means if a wider window is selected for the estimation of scale factors, necessary time to start up the estimation process extends.

Nonetheless, it is possible to estimate also all other parameters accurately via this algorithm. As seen in Figure 16.4 and Figure 16.5, attitude parameter estimation errors are in acceptable limits for a pico satellite which are $\pm 1°$ for angles and $\pm 0.01°$ /s for angular rates. These results verify that proposed algorithm's estimation accuracy for both angles and angular rates is within the desired bounds. Besides, all three bias terms can be estimated correctly.

## 16.3 RECONFIGURABLE UKF FOR SIMULTANEOUS ATTITUDE ESTIMATION AND MAGNETOMETER CALIBRATION

In the previous section, we showed that using the proposed UKF algorithm it is possible to estimate all nine parameters of the state vector (16.3) and also the scale factors precisely. However, nine-dimensional state vector and the additional part to UKF for scale factor estimation demand a high computational effort. In general, Kalman

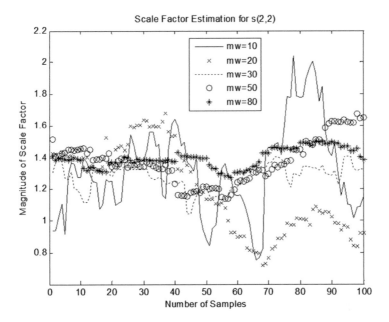

**FIGURE 16.3** Estimation of the scale factor $\hat{s}(2,2)$ for different sizes of the moving window.

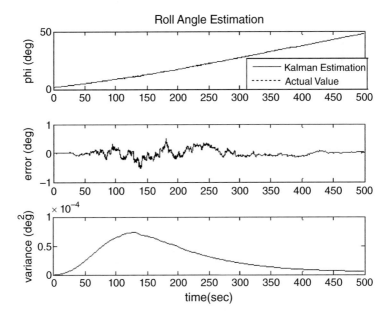

**FIGURE 16.4** Roll angle estimation via UKF without reconfiguration.

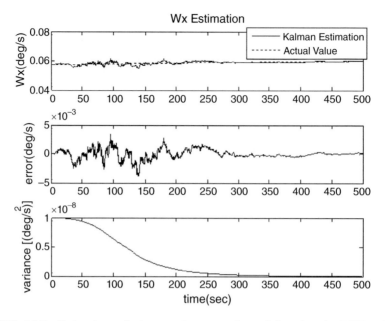

**FIGURE 16.5** Estimation of the angular rate about "x" axis via UKF without reconfiguration.

filter's computational burden increases in a direct relation with the cube of the number of states. Hence, constituting a state vector with nine states means more computational burden than a state vector with six states. That is an important redundancy for considerably limited attitude computer of a pico satellite.

Hence, it is better to reconfigure filter regarding the scale factors and biases of magnetometers and proceed estimation procedure by only estimating Euler angles and angular rates as

$$\bar{x} = \begin{bmatrix} \varphi & \theta & \psi & \omega_x & \omega_y & \omega_z \end{bmatrix}^T, \tag{16.10}$$

A stopping rule is proposed in order to reconfigure UKF after convergence of the bias estimations to the actual values. In that second stage, previously estimated magnetometer bias terms are taken into account by UKF algorithm and number of states to be estimated reduces to 6 as shown with Eq. (16.10). Structural diagram for the proposed method is given in Figure 16.6.

As shown in the structural diagram, initially primary UKF runs. In this case, switches number I are closed. When the bias stopping rule is satisfied such that the biases are converged to the desired actual value, switches number I open and switches number II close. After that instant (indicated with a dashed line in the corresponding figures given hereafter) reconfigured UKF runs taking the incoming initial estimates from the memory block (these values are the latest estimates of primary UKF).

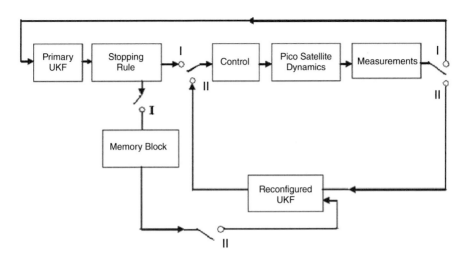

**FIGURE 16.6** Structural diagram of the reconfigurable attitude estimation method.

## 16.3.1 STOPPING RULE FOR BIAS ESTIMATION

### 16.3.1.1 Stopping Rule Formation

The following stopping rule may be introduced for the bias estimation problem [15, 16]:

$$r_k^2 = \left( \hat{b}_k - \hat{b}_{k-1} \right)^T D_{\Delta b_k}^{-1} \left( \hat{b}_k - \hat{b}_{k-1} \right) = \varepsilon, \qquad (16.11)$$

where $D_{\Delta b_k}$ is the covariance matrix of the discrepancy between two successive bias estimates $\hat{b}_k$ and $\hat{b}_{k-1}$, and $\varepsilon$ is a predetermined small number.

It is assumed that the well-developed theory of Kalman filtering is used to estimate the parameters from a sequence of observations with Gaussian tolerances of the measurement errors and system noise. In this situation , the Kalman filter yields an estimate with an expected value equal to the estimated quantity and a Gaussian distribution function. The discrepancy $\hat{b}_k - \hat{b}_{k-1}$ then has a normal distribution as well, since it is a linear combination of two Gaussian random variables [17]. With these considerations in mind it is known that the statistic $r^2$ has a $\chi^2$ distribution with $n$ degrees of freedom ($n$ is the number of dimensions of the bias vector $b$), and the threshold values of $r^2$ can be found by determining the tabulated values of the $\chi^2$ distribution for a given level of significance.

It is evident from relation (16.11) that for smaller the value of $r^2$, the consistency of the estimates will be greater. Usually in the consistency tests in such cases, the lower limit of the confidence interval must be equal to zero, and the upper limit is determined by the level of significance $\alpha$.

To test the consistency of the estimates, the level of significance $\alpha$ is adopted, which corresponds to the confidence coefficient $\beta = 1 - \alpha$. The threshold $\chi_\beta^2$ is specified in terms of this probability, using the distribution of the investigated statistic $r^2$:

$$P\left\{ \chi^2 < \chi_\beta^2 \right\} = \beta, \ 0 < \beta < 1. \qquad (16.12)$$

The estimation process is stopped when $r_k^2 < \chi_\beta^2$, since further observations yield insignificant improvement of the identified model and are deemed impractical in this event. If the quadratic form $r^2$ is larger than or equal to the specified threshold $\chi_\beta^2$, estimation should be continued.

This stopping rule can be used to make a timely decision to stop the bias estimation process and it does not require large computational expenditures.

### 16.3.1.2 Computation of the Covariance Matrix of the Discrepancy between Two Successive Bias Estimates

We investigate the problem of determining the covariance matrix of the discrepancy between successive bias estimates. The estimates $\hat{b}_{k-1}$ and $\hat{b}_k$ are determined via UKF from sets of successive measurements $\{z_1, z_2, \ldots, z_{k-1}\} = Z_1^{k-1}$ and $\{z_1, z_2, \ldots, z_{k-1}, z_k\} = Z_2^k$, respectively:

$$\hat{b}_{k-1} = E\left\{b_{k-1} \big| Z_1^{k-1}\right\}; \quad \hat{b}_k = E\left\{b_k \big| Z_2^k\right\}. \tag{16.13}$$

These estimates of the biases are correlated, since common data are used. The correlation between them is also attributable to shared initial conditions and shared system noise.

We consider the magnitude of the discrepancy between two successive bias estimates:

$$\delta_k = \hat{b}_k - \hat{b}_{k-1}. \tag{16.14}$$

Define the estimation errors $e_{k-1}$ and $e_k$ as

$$\begin{aligned} e_{k-1} &= \hat{b}_{k-1} - b, \\ e_k &= \hat{b}_k - b. \end{aligned} \tag{16.15}$$

Taking into account (16.15) the expression (16.14) can be written as

$$\delta_k = e_k - e_{k-1}. \tag{16.16}$$

The estimates of UKF are unbiased, therefore

$$E\{\delta_k\} = E\{e_k - e_{k-1}\} = 0. \tag{16.17}$$

The covariance for $\delta_k$ is written in the form

$$E\{\delta_k \delta_k^T\} = E\{e_k e_k^T - e_k e_{k-1}^T - e_{k-1} e_k^T + e_{k-1} e_{k-1}^T\} = P_k + P_{k-1} - P_{k,k-1} - P_{k,k-1}^T = D_{\Delta b_k}, \tag{16.18}$$

where

$$P_{k,k-1} = E\{e_k e_{k-1}^T\} = P_{k-1,k}^T. \tag{16.19}$$

$P_k$ and $P_{k-1}$ are the covariance matrices of the errors of two successive bias estimates $\hat{b}_k$ and $\hat{b}_{k-1}$, respectively, and $P_{k,k-1}$ and $P_{k-1,k}$ are the cross covariance matrices between the errors of mentioned estimates.

Consequently, to compute the covariance of the difference between two successive bias estimates of the UKF, it is necessary to obtain the cross terms of the covariance of the errors. Algorithms for determining the cross covariance terms are described in detail in [18, 19].

Once $P_k$, $P_{k-1}$ and $P_{k,k-1}$ have been found, Eq. (16.18) is used to determine the covariance matrix $D_{\Delta b_k}$, and a comparison of the statistic $r_k^2$ computed from this matrix with the confidence limit obtained from the corresponding $\chi^2$ distribution leads to decision whether to stop the bias estimation process or not .

Investigated estimates of Kalman filter $\hat{b}_{k-1}$ and $\hat{b}_k$ are evaluated based on the same system state model, their initial conditions are equal, the initial covariances of errors of mentioned estimates and the initial cross covariance between them are also equal. Then it can be shown that [18]

$$P_{k,k-1} = P_k. \tag{16.20}$$

Substituting (16.20) into (16.18) and taking into account that the covariance matrix $P_k$ is symmetrical, i.e. $P_k = P_k^T$, the covariance matrix $D_{\Delta b_k}$ may be written in the form [16]

$$D_{\Delta b_k} = P_{k-1} - P_k. \tag{16.21}$$

Here, $D_{\Delta b_k}$ is positive definite matrix [20].

## 16.3.2 Demonstration of Reconfigured UKF for Simultaneous Attitude Estimation and Magnetometer Calibration

After stopping the magnetometer bias estimation process using stopping rule (16.11), the UKF is reconfigured; number of states to be estimated reduces to 6 instead of 9. For the reconfigured UKF bias terms are not estimated anymore. However, since the scale factors are found by the use of innovation with (16.9), convergence to an actual value cannot be mentioned for their estimations as proved in Figure 16.2. Therefore, two different scenarios are tried for the reconfigured UKF algorithm in scope of scale factors. In first algorithm, mean of the scale factors estimated till the stopping instant are taken into account by the reconfigured filter and also scale factor estimation is stopped with the biases. In second scenario, scale factor estimation is continued although the biases are not estimated anymore after the stopping criteria are satisfied. Absolute values of errors are compared in Table 16.2.

As seen from Table 16.2, in both scenarios reconfigured filter gives better estimation results than the filter without reconfiguration in terms of absolute value of estimation errors. Reducing the size of the state vector without any change at the measurement vector brings about an increment in the accuracy of estimation of the states. Nonetheless, a smaller state vector means lower computational burden and risk for the divergence of the filter. Moreover, it is apparent that there is a little supremacy of the filter with reconfiguration when the scale factor estimation

**TABLE 16.2**

**Comparison of Absolute Values of Error for UKFs with and without Reconfiguration**

| Parameter | Abs. Values of Err. for UKF with Reconfiguration (mean of scale factors are used) | | Abs. Values of Err. for UKF with Reconfiguration (scale factor estimation continues) | | Abs. Values of Err. for UKF without Reconfiguration | |
|---|---|---|---|---|---|---|
| | 300 s | 400 s | 300 s | 400 s | 300 s | 400 s |
| $\varphi$ (deg) | 0.1212 | 0.0852 | 0.0603 | 0.0751 | 0.1119 | 0.191 |
| $\theta$ (deg) | 0.1041 | 0.1278 | 0.1021 | 0.0661 | 0.1451 | 0.2153 |
| $\psi$ (deg) | 0.1346 | 0.1293 | 0.0998 | 0.1175 | 0.1598 | 0.1837 |
| $\omega_x$ (deg/s) | 0.0004 | 0.0002 | 0.0003 | 0.0002 | 0.0004 | 0.0003 |
| $\omega_y$ (deg/s) | 0.0001 | 0.0004 | 0.0001 | 0.0003 | 0.0002 | 0.0003 |
| $\omega_z$ (deg/s) | 0.001 | 0.0004 | 0.0006 | 0.0004 | 0.0006 | 0.0006 |

continues. As a result, since scale factors are not estimated as a part of the state vector and their estimations do not require a heavy computational burden, maintaining scale factor estimation throughout the whole process is suggested for the sake of estimation accuracy. Illustrative examples for the reconfigured UKF algorithm (scale factor estimation continues) are given in Figure 16.7 and Figure 16.8. Reconfiguration time

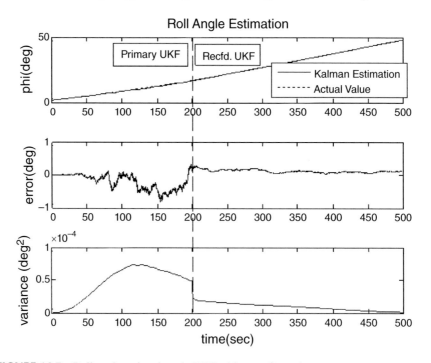

**FIGURE 16.7**  Roll angle estimation via UKF with reconfiguration.

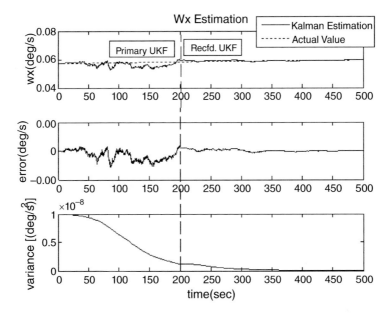

**FIGURE 16.8**   Estimation of the angular rate about "x" axis via UKF with reconfiguration.

is about 200th s and it is marked on figures with dashed line. All the results show that reconfiguration of the filter has a regenerative effect on the estimation accuracy. After reconfiguration, noisy characteristic of the estimation process lessens and the parameters are estimated more accurately.

## 16.4  TWO-STAGE UKF FOR SIMULTANEOUS ATTITUDE ESTIMATION AND MAGNETOMETER CALIBRATION

In this section, an UKF-based procedure for the bias estimation of both the magnetometers and the gyros, which are carried onboard by a pico satellite, is proposed. The filter is a dynamics-based attitude filter and uses the gyro measurements to correct the states at the update stage. At the initial phase, biases of three orthogonally located magnetometers are estimated as well as the attitude and attitude rates of the satellite. During initial period after the orbit injection, gyro measurements are accepted as bias free. At the second phase, estimated constant magnetometer bias components are taken into account and the algorithm is run for the estimation of the gyro biases. As a result, six different bias terms for two different sensors are obtained in two stages where attitude and attitude rates are estimated regularly. For both estimation phases of the procedure, UKF is used as the estimation algorithm [21].

If the kinematics of the pico satellite is derived in the base of Euler angles, then the mathematical model can be expressed with a nine-dimensional system vector which is made of attitude Euler angles vector, the body angular rate vector with respect to the inertial axis frame and the vector formed of bias terms of magnetometer or rate gyros. Hence,

- For the first stage where bias terms of three magnetometers are estimated:

$$\bar{x} = \begin{bmatrix} \varphi & \theta & \psi & \omega_x & \omega_y & \omega_z & b_{m_x} & b_{m_y} & b_{m_z} \end{bmatrix}^T, \qquad (16.22)$$

- For the second stage, where bias terms of three gyros are estimated:

$$\bar{x} = \begin{bmatrix} \varphi & \theta & \psi & \omega_x & \omega_y & \omega_z & b_{g_x} & b_{g_y} & b_{g_z} \end{bmatrix}^T, \qquad (16.23)$$

Here, subscript $m$ or $g$ for bias terms represents either magnetometer or gyro.

### 16.4.1 Two-Stage Estimation Procedure

The proposed estimation procedure for in-flight calibration of magnetometers and rate gyros of pico satellite consists of two stages: Magnetometer bias estimation stage and gyro bias estimation stage. These are examined below.

### 16.4.1.1 Magnetometer Bias Estimation Stage

For the first stage of the estimation, where the magnetometer biases as well as the attitude angles and attitude rates are estimated, estimated state is formed as given in (16.22) and at that stage it is assumed that gyros are bias free. Architectural scheme is given as Figure 16.9.

As seen from Figure 16.9, first-stage Kalman filter takes the measurements of magnetometers and gyros as input on one hand. On the other hand, the Earth Magnetic Field model is also taken into consideration by the filter in line with the common

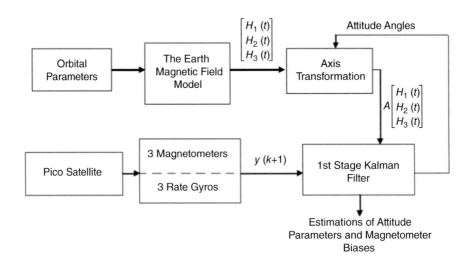

**FIGURE 16.9**  In-flight calibration scheme for magnetometers.

vector measurement model. UKF, which has (16.22) as the state to be estimated, runs and it gives magnetometer bias and attitude parameter estimations as the results.

### 16.4.1.2 Gyro Bias Estimation Stage

As it is aforementioned, at the second stage of the estimation procedure, this time, gyro biases are estimated. At that stage, previously estimated magnetometer bias terms are taken into consideration and the algorithm is run for the estimation of the gyro bias terms. Estimated state is formed as given in (16.23). Figure 16.10 presents the architectural scheme for the gyro bias estimation stage.

Similarly, with the magnetometer bias estimation scheme, second-stage Kalman filter takes the measurements of magnetometers and gyros as input.

### 16.4.1.3 Overall Look to Two-Stage Estimation Scheme

If the proposed method is thought as a part of a prospective attitude determination and control system, a block diagram as given in Figure 16.11 may be presented.

In diagram, initially the first-stage estimation procedure runs (indicated with label 1 for dashed lines). Estimated magnetometer bias vector is used by the filter to correct the magnetometer measurements and improve the attitude estimation results. Filter would diverge by time when the present magnetometer biases are disregarded even though they are small in magnitude [5, 21]. As the next step in attitude determination and control procedure, estimated attitude parameters are taken into consideration by the control algorithm so as to produce a control moment and change the orientation of the satellite in the desired direction.

Nevertheless, when the magnetometer biases are not necessary anymore after they are accurately estimated, first stage stops and the second-stage estimation procedure begins processing (indicated with label 2 for dashed lines) for gyro bias estimation. Remaining procedure repeats in the same way as stated for the first stage.

On the other hand, although magnetometers' constant bias terms are estimated initially, they may vary by time as discussed [22]. As a result of those situations,

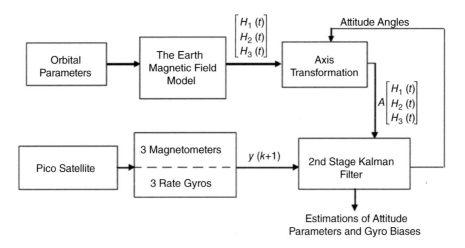

**FIGURE 16.10** In-flight calibration scheme for gyros.

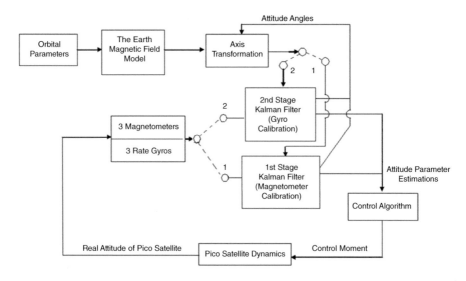

**FIGURE 16.11**   Block diagram of the proposed pico satellite attitude determination and control system.

magnetometers may need to be calibrated again in a period that changes according to the type of mission and pico satellite. Thus, the proposed two-stage procedure may be repeated when it is necessary.

### 16.4.2  DEMONSTRATION OF TWO-STAGE UKF FOR SIMULTANEOUS ATTITUDE ESTIMATION AND MAGNETOMETER CALIBRATION

Simulations are realized for 2000 seconds with a sampling time of $\Delta t = 0.1$ s. As an experimental platform a cubesat model is used and the inertia matrix is taken as

$$
J = \begin{bmatrix}
2.1 \times 10^{-3} & 0 & 0 \\
0 & 2.0 \times 10^{-3} & 0 \\
0 & 0 & 1.9 \times 10^{-3}
\end{bmatrix}
$$

The orbit of the satellite is a circular orbit with an altitude of $r = 550$ km.

First part of figures gives UKF parameter estimation results and the actual values in a comparing way. Second part of the figures shows the error of the estimation process based on the actual estimation values of the satellite. The last part indicates the variance of the estimation.

### 16.4.2.1  Simulation Results for Magnetometer Bias Estimation

A constant bias term that is about %10 of the measurement is implemented to the magnetometer measurements for the first stage of the estimation scenario.

As it is apparent from Figure 16.12, UKF algorithm estimates magnetometer biases with a sufficient precision.

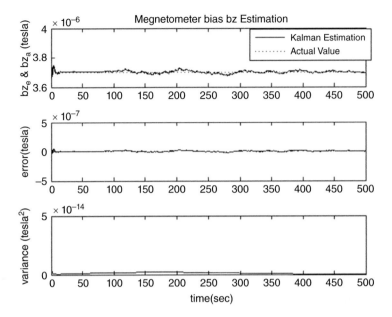

**FIGURE 16.12**    Magnetometer bias estimation for the magnetometer aligned in the "z" axis.

Simulation results have proved that it is possible to estimate the bias terms within bounds of ± % 3, ± % 15 and ± % 2 error percentages with respect to the actual values of the biases when the data are considered for magnetometers aligned in $x$, $y$ and $z$ axis, respectively. Larger error for the estimation of bias of the magnetometer aligned in $y$ axis is caused by the relatively smaller value of the sensed magnetic field in this axis for the simulation duration.

### 16.4.2.2 Simulation Results for Gyro Bias Estimation

This time, a random walk bias term as in Eq. (4.30) is implemented to the gyros assuming that the bias terms which are not estimated in the first stage are not negligible anymore. Estimation results presented in Figure 16.13 show that gyro biases can also be estimated accurately by the use of the proposed two-stage estimation procedure.

Nonetheless, in addition to the bias estimations, algorithms also do estimate the Euler angles and the attitude rates as a part of the state vector. Two examples for these parameters' estimation process are presented in Figure 16.14 and Figure 16.15. As obvious attitude angles and the attitude rates of the satellite can also be accurately estimated by the proposed two-stage estimation scenario. Note that the results are similar for all other attitude angles and the attitude rates.

## 16.5 MAGNETOMETER CALIBRATION WITH KNOWN ATTITUDE

Magnetometer calibration with known attitude is the second type of attitude-dependent algorithms. If spacecraft's attitude, $A_k$, is known accurately, then the

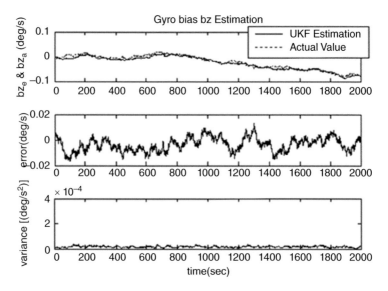

**FIGURE 16.13**    Gyro bias estimation for the gyro aligned in the "z" axis.

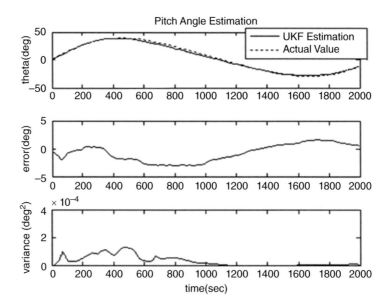

**FIGURE 16.14**    Pitch angle estimation.

magnetometer calibration problem is easy to solve. Regarding the magnetometer measurement model (Eq. 15.5), we can derive the measurement model for the recursive estimator as

$$A_k H_k - B_k = DB_k - b_k. \tag{16.24}$$

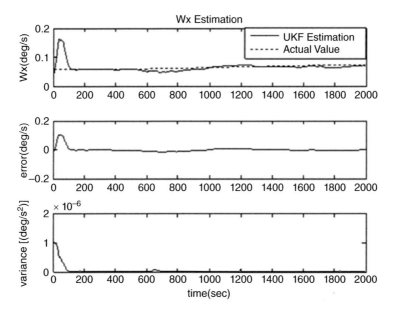

**FIGURE 16.15**   Estimation of the angular rate about "x" axis.

If it is reformulated

$$A_k H_k - B_k = \Phi_k X,$$ (16.25)

where

$$X = \begin{bmatrix} b_k^T & D_{1k}^T & D_{2k}^T & D_{3k}^T \end{bmatrix}^T,$$ (16.26a)

$$D_{ik} = \begin{bmatrix} D_{i1} & D_{i2} & D_{i3} \end{bmatrix}^T \quad \text{for} \quad i = 1, 2, 3,$$ (16.26b)

$$\Phi_k = \begin{bmatrix} -I_{3\times3} & \begin{matrix} B_k^T \\ 0_{2\times3} \end{matrix} & \begin{matrix} 0_{1\times3} \\ B_k^T \\ 0_{1\times3} \end{matrix} & \begin{matrix} 0_{2\times3} \\ B_k^T \end{matrix} \end{bmatrix}.$$ (16.26c)

As we see the model is linear in terms of the state vector $X$. Therefore, if we know spacecraft's attitude we can estimate the error terms using a linear KF.

Spacecraft's attitude information can be obtained using another estimator or directly from the outputs of a star tracker. In [23], the magnetometer bias vector, $b_k$, is estimated using this method. Likewise in [24] this formulation is used to estimate the magnetometer bias vector, $b_k$, and the diagonal terms (scaling terms) of $D$ matrix with a linear KF. In both [23, 24], it is assumed that the spacecraft's attitude is known and only the recursive magnetometer calibration algorithm is discussed.

In [1], an expectation-maximization algorithm is investigated for magnetometer calibration. It is a recursive version of the batch algorithm introduced in [25].

The expectation part is a UKF which estimates the spacecraft's attitude. In the maximization part the magnetometer errors are estimated with the sequential linear least-squares algorithm using the attitude estimations from the UKF.

In [26], the TRIAD algorithm is used in conjunction with the UKF to form an attitude filter. First the vector measurements from the magnetometer and Sun sensor are processed with the TRIAD algorithm to obtain a coarse attitude estimate for the spacecraft. In the second phase, the estimated coarse attitude is used as quaternion measurements for the UKF. The UKF estimates the fine attitude, and the gyro and magnetometer biases. In [27], this approach is extended such that the UKF estimates full error states for the magnetometers. In these studies, it is shown that the magnetometer calibration is possible with the linear model in Eq. (16.25) even if the attitude information is a coarse one provided by the TRIAD algorithm.

### 16.5.1 MAGNETOMETER BIAS AND SCALE FACTOR ESTIMATION USING A LINEAR KF

Here, we present the details of a linear KF (LKF) based calibration algorithm for in-orbit magnetometer bias and scale factor estimation, which is originally published in [24]. This algorithm is suitable for small satellites whose processing capacity is limited. The algorithm assumes that the attitude of the satellite is measured using other available attitude sensors (e.g. star tracker).

The magnetometer measurement model can be expressed in the discrete-time form as follows:

$$B_m(k) = S(k)A(k)B_o(k) + b(k) + v(k) \tag{16.27}$$

Here,

$$S = \begin{bmatrix} S_x & 0 & 0 \\ 0 & S_y & 0 \\ 0 & 0 & S_z \end{bmatrix} = diag(S_x, S_y, S_z) \tag{16.28}$$

is the scale factor.

For a random walk process, the continuous-time state equations of the magnetometer bias and scale factor are given by

$$\dot{b} = u_b \tag{16.29}$$

$$\dot{S} = u_S \tag{16.30}$$

where $u_b$ and $u_S$ are the white noises with zero mean.

The state equations (16.29)–(16.30) can be written in the discrete-time form as

$$b(k) = b(k-1) + T_s u_b(k-1) \tag{16.31}$$

$$S(k) = S(k-1) + T_s u_S(k-1) \tag{16.32}$$

where $T_s$ is the sampling time, $b(k) = \begin{bmatrix} b_x(k) & b_y(k) & b_z(k) \end{bmatrix}^T$ and $S(k) = \begin{bmatrix} S_x(k) & S_y(k) & S_z(k) \end{bmatrix}^T$

Let us form the expanded state vector as

$$\Sigma(k) = \begin{bmatrix} b_x(k) & b_y(k) & b_z(k) & S_x(k) & S_y(k) & S_z(k) \end{bmatrix}^T$$

Then the following equation can be written

$$\Sigma(k) = \Sigma(k-1) + T_s U(k-1) \tag{16.33}$$

where $U(k) = \begin{bmatrix} u_{b_x}(k) & u_{b_y}(k) & u_{b_z}(k) & u_{S_x}(k) & u_{S_y}(k) & u_{S_z}(k) \end{bmatrix}^T$.

We can rewrite the Equations (16.31)–(16.32) in state-space form, yielding

$$\begin{bmatrix} b_x(k) \\ b_y(k) \\ b_z(k) \\ S_x(k) \\ S_y(k) \\ S_z(k) \end{bmatrix} = \begin{bmatrix} 1 & 0 & 0 & 0 & 0 & 0 \\ 0 & 1 & 0 & 0 & 0 & 0 \\ 0 & 0 & 1 & 0 & 0 & 0 \\ 0 & 0 & 0 & 1 & 0 & 0 \\ 0 & 0 & 0 & 0 & 1 & 0 \\ 0 & 0 & 0 & 0 & 0 & 1 \end{bmatrix} \begin{bmatrix} b_x(k-1) \\ b_y(k-1) \\ b_z(k-1) \\ S_x(k-1) \\ S_y(k-1) \\ S_z(k-1) \end{bmatrix} + T_s \begin{bmatrix} u_{b_x} \\ u_{b_y} \\ u_{b_z} \\ u_{S_x} \\ u_{S_y} \\ u_{S_z} \end{bmatrix} \tag{16.34}$$

This means that the systems dynamics matrix is the (6x6) unit matrix.
Let's move linear state model (16.34) into following form:

$$\Sigma(k) = \Sigma(k-1) + T_s U(k-1) \tag{16.35}$$

$$B_m(k) = S(k) A(k) B_o(k) + b(k) + v(k) \tag{16.36}$$

The measurement matrix in this case is [24]

$$H(k) = \begin{bmatrix} 1 & 0 & 0 & B_x^b(k) & 0 & 0 \\ 0 & 1 & 0 & 0 & B_y^b(k) & 0 \\ 0 & 0 & 1 & 0 & 0 & B_z^b(k) \end{bmatrix} \tag{16.37}$$

where

$$B_x^b(k) = a_{11}(k) B_{ox}(k) + a_{12}(k) B_{oy}(k) + a_{13}(k) B_{oz}(k)$$
$$B_y^b(k) = a_{21}(k) B_{ox}(k) + a_{22}(k) B_{oy}(k) + a_{23}(k) B_{oz}(k)$$
$$B_z^b(k) = a_{31}(k) B_{ox}(k) + a_{32}(k) B_{oy}(k) + a_{33}(k) B_{oz}(k)$$

are the components of the Earth magnetic field vector in body frame, $a_{ij}, (i,j = \overline{1,3})$ are the elements of the direction cosine matrix, $A$.

Applying the Optimum Kalman filter equations [28] to model (16.35–16.36) we obtain the following Kalman filter for estimation of magnetometer bias and scale factor

$$\hat{\Sigma}(k) = \hat{\Sigma}(k-1) + K(k)\left[ B_m(k) - \hat{b}(k-1) - \hat{S}(k-1)A(k)B_o(k) \right] \quad (16.38)$$

$$K(k) = P(k/k-1)H^T(k) \times \left[ H(k)P(k/k-1)H^T(k) + R_v(k) \right]^{-1} \quad (16.39)$$

$$P(k/k) = \left[ I - K(k)H(k) \right] P(k/k-1) \quad (16.40)$$

$$P(k/k-1) = P(k-1) + T_s^2 Q_U \quad (16.41)$$

where $\hat{\Sigma}(k)$ is the estimation value, $K(k)$ is the Kalman gain, $P(k/k)$ is the covariance matrix of the estimation error, $P(k/k-1)$ is the covariance matrix of the extrapolation error, $I$ is the identity matrix.

The innovation sequence of an optimal Kalman filter has precisely defined characteristics, which can be compared with the output of an actually implemented Kalman filter. The innovation process contains all information to assess the optimality of filter operations [29].

The innovation sequence is defined as the difference between the actual system output and the predicted output based on the predicted state. This sequence has the property below.

If the system operates normally then innovation sequence

$$\Delta(k) = B_m(k) - \hat{b}(k-1) - \hat{S}(k-1)A(k)B_o(k) \quad (16.42)$$

in Kalman filter which is adjusted according to the model of system dynamic will be white Gaussian noise with zero-mean and covariance matrix [30]:

$$P_\Delta(k) = H(k)P(k/k-1)H^T(k) + R_v(k) \quad (16.43)$$

It is more appropriate to use normalized innovation sequence to supervise the Kalman filter operation:

$$\tilde{\Delta}(k) = \left[ H(k)P(k/k-1)H^T(k) + R_v(k) \right]^{-1/2} \Delta(k) \quad (16.44)$$

Because in this case,

$$E\left[ \tilde{\Delta}(k)\tilde{\Delta}^T(j) \right] = P_{\tilde{\Delta}} = I\delta(kj) \quad (16.45)$$

Since the normalized innovations have the distribution $N(0,1)$, they will be in the confidence interval $[-3,+3]$ with probability 0.9986 in case of optimal operation of Kalman filter.

### 16.5.2 Simulation Results for Magnetometer Calibration via LKF

Simulations are realized with a sampling time of $T_s = 0.1$ s. As an experimental platform a cubesat model is used. For the magnetometer measurements, sensor noise is characterized by zero mean Gaussian white noise with a standard deviation of $\sigma_m = 0.83 \mu T$.

The magnetometer biases and scale factors estimation results are shown in Figures 16.16–16.21. In the proposed algorithm, state vector is formed of six states: three magnetometer biases and three scale factors. Bias and scale factor estimation results are given in Figure 16.16 and Figure 16.17, respectively.

Behaviors of the normalized innovations of LKF are shown in Figure 16.18. The dotted line indicates the confidence interval $[-3, +3]$. As seen from presented graphs, the normalized innovations of all measurement channels lay between the bounds of the confidence interval $[-3, +3]$.

Comparison of the calibrated and uncalibrated TAM results is shown in Figures 16.19–16.21. In these figures the true values of the geomagnetic field magnitudes in body frame and the calibrated and uncalibrated magnetometers measurement results are presented. As seen from presented graphs, the calibrated magnetometer results are very close to the true values of geomagnetic field and considerably better than the uncalibrated magnetometer results. The uncalibrated results are significantly biased and scaled, but calibrated results are free from these components.

Absolute and relative estimation errors of magnetometer biases and scale factors when the proposed LKF is used are presented in Table 16.3 and Table 16.4, respectively. In Table 16.3, the absolute estimation errors of magnetometer biases and scale factors

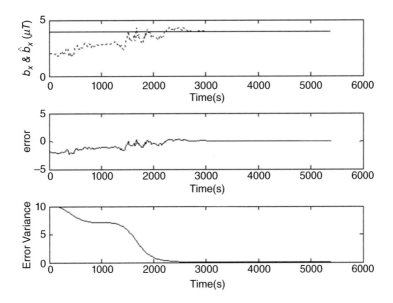

**FIGURE 16.16** Actual ($b_x$: solid line) and estimated ($\hat{b}_x$: dotted line) bias values, difference between them and error variance.

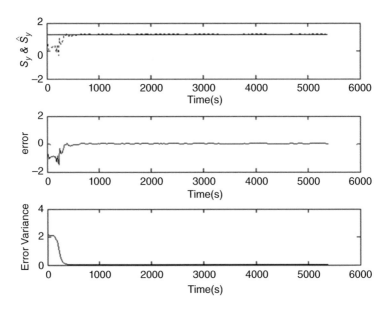

**FIGURE 16.17**   Actual ($S_y$: solid line) and estimated ($\hat{S}_y$: dotted line) scale factor values, difference between them and error variance.

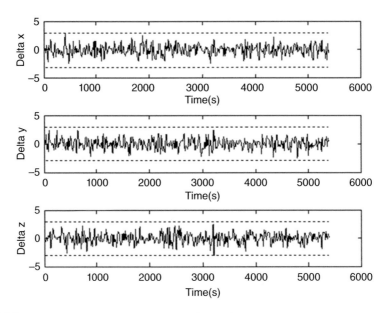

**FIGURE 16.18**   Behaviors of the normalized innovations.

are presented. The results in Table 16.4 correspond to the relative estimation errors of magnetometer biases and scale factors.

As seen from results presented in Table 16.3 and Table 16.4, the bias and scale factor estimation errors are very small after 750 sec of estimation procedure.

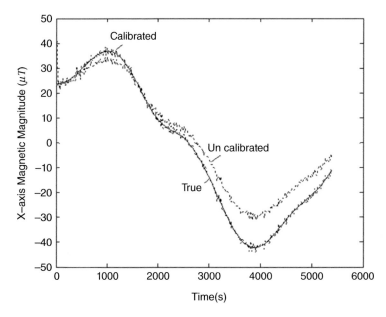

**FIGURE 16.19** X-axis Geomagnetic Field magnitude in body frame of the calibrated and uncalibrated magnetometers.

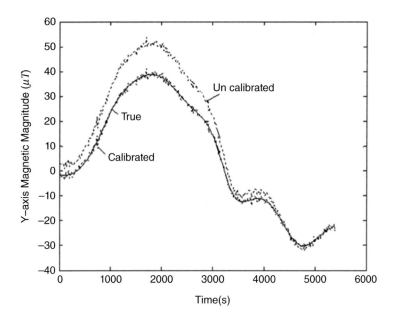

**FIGURE 16.20** Y-axis Geomagnetic Field magnitude in body frame of the calibrated and uncalibrated magnetometers.

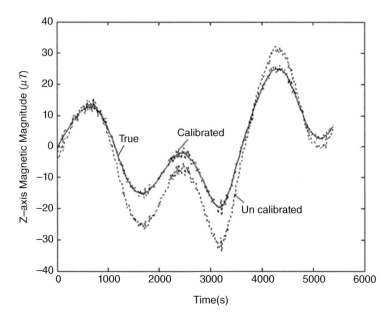

**FIGURE 16.21** Z-axis Geomagnetic Field magnitude in body frame of the calibrated and uncalibrated magnetometers.

**TABLE 16.3**
**Absolute Estimation Errors of Magnetometer Biases and Scale Factors**

| Time (s) | Abs. Error $\Delta b_x(\mu T)$ | Abs. Error $\Delta b_y(\mu T)$ | Abs. Error $\Delta b_z(\mu T)$ | Abs. Error $\Delta S_x$ | Abs. Error $\Delta S_y$ | Abs. Error $\Delta S_z$ |
|---|---|---|---|---|---|---|
| 150 | −1.9864 | −1.5691 | 0.3526 | 0.1123 | −0.9187 | −0.2604 |
| 750 | −1.1442 | −0.0781 | −0.0197 | 0.0195 | −0.0106 | 0.0107 |
| 1500 | −0.6148 | −0.0844 | 0.0035 | 0.0290 | −0.0099 | 0.0165 |
| 2250 | 0.0099 | −0.0836 | 0.0119 | 0.0269 | 0.0150 | 0.0296 |
| 3000 | 0.0038 | −0.0914 | 0.0273 | −0.0361 | 0.0101 | −0.0222 |
| 3750 | −0.0105 | −0.0494 | −0.0234 | −0.0206 | −0.0046 | −0.0210 |
| 4500 | −0.0118 | −0.0613 | −0.0571 | −0.0047 | −0.0113 | 0.0245 |
| 5250 | −0.0048 | −0.0631 | −0.0786 | −0.0108 | 0.0254 | −0.0180 |
| 5400 | 0.0020 | −0.0570 | −0.0648 | −0.0023 | −0.0151 | 0.0369 |

## 16.6 TRIAD+UKF APPROACH FOR ATTITUDE ESTIMATION AND MAGNETOMETER CALIBRATION

Here, details of a two-stage attitude estimation and magnetometer calibration algorithm are presented. The algorithm, which is designed for small satellite applications and originally published in [27] integrates the TRIAD algorithm with an UKF to both estimate the attitude and calibrate the magnetometers. TRIAD and UKF work in a sequential architecture to form the overall algorithm. First the vector measurements

**TABLE 16.4**

**Relative Estimation Errors of Magnetometer Biases and Scale Factors**

| Time (s) | Rel. Error $Rb_x(\%)$ | Rel. Error $Rb_y(\%)$ | Rel. Error $Rb_z(\%)$ | Rel. Error $RS_x(\%)$ | Rel. Error $RS_y(\%)$ | Rel. Error $RS_z(\%)$ |
|---|---|---|---|---|---|---|
| 150 | 49.66 | 31.382 | 8.815 | 14.0375 | 76.5583 | 18.6 |
| 750 | 28.605 | 1.562 | 0.4925 | 2.4375 | 0.8833 | 0.7642 |
| 1500 | 15.37 | 1.688 | 0.0875 | 3.625 | 0.825 | 1.1785 |
| 2250 | 0.2475 | 1.672 | 0.2975 | 3.3625 | 1.25 | 2.1142 |
| 3000 | 0.095 | 1.828 | 0.6825 | 4.5125 | 0.8416 | 1.5857 |
| 3750 | 0.2625 | 0.988 | 0.585 | 2.575 | 0.3833 | 1.5 |
| 4500 | 0.295 | 1.226 | 1.4275 | 0.5875 | 0.9416 | 1.75 |
| 5250 | 0.12 | 1.262 | 1.965 | 1.35 | 2.1166 | 1.2857 |
| 5400 | 0.05 | 1.14 | 1.62 | 0.2875 | 1.2583 | 2.6357 |

from the magnetometers and Sun sensor are used in the TRIAD to estimate a coarse attitude. Then this coarse attitude information is fed into the UKF. The UKF uses the coarse attitude from TRIAD as measurements together with gyro measurements for getting fine attitude estimates and estimating the gyro biases and magnetometer error terms.

In this configuration, the TRIAD algorithm provides a coarse attitude estimate, specifically at the beginning of the estimation process, when the UKF has not converged to the true magnetometer errors and the errors have not been compensated yet. Nonetheless, this attitude information from the TRIAD enables complete magnetometer calibration with a linear model as introduced in Eq. (16.25).

The UKF estimates the attitude quaternion ($q$), gyro biases ($b_g$) and the magnetometer error terms ($\theta_m$) as

$$X = \begin{bmatrix} q^T & b_g^T & \theta_m^T \end{bmatrix}^T, \tag{16.46}$$

where the error parameters for the magnetometer are selected as

$$\theta_m = \begin{bmatrix} b_m^T & D_{11} & D_{22} & D_{33} & D_{12} & D_{13} & D_{23} \end{bmatrix}^T, \tag{16.47}$$

assuming an asymmetric $D$ matrix with six parameters. In this case, $\Phi_k$ in Eq. (16.26c) should be reformulated as

$$\Phi_k = \begin{bmatrix} -1 & 0 & 0 & B_{bx} & 0 & 0 & B_{by} & B_{bz} & 0 \\ 0 & -1 & 0 & 0 & B_{by} & 0 & B_{bx} & 0 & B_{bz} \\ 0 & 0 & -1 & 0 & 0 & B_{bz} & 0 & B_{bx} & B_{by} \end{bmatrix}. \tag{16.48}$$

The scheme for the TRIAD+UKF algorithm for attitude estimation and magnetometer calibration is given in Figure 16.22. The TRIAD algorithm runs at the first stage using magnetometer ($\bar{B}_b$) and Sun sensor ($S_b$) measurements and estimates a coarse attitude ($\hat{q}_{tr}$). The UKF in the second stage uses this coarse attitude information together with the angular rates from gyros ($\bar{\omega}_{BI}$) as measurements. Note that here

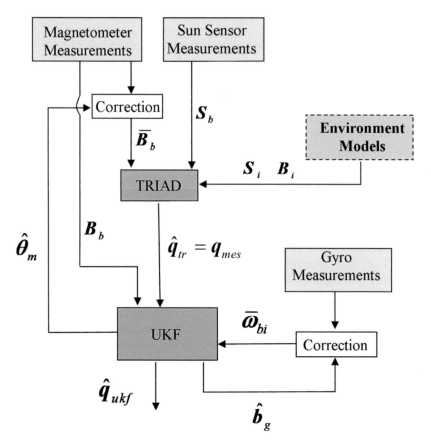

**FIGURE 16.22**    TRIAD+UKF approach for attitude estimation and complete magnetometer calibration.

we use corrected TAM measurements in the TRIAD (represented as $\overline{B}_b$ with an over bar), whereas the UKF still needs raw magnetometer measurements. The magnetometer measurements are corrected using the magnetometer error estimates of the UKF from the previous recursive step as

$$\overline{B}_{b,k+1} = \left( I_{3\times3} + \hat{D}_k \right) B_{b,k+1} - \hat{b}_{m,k}. \tag{16.49}$$

Figure 16.23 presents the attitude estimation error for the TRIAD+UKF approach and the standalone TRIAD algorithm ($\hat{q}_{ukf}$ and $\hat{q}_{tr}$ in Figure 16.22, respectively). Shaded areas in the figure represent the eclipse. As seen the error converges below 0.5° at about 5000th s and this is an acceptable accuracy regarding the used sensors and their error characteristics. Almost periodic increases in the TRIAD estimation errors (e.g. at 9500th s) are mainly due to the two measurement vectors approaching to parallelism. Before the convergence, the TRIAD+UKF attitude estimations are also affected by this phenomenon. Once the algorithm converges, this effect is mitigated in the TRIAD+UKF results by magnetometer error compensation and

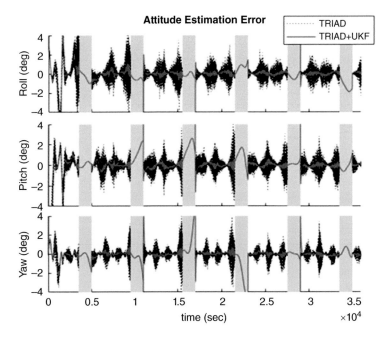

**FIGURE 16.23** Attitude estimation error for TRIAD and TRIAD+UKF algorithms. Presented TRIAD results are for the algorithm when it is run as a part of the proposed scheme.

incorporating the gyro measurements (compare e.g. the estimation results at the 2nd daylight period – from 5000th s to 9500th s – and the last daylight period – from 29000th s to 33500th s).

Figures 16.24–16.26 present the estimation results for the magnetometer error terms. The filter quickly converges to the actual values, within two orbital periods. Convergence is rather slow for the D matrix terms, but for the magnetometer bias estimations, errors converge below 300nT ($1\sigma$) at approximately 3100th s.

Overall, the proposed algorithm can estimate the attitude – in daytime – with an accuracy, and for roll, pitch and yaw angles, respectively, and these values are far better than the desired accuracy. Filter achieves this accuracy level with successful real-time complete calibration of magnetometer errors.

Here, we shall mention that all these algorithms presented so far in scope of recursive magnetometer calibration algorithm are capable of tracking any variation in the estimated error terms. Despite, the assumption of constant error terms as in Eq. (16.1), with proper tuning, the filtering algorithm can quickly converge to the new values of the states. In. Figure 16.27, magnetometer bias estimation results for the TRIAD+UKF algorithm, presented in this example, are given when the bias terms change at 24000th s.

## 16.7  CALIBRATION WITHOUT ATTITUDE

Recursive magnetometer calibration algorithm, presented so far, requires attitude knowledge. In contrast, it is also possible to calibrate the magnetometers without needing attitude information at all. Attitude-independent recursive algorithms

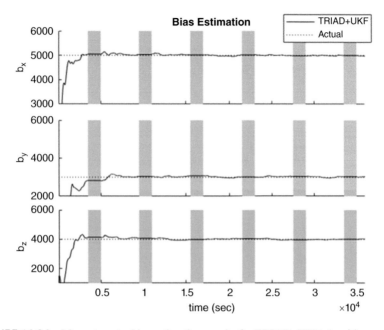

**FIGURE 16.24**    Magnetometer bias estimation results for TRIAD+UKF algorithm.

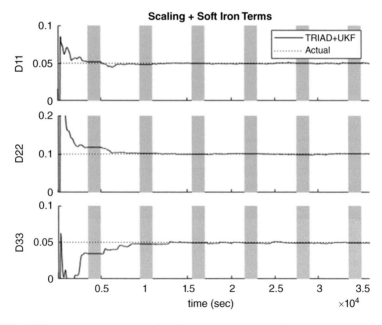

**FIGURE 16.25**    Estimation results for the diagonal terms of the scaling matrix when TRIAD+UKF algorithm is used.

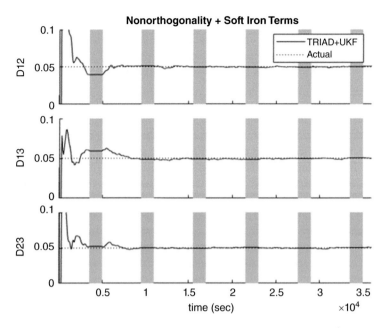

**FIGURE 16.26** Estimation results for the non-diagonal terms of the scaling matrix when TRIAD+UKF algorithm is used.

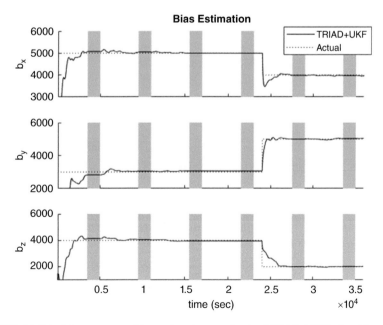

**FIGURE 16.27** Magnetometer bias estimation results for TRIAD+UKF algorithm when magnetometer bias terms change at 24000th s.

reformulate the batch magnetometer calibration algorithm to provide real-time estimation for the magnetometer error terms. Mainly they make use of the scalar attitude-independent measurement equation given in Eq. (15.8) [31]. In recent studies, additional quasi measurements are proposed for improving the accuracy and convergence of the algorithm [32, 33].

An attitude-independent recursive algorithm can be designed to estimate the magnetometer error terms using only the scalar measurement given in Eq. (15.8). In this case, due to the nonlinear nature of the measurement model, a nonlinear version of the KF must be used, unless the equation is linearized using an approach such as centering. In [31], three sequential algorithms are designed for estimating the magnetometer error terms: 1) a sequential centered algorithm that applies linear LS using the centered variables; 2) an EKF; and 3) an UKF. All three algorithms are tested with both simulated and actual data and their results are compared with those of the batch two-step algorithm. In [34], an UKF, which is exactly formulated same as the one in [31], is used to calibrate the magnetometers after the launch.

In recent studies, recursive attitude-independent algorithms that make use of other measurements are introduced. We refer these measurements as quasi-measurements since they are not directly measured but derived from the magnetometer measurements (usually based on the assumptions). These measurements are used either as a replacement or in addition to Eq. (15.8).

In [32], the authors proposed using the fact that the magnitude of the Earth's magnetic field does not change within a short time. Hence,

$$\left|\boldsymbol{H}_k\right|^2 - \left|\boldsymbol{H}_{k+\Delta k}\right|^2 \cong 0, \tag{16.50}$$

where $\Delta k$ is small enough. Then the pseudo-measurement is defined as

$$Z_{2k} = \left|\boldsymbol{B}_k\right|^2 - \left|\boldsymbol{B}_{k+\Delta k}\right|^2 = g\left(\boldsymbol{B}_k, \boldsymbol{\theta}\right) + \eta_{2k}. \tag{16.51}$$

Here,

$$\eta_{2k} = \eta_k - \eta_{k+\Delta k}. \tag{16.52}$$

A sliding window filter, which is designed in the framework of an EKF, is used to estimate the magnetometer error terms using the measurements given in Eqs. (15.8) and (16.51).

In [35], the same quasi-measurement (Eq.16.51) is used for estimating only the bias vector with an algorithm called the differential value iterative algorithm.

In [33], first, the gyro measurements are used to propagate the measured magnetic field in time. Then assuming

$$\boldsymbol{H}_{k+\Delta k} \cong \boldsymbol{H}_k \tag{16.53}$$

for a $\Delta k$, which is small enough, the following linear equation is derived to estimate the bias vector,

$$\boldsymbol{B}_{k+\Delta k} - \boldsymbol{B}_{p,k+\Delta k} = h\left(\Delta k, \omega_k\right)\boldsymbol{b}_k. \tag{16.54}$$

Here, $B_{k+\Delta k}$ is the measured magnetic field vector at time $k + \Delta k$, $B_{p,k+\Delta k}$ is the magnetic field vector, which is measured at time $k$ and propagated to time $k + \Delta k$ and $\omega_k$ is the spacecraft's angular rate vector at $k$.

Finally, a recursive linear least squares algorithm is designed to estimate the bias vector in real time. An extended version of the algorithm estimates also the gyro biases [33].

In [36], a similar approach with that of [37] is used for estimating the bias terms. Assuming the reference magnetic field is not changing at two consecutive measurements (Eq. 16.53), the following equation is derived:

$$\left(A_k^{k+1} - I_{3\times3}\right)b_k = A_k^{k+1}B_k - B_{k+1}. \tag{16.55}$$

Here, $A_k^{k+1}$ is the relative attitude matrix which is defining the rotation in between two consecutive measurements and can be calculated using gyro measurements.

The authors design a linear KF using Eq. (16.55). The state model for the KF in discrete form is given as

$$\begin{bmatrix} B_{k+1} \\ b_{k+1} \end{bmatrix} = \begin{bmatrix} A_k^{k+1} & I_{3\times3} - A_k^{k+1} \\ 0_{3\times3} & I_{3\times3} \end{bmatrix} \begin{bmatrix} B_k \\ b_k \end{bmatrix} \tag{16.56}$$

whereas the measurement equation is

$$Y_k = \begin{bmatrix} I_{3\times3} & | & 0_{3\times3} \end{bmatrix} \begin{bmatrix} B_{k+1} \\ b_{k+1} \end{bmatrix}. \tag{16.57}$$

Their results, which include results for application for a mobile phone magnetometer, show that the algorithm performs better compared to two-step method [38], the attitude-independent EKF in [31], as well as a batch calibration version of their algorithm.

In [37], the magnetometer bias estimation for a spinning spacecraft is investigated. Details and the results for this algorithm will be presented in Section 16.8.

## 16.8 MAGNETOMETER BIAS ESTIMATION FOR A SPINNING SMALL SPACECRAFT

As discussed in [31], the scalar Eq. (15.8) can be sufficient to estimate the complete set of magnetometer errors. However, this algorithm works well only with an UKF, which comes with a computational cost. In this example, we present a simple KF algorithm to estimate the bias terms in real time. The scalar measurement is aided with two quasi-measurements and both the accuracy and the convergence rate of the filter is improved.

The quasi-measurements are derived based on the fact that in an ideal case, where there is no bias, the spin-planar magnetometer measurements oscillate about a zero-mean. In contrast, when there is bias, the mean of oscillations for the spin-planar measurements will shift depending on the magnitude of the bias (Figures 16.28 and 16.29). Thus, if the spin-planar magnetometer measurements are averaged within

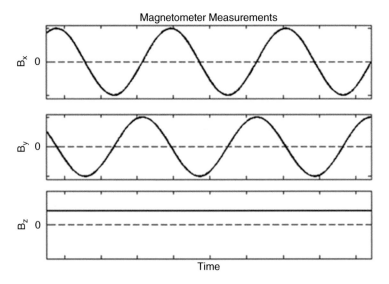

**FIGURE 16.28**  Magnetometer measurements on a spinning spacecraft in ideal case.

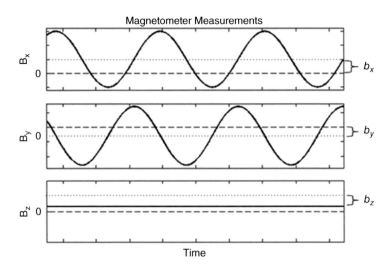

**FIGURE 16.29**  Magnetometer measurements on a spinning spacecraft when there is bias in the measurements. Dashed line is for zero level. Dotted line is for the mean of spin planar measurements and for the non-biased value of spin direction measurements.

a sliding-window, a quasi-measurement for the corresponding bias term can be obtained,

$$\tilde{b}_{p,k} = \frac{1}{N} \sum_{i=k-N+1}^{k} B_{p,i} = b_{p,k} + \underline{v}_{p,k}, \tag{16.58}$$

where

$$\underline{v}_{p,k} = \frac{1}{N} \sum_{i=k-N+1}^{k} v_{p,i}. \tag{16.59}$$

Here, subscript $p$ represents one of the spin-planar body axes for the spacecraft (e.g. $p = x$, $y$ if the spin axis is body $z$ axis). Hence, $B_{p,i}$ is the magnetometer measurements for spin-planar body axis $p$ sampled at $i^{th}$ time-step. $N$ is the size of the sliding window.

Note that such two quasi-equations can be derived for a spinning spacecraft, one for each spin-planar axis. If we use these two quasi-measurements together with the scalar measurement of Eq. (15.8) then we can build a pseudo-linear measurement model for the KF as

$$\begin{bmatrix} \tilde{b}_{x,k} \\ \tilde{b}_{y,k} \\ z_k \end{bmatrix} = \begin{bmatrix} 1 & 0 & 0 \\ 0 & 1 & 0 \\ 2B_{x,k} - b_{x,k} & 2B_{y,k} - b_{y,k} & 2B_{z,k} - b_{z,k} \end{bmatrix} \begin{bmatrix} b_{x,k} \\ b_{y,k} \\ b_{z,k} \end{bmatrix} + \begin{bmatrix} v_{x,k} \\ v_{y,k} \\ \eta_k \end{bmatrix}. \tag{16.60}$$

Table 16.5 summarizes the results for 100 runs executed the pseudo-KF algorithm. The results estimated by an UKF are also given. The UKF is same with the one presented in [31] in essence but estimates only the bias vector, not all the calibration parameters. Comparison with the UKF provides us two opportunities. First, we can compare the accuracy of the proposed algorithms with that of the UKF, which is shown to be the most accurate algorithm for real-time attitude-independent magnetometer calibration in [31]. Second, we can understand the computational efficiency of the proposed algorithms. The estimation results are picked at the end of each simulation run (at $18000^{th}$ s) and Table 16.5 gives mean values of the picked estimation results for 100 runs. The table also includes computed $3\sigma$ bounds for the estimations.

The results show that the pseudo-linear KF and UKF provide bias estimates very close to the actual values. The $3\sigma$ bounds for spin planar bias terms are slightly lower for the pseudo-linear KF, which is showing the advantage of including the spin motion derived measurements. Nonetheless, the UKF demands more than four times

**TABLE 16.5**

**Results for Magnetometer Bias Estimation using Two Attitude-Independent Recursive Algorithms, UKF and Pseudo-linear KF**

| Bias Term | Truth | UKF | Pseudo-Linear KF |
|---|---|---|---|
| $b_x$ | 5000 nT | 4999.22 ± 22.24 | 4997.54 ± 20.03 |
| $b_y$ | 3000 nT | 2996.30 ± 26.15 | 3001.58 ± 21.83 |
| $b_z$ | 4000 nT | 3996.08 ± 32.10 | 3998.31 ± 34.59 |

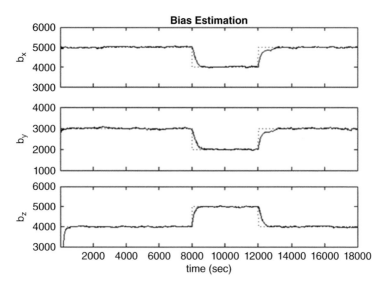

**FIGURE 16.30**  Estimation results for pseudo-linear KF in case of variation in the bias terms.

higher computational load compared to the pseudo-linear KF. The computational loads are calculated based on the required time for running the algorithms on MATLAB®. We do not give exact computation times as they may be misleading and vary depending on the simulation platform. Yet the relative numbers give tips for quantifying the computational load of each algorithm. Such reduced computational load with pseudo-linear KF despite the competitive accuracy makes it ideal for small satellite applications.

Estimation results for a scenario, which includes a change in the bias terms, show the pseudo-linear KF is also capable of quickly tracking the variations in bias terms (Figure 16.30). In this scenario, the bias values change to $b = \begin{bmatrix} 4000 & 2000 & 5000 \end{bmatrix}^T nT$ in between 8000th and 12000th s.

## 16.9 DISCUSSION ON RECURSIVE MAGNETOMETER CALIBRATION

Compared to the batch magnetometer calibration algorithms, the main advantage of the recursive algorithms is being capable of estimating the time-varying error terms even with a simple state model as given in Eq. 16.1. They do not require any extension to the general magnetometer measurement model (Eq. 15.5) to track the changes in the estimated states as examples show.

The biggest challenge while using recursive magnetometer calibration algorithms is to have sufficient number of measurements to assure the observability and accuracy. As long as the assumptions lying on the basis of the algorithms are valid, the recently proposed algorithms in [33, 36, 37] can provide accurate calibration results. Yet these algorithms should be validated for different cases. One of the further investigations might be introducing new quasi-measurements for especially attitude-independent recursive calibration algorithms to improve the algorithms' observability condition as well as the accuracy.

## 16.10 CONCLUSION

In this chapter, we reviewed the recursive magnetometer calibration algorithms which are capable of sensing the time-variation in the estimated error terms. We categorized the recursive magnetometer calibration algorithms in two: attitude-dependent and attitude-independent algorithms. Studies that propose different filtering algorithms (e.g. Extended Kalman Filter, Unscented Kalman Filter) and different approaches for calibration are reviewed.

A reconfigurable UKF-based estimation algorithm for magnetometer biases and scale factors is presented as a part of the attitude estimation scheme of a pico satellite. In this approach, scale factors are not treated together with other parameters as a part of the state vector; they are estimated separately. After satisfying the conditions of the proposed stopping rule for bias estimation, UKF reconfigures itself for estimation of attitude parameters alone.

This approach is advantageous because simulation results prove that reducing the size of the state vector without any change at the measurement vector brings about an increment in the accuracy of estimation of the states. Also, a smaller state vector means lower computational burden and risk of filter divergence.

As another example for simultaneous attitude estimation and magnetometer calibration, a two-stage algorithm is given. At the initial phase, biases of three orthogonally located magnetometers are estimated as well as the attitude and attitude rates of the satellite. During initial period after the orbit injection, gyro measurements are accepted as bias free. At the second phase, estimated constant magnetometer bias components are taken into account and the algorithm is run for the estimation of the gyro biases. As a result, six different bias terms for two different sensors are obtained in two stages where attitude and attitude rates are estimated regularly. For both estimation phases of the procedure, UKF is used as the estimation algorithm.

A Linear Kalman filter-based algorithm for the estimation of magnetometer biases and scale factors is presented as the magnetometer calibration method with known attitude. For this method, the attitude transformation matrix is needed to transform the modeled magnetic field vector in orbit frame to the body frame. Measurements from other attitude sensors (e.g. star trackers) are thus essential for this bias calibration algorithm. In TRIAD+UKF approach for attitude estimation and magnetometer calibration, it is shown that even coarse attitude estimates provided by the TRIAD can be used to form a calibration algorithm with known attitude. The proposed approach is used for concurrent magnetometer calibration and attitude estimation.

Last, attitude-independent recursive calibration algorithms are introduced. These algorithms rely on the scalar check for the magnitude difference for the body frame measurements and the modeled magnetic field. It is shown that including additional quasi-measurements as in case for a spinning spacecraft can be useful for improving both the convergence rate and the accuracy of the algorithm.

Using recursive magnetometer calibration algorithms is more advantageous in case of time-variation in the magnetometer error terms. These algorithms are capable of capturing the time-variation in the error terms even with a simple state model. One essential future discussion might be improving the convergence speed especially for attitude-independent recursive calibration algorithms by introducing new quasi-measurements. Additional measurements are also needed for observability when a fully populated scaling matrix is assumed.

## REFERENCES

1. H. E. Soken and S. Sakai, "Magnetometer calibration for advanced small satellite missions," in *30th International Symposium on Space Technology and Science*, 2015, pp. 10–12.
2. H. E. Soken, S. I. Sakai, and R. Wisniewski, "In-orbit estimation of time-varying residual magnetic moment," *IEEE Trans. Aerosp. Electron. Syst.*, vol. 50, no. 4, pp. 3126–3136, 2014.
3. R. Da Forno, F. Reali, S. Bristot, and S. Debei, "Autonomous navigation of MegSat1: Attitude, sensor bias and scale factor estimation by EKF and magnetometer-only measurement," in *22nd AIAA International Communications Satellite Systems Conference & Exhibit 2004 (ICSSC)*, Monterey, USA, 2004, pp. 1–14.
4. H. E. Soken and C. Hajiyev, "UKF-based reconfigurable attitude parameters estimation and magnetometer calibration," *IEEE Trans. Aerosp. Electron. Syst.*, vol. 48, no. 3, pp. 2614–2627, Jul. 2012.
5. H. E. Soken and C. Hajiyev, "Reconfigurable UKF for in-flight magnetometer calibration and attitude parameter estimation," in *IFAC Proceedings Volumes (IFAC-PapersOnline)*, Milano, Italy, 2011, vol. 18, no. PART 1.
6. H. E. Söken and C. Hajiyev, "UKF based in-flight calibration of magnetometers and rate gyros for pico satellite attitude determination," *Asian J. Control*, vol. 14, no. 3, 2012.
7. T. Inamori, N. Sako, and S. Nakasuka, "Strategy of magnetometer calibration for nano-satellite missions and in-orbit performance," in *AIAA Guidance, Navigation, and Control Conference*, Toronto, Canada, 2010.
8. T. Inamori, R. Hamaguchi, K. Ozawa, P. Saisutjarit, N. Sako, and S. Nakasuka, "Online magnetometer calibration in consideration of geomagnetic anomalies using Kalman filters in nanosatellites and microsatellites," *J. Aerosp. Eng.*, vol. 29, no. 6, p. 04016046, 2016.
9. H. E. Soken, "An Attitude Filtering and Magnetometer Calibration Approach for Nanosatellites," *Int. J. Aeronaut. Space Sci.*, vol. 19, pp. 164–171, 2018.
10. M. Kiani and S. H. Pourtakdoust, "Spacecraft attitude and system identification via marginal modi ed unscented Kalman lter utilizing the sun and calibrated three-axis-magnetometer sensors," *Sci. Iran. B*, vol. 21, no. 4, pp. 1451–1460, 2014.
11. H. E. Soken and C. Hajiyev, "Pico satellite attitude estimation via robust unscented Kalman filter in the presence of measurement faults," *ISA Trans.*, vol. 49, no.3, pp. 249–256, 2010.
12. Ch. Hajiyev, "Adaptive filtration algorithm with the filter gain correction applied to integrated INS/Radar altimeter," *Proc. Inst. Mech. Eng., Part G: J. Aerosp. Eng.*, vol. 221, no. 5, pp. 847–885, 2007.
13. H.E. Soken and C. Hajiyev. "In flight magnetometer calibration via unscented Kalman filter. In *Proc. 5th International Conference on Recent Advances in Space Technologies*, Istanbul, Turkey, 2011, pp. 885–890.
14. D. J. Jwo and S. C. Chang, "Particle swarm optimization for GPS navigation Kalman filter adaptation," vol. 81, no. 4, pp. 343–352, 2009.

15. Ch. M. Hajiyev, "Some problems of dynamic systems parametrical identification," in *Digest of First Asian Control Conference*, Japan, July 27–30, 1994, vol. 2, pp. 921–924.
16. Ch. M. Hajiyev, "Stopping rules formation and faults detection in parametric identification problems," *IMechE Proceedings , Part I, J.Systems & Control Engineering*, vol. 215 (I4), pp. 357–364, 2001.
17. V.S. Pugachev, *Probability Theory and Mathematical Statistics (in Russian)*. Moscow: Nauka, 1979.
18. B. D. Brumback and M. D. Srinath, "A fault tolerant multisensor navigation system design," *IEEE Trans. Aerosp. Electron. Syst.*, vol.AES-23, no. 6, pp. 738–755, 1987.
19. P. J. Wolff et al., "Computation of the factorized error covariance of the difference between correlated estimators," *IEEE Trans. Automat. Control*, vol. AC-35, no.12, pp. 1284–1292, 1990.
20. Ch. M. Gadzhiev, "New control conditions for diagnostics of Kalman filter," *Metrologiya*, 5, pp. 3–15, 2001 (in Russian).
21. H. E. Soken and Ch. Hajiyev, "In flight calibration of pico satellite attitude sensors via unscented Kalman filter," *Gyroscopy and Navigation*, vol. 2, no. 3, pp. 156–163, 2011.
22. H. E. Soken, "A survey of calibration algorithms for small satellite magnetometers," *Meas. J. Int. Meas. Confed.*, vol. 122, no. September 2017, pp. 417–423, 2018.
23. C. Hajiyev, "Orbital calibration of microsatellite magnetometers using a linear Kalman filter," *Meas. Tech.*, vol. 58, no. 9, pp. 1037–1043, 2015.
24. C. Hajiyev, "In-orbit magnetometer bias and scale factor calibration," *Int. J. Metrol. Qual. Eng.*, vol. 7, no. 1, p. 104, 2016.
25. Y. Cheng and J. L. Crassidis, "An expectation-maximization approach to attitude sensor calibration," *Adv. Astronaut. Sci.*, vol. 130 PART 2, pp. 1749–1764, 2008.
26. H. E. Söken, "An attitude filtering and magnetometer calibration approach for nanosatellites," *Int. J. Aeronaut. Sp. Sci.*, vol. 19, no. 1, pp. 164–171, 2018.
27. H. E. Soken and S. Sakai, "Attitude estimation and magnetometer calibration using reconfigurable TRIAD+filtering approach," *Aerosp. Sci. Technol.*, vol. 99, p. 105754, Apr. 2020.
28. R.E. Kalman, A new approach to linear fltering and prediction problems: ASME *J. Basic Eng.*, 1960, 35–45.
29. C. Hajiyev and F. Caliskan, *Fault Diagnosis and Reconfiguration in Flight Control Systems*. Boston, USA: Kluwer Academic Publishers, 2003.
30. R. K. Mehra and J. Peschon, "An innovations approach to fault detection and diagnosis in dynamic systems," *Automatica*, vol. 7, pp. 637–640, 1971.
31. J. L. Crassidis, K.-L. Lai, and R. R. Harman, "Real-time attitude-independent three-axis magnetometer calibration," *J. Guid. Control. Dyn.*, vol. 28, no. 1, pp. 115–120, 2005.
32. B. Grandvallet, A. Zemouche, M. Boutayeb, and S. Changey, "Real-time attitude-independent three-axis magnetometer calibration for spinning projectiles: A sliding window approach," *IEEE Trans. Control Syst. Technol.*, vol. 22, no. 1, pp. 255–264, 2014.
33. S. Sakai, N. Bando, S. Shimizu, Y. Maru, and H. Fuke, "Real-time estimation of the bias error of the magnetometer only with the gyro sensors," *28th International Symposium on Space Technology and Science*, Okinawa, Japan, 2011, pp. 1–6.
34. K. Han, H. Wang, T. Xiang, and Z. Jin, "Magnetometer compensation scheme and experimental results on ZDPS-1A pico-satellite," *Chinese J. Aeronaut.*, vol. 25, no. 3, pp. 430–436, 2012.
35. Z. Zhang, J. Xiong, and J. Jin, "On-orbit real-time magnetometer bias determination for micro-satellites without attitude information," *Chinese J. Aeronaut.*, vol. 28, no. 5, pp. 1503–1509, 2014.

36. G. Troni and R. M. Eustice, "Magnetometer bias calibration based on relative angular position: Theory and experimental comparative evaluation," in *2014 IEEE/RSJ International Conference on Intelligent Robots and Systems*, Chicago, USA, 2014, pp. 444–450.
37. H. E. Soken and S. Sakai, "Real-time attitude-independent magnetometer bias estimation for spinning spacecraft," *J. Guid. Control. Dyn.*, 2017.
38. R. Alonso and M. D. Shuster, "Complete linear attitude-independent magnetometer calibration," *J. Astronaut. Sci.*, vol. 50, no. 4, pp. 477–490, 2002.

# Index

Page numbers in **bold** refer to content in **tables**.

## A

abnormal measurements, 136, 179, 199, 200
abrupt changes, 158, 180, 191
absolute estimation errors, 209, 210, 212, 214, 215, 299, 302
absolute values of error, **159**, 208, 248, 287, 288
active fault tolerant attitude estimation, 175, 194, 195
active fault tolerant attitude estimation system, 194
active magnetic controller, 18
active-pixel sensor, 36
actuator faults, 117, 119, 231, 241
actuators, 15, 23, 117, 119, 175, 221
adaptation, 117, 118, 119, 120, 121, 124, 128, 129, 130, 131, 132, 136, 137, 158, 199, 200, 201, 203, 206, 215, 217, 221, 222, 225, 227, 228, 231, 232, 239, 241, 242, 244, 246, 247, 248, 251, 253, 254, 255, 256, 258, 259, 262, 263
adaptation of the process noise covariance, 124
adaptation of the R matrix, 132
adaptation problem, 120
adaptation techniques, 121, 129, 136, 215, 241
adaptation with multiple scale factors, 242
adaptive fading UKF algorithm, 234, 263
adaptive Fading Unscented Kalman Filter, 231
adaptive estimation, 117, 120, 121, 124, 128, 136, 175, 226, 228
adaptive estimator, 175, 200
adaptive factor, 217, 233
adaptive fading factor, 232, 233, 237, 242
adaptive fading Kalman filter, 118
adaptive fading UKF, 234, 263
adaptive filters, 100
adaptive filtration algorithm, 125, 127
adaptive Kalman filter, 117, 118, 136
adaptive Kalman filtering, 117, 118, 119, 124, 129, 155, 221, 226
adaptive LS method, 273
adaptively tuned MEKF, 226
adaptive noise covariance estimation methods, 128, 136
adaptive system identification, 94
adaptive tuning, 226, 231

adaptive UKF, 228, 229, 243, 254
additive error model, 62
additive process noise, 232, 254
aerodynamic disturbance torque, 16, 18
aerodynamic torque, 18
aerodynamic drag, 17, 20
aerodynamic pressure, 18
aerodynamic pressure vector, 18
albedo, 32, 35, 36, 38, 42, 57, 58
algebraic method, 164
alignment of the TAM, 269
ambiguity, 2, 40, 54, 71, 79, 80, 89
analog cubesat Sun sensor, 31
analog quadrant Sun sensor, 36
analog sensors, 30, 31, 35, 58
analog Sun sensors, 31
analytical approximation, 221, 223, 228, 231
analytical redundancy-based approach, 175
angular momentum vector, 13, 16
angular motion, 26, 164
angular rate, 45, 47, 63, 140, 141, 152, 156, 169, 206, 259, 260, 261, 262, 278, 279, 284, 289, 295, 309
angular rate estimation, 259, 260, 261, 262
angular velocities, 3, 44, 47, 164, 191, 260, 262
angular velocity estimates, 259
angular velocity vector, 16, 278
a posterior possibility density, 97
a posterior probability distribution density, 95
a priori statistical characteristics, 120, 176
a priori statistical uncertainty, 119
a priori uncertainty, 119, 120
a priori uncertainty degree, 119
argument of latitude, 14
argument of perigee, 14, 58
ascending node, 14, 58, 75
artificial neural networks, 176
atmospheric density, 18
attitude accuracy, 75, 78, 79, 90, 210
attitude angle error, 166
attitude angles, 70, 73, 76, 79, 164, 166, 290, 291, 292, 293
attitude control, 13, 18, 21, 152, 158, 176
attitude-dependent algorithms, 278
attitude-dependent magnetometer calibration algorithms, 277

attitude determination, 21, 25, 48, 49, 67, 73,
        74, 75, 77, 78, 79, 80, 81, 82, 88,
        89, 90, 164, 175, 176, 177, 188,
        196, 221, 239, 265, 266, 291, 292
attitude determination and control, 21, 25,
        175, 176, 291, 292
attitude determination and control system,
        25, 175, 176, 291, 292
attitude determination system, 164, 175,
        177, 188, 196, 221, 239, 265
attitude dynamics, 8, 9, 16, 22, 152, 153,
        160, 163
attitude error, 10, 40, 62, 139, 142, 144, 148,
        150, 168, 210, 218
attitude estimation, 25, 42, 44, 48, 49, 53,
        55, 57, 60, 67, 74, 79, 81, 85, 86,
        87, 88, 89, 90, 94, 113, 114, 117,
        118, 119, 123, 136, 139, 141, 145,
        148, 149, 152, 153, 160, 163, 164,
        165, 166, 168, 170, 171, 172, 175,
        179, 191, 193, 194, 195, 196, 199,
        200, 206, 207, 208, 209, 210, 212,
        213, 215, 217, 223, 224, 225, 226,
        228, 229, 230, 232, 239, 242, 243,
        244, 245, 248, 249, 250, 251, 252,
        253, 255, 256, 257, 258, 266, 277,
        278, 280, 282, 285, 287, 289, 291,
        292, 302, 303, 304, 305, 313
attitude estimation algorithm, 44, 60, 81,
        113, 165, 232, 248
attitude estimation accuracy, 55, 152, 194,
        215, 250, 253, 266
attitude estimation filters, 200
attitude estimation errors, 85, 193, 194, 208,
        209, 210, 212, 213, 215, 229,
        230, 258
attitude estimation method, 67, 81, 251, 255
attitude estimator reconfiguration, 175, 196
attitude filter, 200, 202, 221, 222, 226, 231,
        239, 241, 242, 248, 253, 255, 258,
        259, 262, 263, 278, 289, 296
attitude filtering, 1, 45, 64, 67, 139, 140,
        141, 142, 147, 148, 149, 152, 159,
        160, 164, 169, 177, 222, 239, 241,
        251, 253, 262, 263
attitude filtering algorithm, 64, 67, 140, 177,
        222, 253
attitude-independent EKF, 309
attitude-independent observation, 272
attitude-independent recursive algorithms,
        305, 311
attitude information, 25, 37, 41, 45, 63, 165,
        277, 295, 296, 303, 305
attitude kinematics, 11
attitude matrix, 67, 68, 69, 70, 79, 81, 82,
        83, 88, 141, 270, 309

attitude parameters, 1, 10, 21, 140, 141, 191,
        235, 290, 291, 313
attitude parameter estimation, 282, 291, 292
attitude parameterization technique, 79, 80
attitude quaternions, 22
attitude rate, 19, 25
attitude representation, 139, 141, 143, 148, 160
attitude sensors, 15, 25, 30, 41, 45, 48, 49,
        64, 67, 75, 80, 88, 90, 177, 199,
        217, 239, 266, 296, 313
attitude sensors' accuracy, 75
attitude stabilization, 18
attitude transformation matrix, 313
Auriga star tracker, 41, 42

**B**

bank of Kalman filters, 117
batch algorithm, 295
batch calibration, 265, 272, 273, 274,
        277, 309
batch calibration method, 273
batch magnetometer calibration, 273, 274,
        275, 308, 312
batch magnetometer calibration algorithms,
        271, 273, 274, 312
batch magnetometer calibration methods,
        271, 272, 273, 274, 275
batch methods, 271, 272, 274
batch methods for magnetometer error
        estimation, 271
batch two-step algorithm, 308
Bayes formula, 95, 166
Bayes' method, 165
bias estimation, 226, 228, 229, 250, 251,
        252, 280, 282, 284, 285, 286, 287,
        289, 290, 291, 292, 293, 294, 305,
        306, 307, 309, 311, 312, 313
bias estimation error, 226, 250, 251, 252
bias estimation stopping, 282
bias estimation stopping instant, 282
binary type Sun presence sensor, 34
body angular rate vector, 63, 141, 278, 289
body frame, 3, 4, 13, 14, 15, 16, 17, 20, 30,
        35, 37, 53, 57, 60, 62, 64, 67, 68,
        69, 73, 86, 88, 90, 146, 156, 269,
        270, 272, 297, 299, 301, 302, 313

**C**

calibration, 34, 54, 57, 64, 90, 164, 250,
        265, 266, 271, 272, 273, 274, 275,
        277, 278, 280, 282, 287, 289, 290,
        291, 292, 293, 294, 295, 296, 299,
        302, 303, 305, 308, 309, 311, 312,
        313, 314

calibration methods, 265, 271, 272, 273, 274, 275
calibration parameters, 271, 274, 311
calibration without attitude, 305
calibrated magnetometer measurements, 270
calibrated measurements, 54
change detection, 158, 176, 254
change detection technique, 176
change in the predicted covariance, 232, 254
characteristic polynomial, 91, 101, 205
chi-square distribution, 207, 235, 244
center of mass, 15, 56, 165, 185, 207
center of pressure, 18
centroiding algorithm, 44
circular orbit, 14, 22, 55, 223, 234, 244, 280, 292
coarse estimate, 273
coarse Sun sensor, 56, 57, 64
complementary metal–oxide–semiconductor, 36
complex dynamical systems, 175
computational burden, 120, 176, 195, 218, 284, 287, 288, 313
conditional distribution density, 96
conditional mathematical expectation, 166
conditional mean, 95
conditional probability density, 166
confidence coefficient, 285
confidence interval, 285, 298, 299
confidence probability, 183
constant magnetometer bias, 229, 280, 289, 313
constraint of quaternion unity, 142
continuous bias failure, 250
continuous model, 102
continuous-time state equations, 296
control algorithm, 291, 292
control distribution matrix, 131
control input, 105
control moment, 291, 292
control torque, 16
convergence characteristic, 114
convergence of estimated values, 114
convergence speed, 210, 314
coordinate frames, 1, 2, 13, 14, 22
Coriolis force, 45, 47
correction term, 99, 106
correction term of the Kalman filter, 106
correlated system and measurement noise, 101, 103
cosine Sun sensors, 31, 32
cost function, 81, 82, 88, 90, 272, 273, 274
covariance estimation, 117, 118, 124, 128, 136, 229
covariance of the difference between two successive bias estimates, 287

covariance matching, 117, 129, 201, 224, 225, 226, 253, 278
covariance matching method, 226, 278
covariance matrix, 87, 94, 95, 96, 97, 98, 99, 100, 102, 103, 112, 113, 118, 119, 122, 129, 130, 131, 133, 136, 142, 145, 149, 166, 169, 170, 176, 178, 179, 180, 182, 184, 188, 190, 195, 199, 200, 201, 202, 203, 204, 207, 217, 222, 224, 225, 226, 227, 228, 229, 231, 232, 233, 234, 235, 237, 239, 244, 249, 253, 254, 258, 285, 286, 287, 298
covariance matrix of the filtering error, 96, 97, 98, 100
covariance matrix of the estimation error, 166, 178, 298
covariance matrix of the extrapolation error, 96, 98, 166, 298
covariance matrix of the measurement noise, 99, 190
covariance matrix of the system noise, 99, 228
covariance matrix tuning process, 199
covariance scaling, 120, 129, 137, 200, 253
covariance tuning, 222
cross covariance, 287
cross covariance terms, 287
cross correlation, 93, 101, 102, 103, 143, 151, 202, 227
cross correlation matrix, 143, 202, 227
cube-satellite, 18
cubesat model, 292, 299
cumulative effect, 102

**D**

damping coefficients, 47
data saturation, 105
Davenport's q-method, 81, 82, 90
data processing, 75, 77, 78, 79, 80, 166
decision making, 185
decision statistics, 119
degree of freedom, 183, 205, 207, 234, 235, 242, 244
deterministic control input, 105
deterministic input, 105
deterministic methods, 164
differential value iterative algorithm, 308
digital fine Sun sensor, 31
digital sensors, 31
digital two-axis Sun sensor, 31
direction cosine matrix, 2, 3, 6, 10, 68, 297
distribution density, 95, 96

discrepancy between two successive bias
  estimates, 285, 286
discrete dynamic system, 94, 97
discrete-time nonlinear equations, 111
disturbance torques, 140, 148, 153, 154,
  170, 175, 221, 231
divergence, 105, 106, 114, 117, 180, 195,
  259, 287, 313
divergence of estimations, 177
divergence of the filter, 195, 287
Doyle-Stein condition, 119, 242
drag coefficient, 18
dummy measurement, 142
dynamic system, 94, 97, 100, 125, 127,
  132, 177

**E**

Earth albedo, 36
Earth angular velocity, 75
Earth aspect angle, 39
Earth's horizon, 37, 38
Earth horizon sensors, 38, 40, 49, 61
Earth Gravitational constant, 207
Earth imaging satellite, 37
Earth's magnetic field, 54, 55, 68, 69, 73,
  74, 80, 164, 166, 270, 278, 308
Earth magnetic field model, 290, 291, 292
Earth's magnetic field vector, 20, 55, 68,
  80, 166
Earth pointing spacecraft, 13, 17
Earth's radiation, 60
Earth radius angle, 39
Earth sensors, 37, 38, 40, 61
Earth sensor measurements, 60, 72
Earth sensor noise, 60
eccentric anomaly, 75
eccentric orbit, 59
eclipse, 19, 21, 58, 59, 60, 72, 148, 156, 163,
  171, 172, 191, 194, 253, 255, 258,
  259, 260, 261, 262, 304
eclipse period, 255, 258
ecliptic longitude of the Sun, 59
ecliptic of the Sun, 59
eigenvalues, 84, 101, 180, 182, 205
eigenvalue/eigenvector problem, 84
eigenvectors, 84
eigenvector of Davenport's K matrix, 84
EKF, 94, 109, 110, 113, 114, 139, 140, 143,
  155, 156, 164, 165, 166, 167, 168,
  170, 171, 179, 180, 182, 185, 188,
  191, 192, 193, 196, 200, 203, 204,
  206, 207, 209, 210, 212, 214, 215,
  216, 217, 218, 221, 228, 239, 273,
  278, 308, 309

electromagnetic interference, 266
ellipse equation, 134
ellipsoid equation, 134
entropy based methods, 200
environmental effects, 199
environmental influences, 268
equatorial plane, 13
error covariance, 93, 103, 106, 225, 258
error covariance matrix, 102, 103, 258
error quaternion, 143, 144, 146
error variance, 93, 105, 106, 299, 300
error's mean quadratic value, 95
ESOQ, 81, 88, 90
estimation accuracy, 113, 114, 131, 139,
  152, 156, 158, 191, 194, 199, 203,
  207, 208, 211, 213, 214, 215, 216,
  217, 222, 237, 241, 250, 253, 266,
  282, 288
estimation algorithms, 25, 81, 93, 119, 122,
  126, 165, 277
estimation equation, 104
estimation error, 102, 109, 110, 148, 149,
  152, 153, 155, 157, **159**, 166, 170,
  171, 178, 191, 192, 193, 206, 207,
  208, 209, 210, 212, 213, 215, 222,
  223, 224, 225, 226, 235, 237, 249,
  250, 251, 252, 257, 281, 298,
  304, 305
estimation of the covariance matrices,
  118, 120
estimation of the roll angle, 252
estimation theory, 94, 95
estimation value, 95, 97, 99, 298
estimation value equation, 97
Estimator of the Optimal Quaternion, 81
Euler angles, 141, 142, 165, 166, 278, 284,
  289, 293
Euclid vector norms, 181
expectation-maximization algorithm, 295
experimental attitude sensors, 199
extrapolation value, 99, 105, 106, 127
extrapolator, 133
extended Kalman filter, 94, 109, 110, 114,
  139, 142, 143, 155, 159, 160, 166,
  200, 203, 222, 273, 313
external torques, 153

**F**

fading matrix, 131, 132
fading memory algorithms, 118
Fast Optimal Attitude Matrix, 81
fault detection, 93, 119, 122, 175, 176, 179,
  194, 195, 196, 204, 207, 216,
  234, 244

fault detection and diagnosis, 93
fault detection and isolation, 122, 175, 194,
    196, 244
fault detection statistics, 216
fault isolation, 182, 194, 216, 241, 242,
    245, 262
fault tolerance, 175, 242, 263
fault tolerant attitude control system, 176
fault tolerant attitude estimation, 194, 195,
    196, 199, 200, 221
fault tolerant control, 122, 175
fault-tolerant estimation, 124, 194, 200
feedback controller, 21
Fiber optic gyros, 45
filter adaptation, 124, 137
filter's estimation performance, 224
filter-gain, 108, 110
filter's measurement noise covariance, 253
filter's performance, 199, 221, 239
filter's stability, 223
filtering algorithm, 64, 94, 109, 120, 127,
    139, 140, 141, 144, 145, 159, 164,
    165, 168, 172, 177, 199, 222, 231,
    239, 251, 253, 256, 278, 305
filtering equations, 94, 109, 143
filtering problem, 95, 226
finite memory filter, 106
fluxgate magnetometers, 25, 26, 27
FOAM, 81
frequency domain, 100
Fresnel reflection, 32
fully calibrated magnetometer, 54

**G**

gain matrix, 98, 106, 110, 130, 131, 133,
    166, 178, 217, 232
gain matrix of the filter, 98, 133
Gain-scheduled Kalman filter, 139, 159
Gauss coefficients, 54, 55
Gauss distribution, 166
Gauss Hermite Quadrature Filter, 139
Gaussian distribution, 95, 97, 285
Gaussian distribution function, 285
Gaussian measurement noise, 133
Gaussian noise, 64, 94, 96, 176, 177, 178,
    180, 298
Gaussian white noise, 53, 57, 60, 61, 62, 63,
    89, 111, 148, 154, 170, 207, 229,
    234, 244, 248, 280, 299
Generalized Rodrigues Parameters, 149
genetic algorithm, 273
Gibbs vector, 1, 2, 9, 10, 11
Gibbs vector representation, 9
gimbaled platform, 45

global minimum in multimodal optimization
    problems, 273
Global Navigation Satellite System, 49, 55,
    73, 88
GLR test, 176
GNSS, 49, 55, 73, 88, 89, 90, 91
GPS, 128, 136
gradient-based optimizations, 273
gravitational constant, 22, 56, 207
gravity gradient, 16, 17, 18, 19, 140, 148, 170
gravity gradient torque, 16, 17, 18
ground observation, 48
guaranteed approach, 119
gyro angular random walk value, 63
gyro-based attitude filter, 139, 140, 141,
    152, 160, 222, 239
gyro-based filtering, 140
gyro-based filtering algorithm, 140
gyro biases, 140, 141, 145, 152, 165, 191,
    222, 223, 229, 231, 239, 278, 289,
    291, 293, 303, 309, 313
gyro bias estimation, 226, 290, 291, 293, 294
gyro bias errors, 226
gyro bias terms, 144, 291
gyro bias vector, 63, 140, 278
gyro measurements, 140, 141, 146, 148,
    163, 194, 199, 289, 303, 305, 308,
    309, 313
gyro measurement model, 140
gyro rate random walk value, 63
gyros, 21, 25, 45, 46, 47, 48, 63, 63, 140,
    141, 148, 152, 164, 170, 194, 223,
    289, 290, 291, 292, 293, 303
gyroscopes, 45, 47, 49

**H**

hard iron, 267, 268
hardware redundancy, 176, 195
Hessian, 273
Hessian matrix, 273
horizon sensor, 37, 38, 59, 60, 61, 67, 68,
    75, 79
Huber based methods, 200
hyper-imaginary numbers, 5, 7
Hyper LS method, 273

**I**

IGRF model, 54, 55, 56, 74
inclination angle, 14
inertial axis frame, 278, 279, 290
inertial attitude, 42, 44
inertial frame, 1, 13, 15, 16, 25, 58, 59, 62,
    90, 156, 270, 272

inertia matrix, 16, 154, 157, 207, 234, 244, 248, 280, 292
in-flight calibration, 290, 291
innovation approach, 179, 195, 196
innovation-based adaptive estimation, 117, 136
innovation based covariance scaling, 200
innovation based KF adaptation, 121
innovation based R-adaptation, 129
innovation covariance, 112, 122, 125, 129, 130, 131, 134, 143, 151, 177, 179, 201, 202, 204, 227, 241, 247, 254, 263, 279, 282
innovation covariance matrix, 129, 130, 131, 201, 204
innovation matrix, 179, 180, 181, 182, 195, 196
innovation matrix of EKF, 180, 182
innovation process, 97, 125, 194, 298
innovation sequence, 99, 112, 124, 130, 179, 180, 182, 183, 184, 185, 186, 187, 188, 201, 227, 253, 254, 263, 279, 298
innovation sequence of an optimal Kalman filter, 298
integrated INS/GPS system, 128, 136
integration of the Q and R-adaptation techniques, 241
intuitively tuned MEKF, 225
intuitive tuning of an attitude filter, 222
in-orbit calibration, 57, 265, 277, 279
in-orbit magnetometer bias and scale factor estimation, 296
invariant principle based approach, 120
iterative nonlinear LS method, 273

**J**

Jacobians, 113
Julian date, 59
Julian Day, 59

**K**

Kalman Bucy filters, 93
Kalman filter, 73, 93, 94, 95, 97, 98, 99, 100, 101, 104, 105, 106, 107, 109, 110, 111, 113, 114, 117, 118, 119, 124, 133, 136, 137, 139, 142, 143, 155, 159, 160, 164, 166, 172, 176, 177, 178, 179, 180, 188, 194, 195, 200, 203, 205, 206, 222, 231, 242, 273, 278, 285, 287, 290, 291, 292, 298, 313

Kalman filter algorithms, 93, 106, 109, 118, 136, 217
Kalman filter equations, 95, 101, 104, 298
Kalman filtering, 93, 104, 109, 114, 117, 118, 119, 124, 129, 139, 143, 155, 160, 206, 211, 214, 216, 217, 221, 226, 285
Kalman filter theory, 97
Kalman gain, 99, 112, 117, 129, 130, 131, 143, 202, 203, 204, 209, 217, 298
Kalman gain matrix, 130, 131, 217
Kalman's method, 93
KF adaptation, 118, 120, 121, 128, 129, 158
Kinematics equation, 21, 22,145
Kinematics equation of motion, 21
Kinematics equation of motion of the satellite, 21
Kronecker symbol, 94, 178

**L**

Lagrange multipliers, 83
Laplace function, 135
largest eigenvalue, 84
least squares algorithm, 309
Legendre functions, 55
LEO satellite, 20, 56, 57, 60
level of significance, 183, 205, 234, 242, 285
Likelihood function, 77
linear filters, 95
linear discrete time Kalman filter, 94
linear dynamic system, 132, 177
linear Kalman Filter, 94, 109, 119, 143, 164, 205, 313
linear least squares algorithm, 309
linear LS using the centered variables, 308
linear measurements, 164, 165, 170, 179
linear model, 59, 94, 109, 296, 303
linear state model, 297
linear system, 94, 101
linear Taylor approximation, 94, 109
linearization, 75, 94, 106, 107, 109, 113, 114, 165
linearized discrete Kalman filter, 109
linearized filter algorithm, 108
linearized estimation error, 109
linearized extrapolation error, 108
linearized Kalman filter, 107, 114
linearized one-stage prediction, 108
LKF, 94, 100, 101, 103, 105, 178, 296, 299
local error quaternion, 144
loss function, 82, 83, 84, 87, 89, 90
low-cost nanosatellite, 45
low-cost small satellite missions, 199
low Earth orbit, 17, 25

low pass filter, 225
low-pass scale factor, 225

**M**

magnetic dipole, 56, 64, 207, 271
magnetic dipole moment, 56, 207, 271
magnetic dipole moment of the Earth, 207
magnetic dipole moment vector, 271
magnetic dipole tilt, 56, 64, 207
magnetic disturbance, 20, 28, 152, 155,
        156, 158
magnetic disturbance torque, 20, 28, 152,
        155, 156
magnetic field, 16, 20, 25, 26, 27, 28, 30, 53,
        54, 55, 56, 57, 59, 64, 67, 68, 69,
        73, 74, 75, 80, 156, 164, 166, 257,
        266, 267, 268, 270, 271, 272, 277,
        278, 290, 291, 292, 293, 297, 299,
        301, 302, 308, 309, 313
magnetic field direction, 56, 271, 272
magnetic torque, 16, 140
magnetic torquers, 21, 266
magnetometer, 25, 26, 27, 28, 29, 30, 53, 54,
        55, 64, 67, 68, 69, 71, 72, 74, 75,
        76, 79, 80, 84, 85, 88, 140, 148,
        154, 156, 157, 164, 169, 170, 171,
        172, 206, 207, 209, 210, 213, 215,
        217, 221, 228, 229, 230, 231, 234,
        244, 248, 249, 250, 251, 252, 255,
        256, 257, 258, 259, 265, 266, 267,
        268, 269, 270, 271, 272, 273, 274,
        275, 277, 278, 279, 280, 281, 282,
        284, 287, 289, 290, 291, 292, 293,
        294, 295, 296, 299, 300, 302, 303,
        304, 305, 306, 307, 308, 309, 310,
        311, 312, 313, 314
magnetometer biases, 221, 228, 229, 231,
        249, 250, 273, 290, 291, 292, 296,
        299, 300, 302, 303, 313
magnetometer bias estimation, 228, 229,
        250, 251, 287, 290, 292, 293, 305,
        306, 307, 309, 311
magnetometer bias estimation error, 250
magnetometer bias terms, 229, 248, 249,
        278, 280, 284, 291, 307
magnetometer bias vector, 278, 279,
        291, 295
magnetometer calibration, 54, 164, 265, 271,
        272, 273, 274, 275, 277, 278, 280,
        282, 287, 289, 292, 293, 294, 295,
        296, 299, 302, 303, 305, 308, 311,
        312, 313, 314
magnetometer calibration algorithms, 54,
        271, 273, 274, 277, 312, 313, 314

magnetometer calibration methods, 265,
        271, 272, 273, 274, 275
magnetometer calibration via LKF, 299
magnetometer calibration with known
        attitude, 293
magnetometer error, 53, 271, 272, 273, 277,
        278, 303, 304, 305, 308, 314
magnetometer misalignment, 272
magnetometer model, 269, 274
magnetometer measurement model, 53, 270,
        271, 294, 296, 312
magnetometer measurement vector, 270
magnetometer sensor noise, 55, 154
magneto-inductive magnetometer, 26, 28
magneto-resistive magnetometer, 26
mathematical expectation, 77, 166, 177,
        179, 180, 181, 182, 183, 185,
        187, 194, 196
mathematical expectation of the spectral
        norm, 181, 196
mathematical expectation statistics, 179, 194
mathematical models, 13, 22, 67
matrix algebra, 83
matrix of scale factor and misalignment, 63
maximum eigenvalue, 84
maximum likelihood criterion, 272
Maximum Likelihood Method, 77, 80
maximum likelihood problem, 273
maximum singular value, 180
mean anomaly, 59, 75
mean anomaly of the Sun, 59
mean longitude of the Sun, 59
measurement covariance, 169, 170, 258
measurement delay, 199, 224
measurement equation, 53, 62, 77, 94, 177,
        308, 309
measurement malfunction, 130, 242, 252,
        257, 262
measurement matrix, 95, 122, 133, 177, 189,
        204, 227, 232, 254, 297
measurement matrix of the system, 95,
        133, 177
measurement models, 53, 64, 101, 269
measurement noise, 95, 99, 101, 102, 103,
        104, 106, 107, 112, 114, 118,
        119, 120, 121, 122, 124, 129,
        131, 133, 136, 142, 166, 172,
        177, 188, 190, 199, 200, 201,
        202, 203, 206, 207, 210, 211,
        212, 216, 217, 227, 228, 246,
        253, 254, 257, 258, 262, 270, 272
measurement noise covariance, 112, 118,
        119, 120, 122, 124, 131, 136, 142,
        166, 172, 199, 200, 201, 202, 203,
        206, 216, 217, 228, 246, 253, 262

measurement noise covariance matrices, 118, 136, 206, 253, 262

measurement noise increment, 207, 210, 211, 212, 217, 257, 258

measurement noise increment failure, 217

measurement noise scale factor, 129

measurement system, 129, 130, 200, 201, 245, 279

measurement vector, 95, 101, 111, 133, 142, 166, 177, 188, 201, 233, 270, 271, 287, 313

mechanical gyros, 45

Mehra algorithm, 126

MEMS gyros, 45, 47, 48

MEMS magnetometers, 28

MEMS vibratory gyroscope, 48

methods of linearization, 94

misalignment, 15, 54, 63, 72, 223, 269, 270, 271, 273, 279, 280

misalignment error, 269, 271, 273

microprocessor, 53

minimum variance filter, 107

mobile phone magnetometer, 309

model accuracy, 21

modified loss function, 83, 84

modified Rodrigues parameters, 1, 2, 10, 149

moment of inertia matrix, 16

monitoring algorithm, 182

Monte Carlo, 123, 274

moving window, 118, 124, 125, 128, 202, 207, 227, 279, 281, 282, 283

multidimensional dynamic systems, 176

multi-dimensional functions, 133

multidimensional normal distribution, 134

multiple fading factors, 131

multiple-model-based adaptive estimation, 117

multiple model based methods, 200

multiple measurement noise scale factors, 119, 217

multiple scale factors, 119, 129, 200, 202, 204, 242

multiplicative EKF, 143

multiplicative error model, 62

**N**

nadir, 13, 17, 37, 40, 59, 60, 61, 67, 68, 141, 164

nadir direction, 13

nadir vector, 37, 40, 59, 60, 61, 68

nanosatellite, 18, 20, 25, 28, 30, 33, 34, 40, 41, 45, 47, 49, 54, 148, 155, 164, 170, 223, 248, 255, 266

nanosatellite attitude determination, 164

nanosatellite missions, 28, 34, 40, 41, 45, 49, 54, 155, 223, 266

navigation, 49, 55, 73, 88, 93, 94, 106

Nearest Neighboring Method, 71

neural network, 176

Newton-Raphson iteration, 86

Newton-Raphson method, 84, 85

Newton's laws, 13, 17

noise covariance, 112, 118, 119, 120, 121, 122, 124, 128, 129, 130, 131, 132, 136, 137, 142, 145, 156, 158, 164, 166, 172, 199, 200, 201, 202, 203, 206, 207, 208, 216, 217, 221, 222, 224, 225, 226, 227, 228, 229, 231, 232, 233, 234, 235, 237, 239, 241, 244, 246, 249, 253, 255, 262

noise covariance estimation, 124, 128, 136

noise covariance matrix, 112, 118, 119, 120, 122, 131, 145, 199, 200, 202, 203, 207, 217, 222, 224, 225, 226, 228, 229, 231, 233, 234, 235, 237, 239, 244, 249, 253

noise covariance scaling, 129, 137

noise estimation, 120

noisy scale factor estimation, 281

noisy scale factor estimation process, 281

noise variances, 106

nonlinear distribution, 94, 114, 148

nonlinear dynamic model, 196

nonlinear function, 74, 114

nonlinear filtering algorithms, 113, 242

nonlinear filters, 113, 139

nonlinear equations, 100, 111

nonlinear estimation, 16, 94, 114

nonlinear estimation problems, 16, 114

nonlinear least squares, 273

nonlinear measurement, 107

nonlinear observation models, 94

nonlinear systems, 107, 111, 113, 114, 118, 136, 231

nonlinear transformation, 111

nonorthogonality, 53, 266, 268, 269, 270, 271, 273, 307

nonorthogonality errors, 269

nonorthogonality for TAM, 269

nontraditional attitude filter, 255, 258, 263

nontraditional attitude filtering, 251, 253, 262

nominal measurement, 107

nominal trajectory, 94, 107

nonstationary noises, 127

normalization constraint, 83

normalized innovation, 143, 176, 179, 180, 182, 183, 185, 186, 187, 188, 195, 196, 298, 299, 300

normalized innovation matrix, 179, 180, 181, 182, 195, 196
normalized innovation sequence, 180, 182, 183, 185, 186, 187, 188, 298
normalized innovation vectors, 180
North Pole, 13
norm of attitude estimation errors, 230
null-shift error, 267, 268

**O**

observation covariance matrix, 112
observation equation, 132
observation function, 94, 109
observation models, 94, 104, 109
observed stars, 42
onboard attitude determination, 67
onboard gyros, 21
optimal attitude matrix, 81, 82
optimal gain matrix, 98
optimality criteria, 120
optimal three-axis attitude, 81, 90
optimized loss function, 84
optimal discrete Kalman filter, 16, 100, 114
optimal discrete LKF, 94
optimal filter, 98
optimal Kalman filter, 100, 298
optimal LKF, 100, 105
optimal quaternion, 81, 85
optimum estimation, 95
Optimum Kalman filter equations, 101, 298
optical gyros, 45
orbital angular momentum vector, 13
orbital angular velocity, 22, 56, 165
orbital angular velocity of the spacecraft, 56
orbital motion, 26
orbital parameters, 73, 75, 290, 291, 292
orbital period, 75, 158
orbit determination, 93, 94, 106, 266
orbit frame, 1, 13, 14, 17, 22, 40, 55, 56, 57, 59, 60, 61, 74, 279, 313
orbit inclination, 56, 207
orbit plane, 13

**P**

parametric approach, 120
parametric adaptive estimation methods, 120
partial derivatives, 204, 232
particle filter, 139
particle swarm optimization, 273
perigee, 14, 58
permanent hard iron bias, 268
permanent magnetic field, 267
permissible ellipsoid, 134, 135

pico satellite, 232, 234, 244, 278, 280, 282, 284, 285, 289, 290, 291, 292, 313
pico satellite attitude determination and control system, 292
pitch angle, 61, 75, 79, 80, 167, 194, 207, 208, 209, 210, 212, 213, 215, 249, 278, 294
pitch angle estimation, 208, 209, 210, 212, 213, 215, 249, 294
pitch rate, 186, 187, 188, 189, 190, 191
pitch rate gyroscope, 186, 187, 188, 189, 190
point methods, 67
Poisson distribution, 134
principal moments of inertia, 255
pre-launch calibration process, 271
prediction, 93, 96, 99, 108, 109, 133
predicted mean, 111, 151, 168
predicted covariance, 122, 204, 225, 227, 232, 254
predicted covariance of UKF, 254
predicted observation, 112, 129, 151, 201, 254
predicted observation vector, 112, 151, 201
predicted quaternion vector, 146
predicted trajectory, 110, 136
probability distribution, 95, 119, 120
probability distribution rule, 119, 120
probability of a fault, 133
process/measurement noise covariances, 119
process noise, 18, 118, 120, 124, 128, 132, 145, 156, 158, 199, 201, 207, 208, 221, 222, 224, 225, 226, 227, 228, 229, 231, 232, 233, 234, 235, 237, 239, 241, 244, 246, 249, 253, 254, 255, 258, 259, 260, 261, 262
process noise covariance, 18, 118, 120, 124, 132, 145, 156, 158, 199, 201, 207, 208, 221, 222, 224, 225, 226, 227, 228, 229, 231, 232, 233, 234, 235, 237, 239, 241, 244, 246, 249, 253
process noise covariance adaptation, 132, 227
process noise covariance matrix, 118, 120, 145, 207, 222, 224, 225, 226, 228, 229, 231, 233, 234, 235, 237, 239, 244, 249
process noise covariance matrix estimation, 224, 226
process noise increment, 255, 259, 260, 261, 262
projected area, 18, 19
propagation of Euler angles, 11
propagation of the mean of the state vector, 104

pseudo- KF algorithm, 311
pseudo-linear KF, 311, 312
pseudo-linear measurement, 311
pseudo-linear measurement model, 311
pseudo-measurement, 308

**Q**

Q adaptation, 118, 131, 221, 225, 228, 231,
      232, 241, 242, 244, 246, 247, 248,
      251, 253, 254, 259, 262, 263
Q-adaptation algorithm, 225, 259
Q-adaptation method, 221, 241, 242, 244,
      247, 254, 262, 263
Q adaptation methods, 221, 241, 242,
      244, 262
Q-adaptive filter, 262
Q-adaptive Kalman filtering methods, 221
Q and R adaptation techniques, 241
Q estimation, 126, 248
*Q* matrix, 61, 62, 63, 126, 132, 144, 213,
      221, 222, 223, 224, 225, 226, 229,
      231, 239, 241, 242, 247, 249, 253
Q matrix estimation algorithm, 126
q-method, 81, 82, 84, 87, 90, 164, 168, 170
quadrant Sun sensor, 32, 33, 36
quaternion attitude representation, 4, 21
quaternion attitude vector, 44
quaternion component, 72
quaternion estimation, 71, 72, 257, 259,
      260, 261
quaternion estimation error, 257, 259,
      260, 261
Quaternion estimator, 81
quaternion inverse, 62
quaternion measurements, 17, 62, 81, 88, 91,
      164, 168, 171, 296
quaternion multiplication, 61, 62, 143, 146
quaternion norm constraint, 142
quaternion parameters, 5
quaternion propagation, 140
quaternions, 1, 2, 4, 5, 6, 8, 9, 10, 11, 22, 62,
      69, 70, 71, 79, 80, 139, 140, 141,
      142, 143, 145, 148, 150, 151, 160,
      168, 223, 278
quaternion vector, 8, 9, 10, 42, 61, 70, 71,
      87, 140, 141, 144, 146, 149,
      151, 222
quasi-equations, 311
quasi-linearization method, 165
quasi measurements, 308, 309, 311,
      312, 313
QUEST, 44, 81, 84, 85, 86, 87, 88, 90, 91,
      164, 168, 169, 170, 171, 191, 192,
      193, 194

QUEST algorithm, 81, 84, 85, 86, 91,
      169, 170
QUEST method, 84

**R**

R adaptation, 119, 129, 131, 132, 199,
      200, 215, 241, 242, 246, 247, 248,
      258, 263
R adaptation methods, 199, 200, 242
R adaptation techniques, 215, 241
R and Q adaptations, 119
R matrix, 126, 129, 130, 131, 132, 199, 200,
      201, 202, 203, 204, 207, 209, 215,
      216, 241, 242, 247, 253, 256, 257,
      258, 262, 263
R matrix estimation algorithm, 126
R-scaling, 132
radiative torque, 19
radiation forces, 19
radiation intensity, 19
radiation pressure, 18, 19
radiation torque, 18, 19
radiation pressure coefficient, 19
random Gaussian noise vector, 94
random Gaussian values, 181
randomly delayed measurements, 199
random matrix, 181
random processes, 119, 120
random walk, 46, 63, 148, 170, 223,
      293, 296
random walk process, 63, 296
rate gyros, 45, 290, 291, 292
rate-integrating gyros, 45
real-time attitude independent magnetometer
      calibration, 311
real-time estimation, 308
real-time estimation for the magnetometer
      error terms, 308
reconfigurable attitude estimation
      method, 285
reconfigured filter, 193, 287
reconfigured UKF, 216, 217, 284, 285,
      287, 288
recursive algorithm, 166, 278, 305, 308,
      311, 312
recursive estimation, 166, 277
recursive estimator, 294
recursive estimation algorithms, 277
recursive linear least squares algorithm, 309
recursive magnetometer calibration, 277,
      295, 305, 312, 313, 314
recursive magnetometer calibration
      algorithms, 277, 312, 313, 314
redundancy techniques, 77, 78

redundant data processing, 75, 77, 78, 79, 80, 166
redundant data processing algorithm, 77, 78, 79, 80, 166
redundant data processing methods, 75
reference directions, 53, 68
reference frame, 1, 3, 4, 6, 13, 15, 17, 21, 22, 42, 53, 54, 55, 57, 58, 59, 60, 61, 64, 68, 69, 86, 141, 156, 165
reference magnetic field vector, 53
REKF, 205, 206, 207, 209, 210, 211, 212, 214, 215, 216, 217, 218
relative estimation errors, 299, 300, 303
relative estimation errors of magnetometer biases and scale factors, 299, 300, 303
relative magnetic permeability, 26
Residual-Based Adaptive Estimation, 117, 136
residual-based adaptive scaling, 132
residual based scaling, 132, 200
residual based KF adaptation, 121
residual covariance, 128, 241, 247, 262
residual error, 120
residual magnetic moment, 16, 28, 148, 152, 155, 170
residual magnetic torque, 16
residual sequence, 128, 263
residual series, 120, 121
residual term, 228
residual vectors, 118, 129
right ascension, 14, 58, 75
right ascension of ascending node, 14
Ring Laser Gyros, 47
RKF against measurement faults, 132
Robust Adaptive UKF, 243
Robust Extended Kalman Filter, 203
Robust Kalman filters, 117, 136, 204, 205, 206, 215, 217
Robust KF insensitive to the actuator failures, 242
Robust Kalman filtering methods, 216, 217
robustness of the filter, 200, 201
Robust Unscented Kalman Filter, 119, 200
Robust UKF insensitive to the system failures, 245
Rodrigues parameters, 1, 2, 10, 149
roll angle, 168, 235, 236, 238, 246, 251, 252, 278, 283, 288
roll angle estimation, 235, 236, 238, 246, 251, 283, 288
rotated ellipsoid, 270
root mean square error, 155, 225
rotation angle, 2, 10, 40, 86, 87, 170, 269
rotation angle error, 86, 87, 170

rotation matrix, 2, 3, 67, 87
RUKF, 205, 206, 207, 208, 209, 210, 211, 212, 213, 214, 216, 217, 218, 253
RUKF algorithms, 216, 217

**S**

Sagnac effect, 45
sample covariance matrix, 184
sample variance, 184
sampling frequency, 63
sampling time, 201, 207, 223, 234, 244, 255, 280, 292, 297, 299
satellite attitude dynamics, 13, 16, 22
satellite body frame, 13, 15, 17, 53, 64
satellite dynamics, 21, 145, 153, 156, 217, 221, 231
scalar attitude-independent measurement, 308
scalar measure for the spectral norm, 180
scale factor, 30, 63, 118, 119, 129, 130, 137, 149, 200, 201, 202, 203, 204, 207, 208, 209, 210, 211, 212, 214, 215, 217, 225, 242, 248, 253, 266, 268, 278, 279, 280, 281, 282, 283, 284, 287, 288, 296, 299, 300, 302, 303, 313
scale factor estimation, 278, 279, 281, 282, 287, 288, 296, 299, 300
scale factor for low-pass effect, 225
scaling and misalignment matrix, 279
scaling matrix, 53, 270, 274, 275, 306, 307, 314
scaling terms/parameter, 111, 270, 295
scanning Earth horizon sensors, 38
Schmitt trigger oscillator, 30
search-coil magnetometers, 25, 26, 27
self tuning, 120
sensor/actuator faults, 117, 119
sensor calibration, 164, 250
sensor failure, 187, 189, 190, 194, 195, 245
sensor failure isolation, 187, 189, 190
sensor fault, 119, 131, 176, 182, 185, 187, 188, 191, 194, 196, 199, 200, 201, 202, 203, 206, 217, 251, 253, 255, 261
sensor fault detection, 196
sensor fault detection and isolation, 196
sensor fusion, 93
sensor measurement models, 53, 56, 59, 64
sensor misalignments, 54, 223, 271
sensor model, 56, 59
sensor noise, 21, 55, 57, 60, 61, 148, 154, 157, 170, 195, 199, 207, 229, 234, 244, 248, 255, 280, 299

sequential algorithm/approach/architecture/ state estimation, 93, 139, 159, 302, 308
sequential centered algorithm, 308
shadow function, 19, 21, 59
short term fault, 215
sigma points, 94, 111, 114, 139, 148, 149, 150, 151, 159, 164
sigma point filters, 139
simultaneous attitude estimation and magnetometer calibration, 277, 278, 280, 282, 287, 289, 292, 313
simultaneous R and Q adaptations, 119, 247, 258
single-frame algorithms, 140, 163, 168, 172
single-frame attitude determination/ estimation method, 67, 81, 90, 251, 255
single-frame attitude estimators, 81, 139, 159, 165
single-frame methods, 139, 159, 163, 165, 168, 170, 171, 172, 255
single scale factor, 119, 129, 200, 201, 203, 208, 209, 210, 211, 212, 214, 215, 242, 279
Singular Value Decomposition (SVD) method, 81, 87, 88, 90, 168, 253
singular values of the matrix, 180
sliding-window, 308, 310, 311
sliding window filter, 308
small satellite magnetometers, 26, 53, 64, 265, 266, 274, 277
small satellite magnetometer calibration, 265
smoothing, 93
soft iron coefficients, 267, 271
soft iron error, 54, 266, 270, 271
solar constant, 19
solar panel currents, 271, 274
solar pressure, 16
solar radiation, 18, 19, 32
solar sail nanosatellite, 18
solar wind, 19
space environment, 221, 239
space guidance, 93, 94
Sparse Grid Quadrature Filter, 139
spectral norm, 179, 180, 181, 182, 185, 195, 196
spherical polar coordinates, 54, 55
spin motion, 27, 38, 311
spin-planar body axes, 311
spin-planar magnetometer measurements, 309
spin-planar measurements, 309, 310
spin rate, 26, 27, 34, 35, 39, 56, 207
spin rate of the Earth, 56, 207

spinning small satellite/spacecraft, 38, 274, 309
spinning spacecraft, 1, 26, 34, 35, 38, 309, 310, 311, 313
spin-type Sun aspect sensor, 34, 35
stability condition of filter, 100
standard deviations, 53, 57, 60, 62, 72, 89, 134, 148, 254, 157, 165, 166, 170, 181, 207, 210, 217, 223, 229, 234, 244, 248, 255, 257, 274, 280, 299
star catalog, 42, 62, 68
star identification, 42, 44, 199
star image sampling, 199
star tracker, 25, 30, 36, 41, 42, 43, 44, 45, 48, 49, 54, 61, 62, 68, 81, 87, 88, 90, 91, 160, 164, 168, 199, 223, 272, 277, 295, 296, 313
star tracker measurement, 43, 61, 90, 160, 164, 168, 199
star tracker measurement model, 61
state covariance, 99, 195, 222
state equation, 94, 132, 177, 296
state estimation, 93, 94, 109, 117, 121, 123, 130, 136, 143, 156, 175
state vector, 93, 94, 95, 97, 104, 106, 107, 109, 111, 113, 125, 128, 133, 136, 137, 140, 141, 143, 144, 148, 150, 151, 156, 157, 165, 177, 178, 182, 185, 191, 222, 228, 231, 232, 275, 278, 279, 282, 284, 287, 288, 293, 295, 297, 299, 313
stationary filter, 125
stationary dynamic filters, 100
stationary systems, 125
statistical analysis, 120, 121
statistical function, 205, 234, 242
statistical filter information matrices, 120
statistical methods, 67, 75, 81, 87, 90, 91
statistical modelling, 106
statistic average operator, 94, 125
static Earth horizon sensors, 38, 40, 41, 61
steady-state Extended Kalman Filter, 139, 159
step-like changes, 200
stochastic, 60, 105
stochastic control problems, 105
stopping criteria, 287
stopping rule, 284, 285, 286, 287, 313
strapdown mode, 45
structural approach, 120
suboptimum filter, 133
Sun aspect angle, 34, 35, 36, 37
Sun angle, 31, 36
Sun direction vector, 34, 35, 56, 57, 58, 59, 60, 68

Sun model, 58
Sun pressure, 17, 19, 140
Sun presence sensors, 31, 34
Sun sensor, 25, 30, 31, 32, 33, 34, 35, 36,
    37, 40, 49, 56, 57, 59, 64, 67, 68,
    69, 71, 72, 74, 79, 84, 85, 88, 148,
    163, 164, 170, 255, 296, 303
Sun sensor measurement, 57, 69, 71, 72, 74,
    163, 164
Sun sensor model, 56
Sun sensor noise, 57, 148, 170
Sun vector, 30, 34, 59, 73, 80
SVD algorithm, 81, 256, 257
SVD method, 81, 87, 88, 90, 168, 253
systematic errors, 60, 61, 72
system dynamics, 178, 242, 298
system failure, 243, 244, 245
system fault, 185, 242, 246, 263
system function, 94, 109
system identification, 94
system model, 93, 104, 105, 108, 111, 120,
    231, 232, 235, 236, 237, 238,
    241, 253
system noise, 94, 95, 97, 99, 104, 106, 107,
    117, 136, 166, 177, 228, 285, 286

T

TAM misalignment, 273
Taylor approximation, 94, 109
theoretical innovation covariance, 202, 279
theoretical variance, 183
thermal stress, 269
three-axis attitude determination, 81, 83,
    84, 90
three-axis magnetometer, 25, 28, 53, 265
thresholding, 43
time-domain filter, 100
time-variable parameters, 100
time-varying bias, 268
time-varying errors, 265, 266, 271, 272, 273,
    274, 312
time-varying magnetometer errors, 265, 266
time-varying parameter estimation, 266
torque estimation, 153, 155
torque vector, 16, 18
traditional approach, 139, 160, 241, 251,
    253, 256, 263
traditional filtering, 256
transition matrix, 94, 99, 107, 132, 140, 145,
    146, 177
transformation matrix, 9, 14, 313
TRIAD algorithm, 67, 69, 71, 72, 73, 74, 75,
    79, 81, 82, 166, 296, 302, 303, 304
TRIAD estimation, 70, 79, 304

TRIAD method, 67, 69, 73, 77, 79, 80, 81
TRIAD+UKF algorithm, 303, 305, 306, 307
TRIAD+UKF approach, 164, 302, 304, 313
trial-error method, 231, 249, 250
true anomaly, 14
truncation error, 106
tuning the attitude filter, 239
tuning the R, Q matrix / covariance matrices,
    199, 221, 225, 262
two-axis attitude, 37
two-stage attitude estimation, 302
two-stage estimation, 291, 293
two-step algorithm, 273, 274, 308
two-step method, 273, 309
two-vector algorithm, 67, 73, 79, 80, 164

U

1U cubesat, 21, 280
3U cubesat, 248
3U nanosatellite, 18, 248
UKF algorithm, 111, 112, 114, 164, 170,
    207, 210, 216, 229, 231, 234, 245,
    251, 253, 254, 263, 282, 284, 287,
    288, 292, 303, 305, 306, 307
UKF bias estimations, 251, 287
UKF for attitude estimation, 148
UKF for scale factor estimation, 282
UKF with reconfiguration, 288, 289
UKF without reconfiguration, 283, 284, 288
uncalibrated magnetometer, 299, 301, 302
unit circle, 101, 205
unit covariance matrix, 176, 179
unit Sun vector, 59
unit velocity vector, 18
unscented Kalman filter (UKF), 94, 111,
    114, 119, 139, 160, 164, 172, 200,
    231, 242, 278, 313
unscented transform, 111, 114

V

variance, 40, 75, 76, 77, 78, 79, 80, 104,
    105, 106, 107, 133, 166, 180, 181,
    182, 183, 185, 187, 188, 190, 194,
    195, 196, 235, 245, 256, 282, 292,
    299, 300
variance of normalized innovation
    sequence, 187
vector magnetometers, 27
vector measurements, 67, 72, 80, 81, 82, 86,
    88, 90, 91, 133, 140, 146, 159,
    163, 164, 165, 170, 171, 172, 253,
    255, 296, 302
vector observations, 67, 81

vector transformation, 2, 5
vehicle-guidance, 94
velocity vector, 15, 18, 26
Vernal Equinox, 13
vibratory driving force, 47

**W**

Wahba's loss function, 90
Wahba's problem, 82, 84, 90, 165
wavelength, 89
white forcing function, 102
whiteness property, 103
white noise, 53, 57, 60, 61, 62, 63, 89, 107,
    111, 124, 148, 154, 170, 177, 179,
    180, 182, 207, 229, 234, 244, 248,
    280, 296, 299
Wheatstone bridge, 28, 29, 30
weighted sum, 82, 121
width of the moving window, 202, 279
Wiener filter, 100
Wiener-Kalman filters, 93

**X**

X-axis Geomagnetic Field, 301

**Y**

yaw angle, 76, 78, 141, 167, 223, 278,
    279, 305
Y-axis Geomagnetic Field, 301

**Z**

Z-axis Geomagnetic Field, 302
z-domain, 101
zero mean Gaussian white noise, 53, 57, 60,
    62, 63, 148, 154, 157, 170, 179,
    207, 229, 234, 244, 248, 280, 299
zero-mean processes, 104
zero-mean system, 104
zero output failure, 211, 213, 214, 215
z-plane, 101, 205
z-transform, 100, 101, 205